Engineering Embedded Systems

Peter Hintenaus

Engineering Embedded Systems

Physics, Programs, Circuits

 Springer

Peter Hintenaus
Embedded Hard and Software
Linz
Austria

ISBN 978-3-319-10679-3   ISBN 978-3-319-10680-9   (eBook)
DOI 10.1007/978-3-319-10680-9

Library of Congress Control Number: 2014947129

Springer Cham Heidelberg New York Dordrecht London

MATLAB® and Simulink® are registered trademarks of The MathWorks, Inc. CrossCore® and SHARC® are registered trademarks of Analog Devices Inc. Maple™ is a trademark of Waterloo Maple Inc. Mathematica® is a registered trademark of Wolfram Research, Inc. ARM® and Cortex® are registered trademarks of Arm Ltd.

Printed on acid-free paper

Springer is part of Springer Science+Business Media (www.springer.com)

Preface

Embedded systems acquire data directly from the physical world and directly act on the physical world. Building such systems requires not only elaborate skills when it comes to designing hardware and software, a solid understanding of the physics of the application and of the physical behavior of the devices used for building the system is at least equally important. In my experience it is the physics which to a large extent determines the architecture of the finished system.

The book grew out of the courses I taught at the University of Applied Sciences FH-Joanneum and at the University of Salzburg. Its main audience are students of computer science and related disciplines in the last year undergraduate and first year graduate levels, who are adventurous enough to pack their computers and leave cyberspace for the physical world, where nature decides whether an idea works or not. A practicing engineer may also find entertainment in the more advanced chapters of the book. A reader should be familiar with engineering mathematics and with signals and systems in particular. Familiarity with a simulation tool such as the Simulink® system helps as does familiarity with computer algebra software like Maxima, Maple™, or Mathematica®. No knowledge of physics beyond what is taught in high school is necessary.

I strongly believe mathematical reasoning about a physical situation is a crucial part of understanding the situation. Therefore, I emphasize mathematical modeling. Whenever possible I derive analytical models of the reality at hand from first principles. Whenever I need a fact whose derivation is beyond the scope of the book, I state this clearly. The book contains a list of acronyms. In the text however, I try to avoid jargon and acronyms as much as possible.

Most chapters contain exercises. I answered all exercises in Appendix D. I moved some arguments into exercises which I felt do not fit into the main development. The provided answers close some gaps the reader might feel the main treatment has. Most chapters contain suggestions for lab exercises too. Naturally, I provide no answers for these.

The first four chapters cover the basics necessary for designing embedded systems. Chapter 1 defines the term embedded system. It motivates how these systems form a natural hierarchy with measurement and control systems at the bottom. The material in this book focuses on this bottom layer. The chapter goes on to introduce models for physical reality and models of computation.

Chapter 2 summarizes the mathematical tools we need in the rest of the book. A well-prepared reader may skim over the chapter and use it later as reference. For the not so well-prepared reader the material together with the given references to the textbook literature can act as guide to the subjects to be learned.

Chapter 3 introduces some basic electrical engineering. It discusses electrical charge, voltage and current, and the associated measurements. It treats resistors, capacitors, and inductors with emphasis on the mathematical description of their behavior. An experienced reader may skip this chapter. Readers without prior exposure to the basics of electrical engineering are encouraged to study the chapter.

Chapter 4 is a compact introduction to digital electronics. It analyzes the properties of the different types of outputs motivating the need for more than two logic levels. Next, it focuses on the temporal behavior of combinatorial circuits including logic races. In the discussion of sequential circuits it again emphasizes the temporal behavior of flip-flops; it covers metastability and motivates the synchronous design discipline. The chapter next describes the realization of state machines in synchronous logic. Systems comprising several clock domains and synchronization follow. After some words on the realization of digital circuits the chapter ends with a first view of the analog problems plaguing fast logic circuits. Readers not familiar with metastability are encouraged to read the chapter.

Each of the following chapters comprises what I call a technical essay. In a technical essay I try to tell a complete story along a well-chosen line of development. The story usually evolves across boundaries between disciplines. To a certain extent such an essay is my personal take on the subject, reflecting my personal experiences and opinions as much as the accepted knowledge in the areas the story touches.

The first two essays can be read in arbitrary order. Both should be read before venturing into one of the last four essays. Chapter 5 contains an essay on programmable devices. The line of development I have chosen for the story follows the sequence of decisions one has to make when designing the digital part of an embedded system. The story starts out with the three architectures for handling external events—event driven, time driven, and polling. It introduces peripheral devices as coprocessors relieving the main processors from mundane but time-critical tasks. After looking at static and dynamic RAM the story goes on to look at ways to deal with memory hierarchies. Finally, it is time to introduce the three possible types of execution engines—embedded microprocessors, digital signal processors, and programmable logic. Two case studies, one signal processor based and the other using programmable logic apply some of the discussed techniques to real designs.

Chapter 6 is an essay titled analog electronics—signal conditioning and conversion. The line of development follows the steps we take when designing the analog part of a system. Starting with the block diagrams of three sampled data systems—a control system, a measurement system, and a digital radio—the story dives into filters and sampling. After stating design rules for anti-aliasing and reconstruction filters it touches upon data converters. The converters' sampling clock gets a special treatment. The second part of the story is devoted to analog building blocks, to active discrete components and to circuits. At first reading, the second part of the story may be skipped.

Chapter 7 tells similar stories of six siblings—six switchmode power converters. The parallel treatment tries to emphasize the importance of switching when we want to achieve high efficiency in energy conversion processes.

The last four essays may be read in any order. These essays are case studies describing systems I have built. The essay in Chap. 8 continues treating energy conversion, this time looking at electrical motors. The story develops along the line physics, programs, circuits. It starts out with a somewhat detailed look at electromagnetism and magnetic materials. Armed with this knowledge it develops the classical model of a permanent magnet synchronous motor—a type of motor finding use in applications from toys to electric vehicles. The story continues with applying control strategies to the model with mixed results. The simple strategy does not work—but close examination of the model gives us clues for a strategy that works—field-oriented control. With this strategy as plan of attack the essay continues with issues arising in the software for such a control scheme in order to end with the circuit of a low-voltage power stage and some measurements produced with this power stage.

The last three essays use Fourier transform methods. The essay in Chap. 9 is the least challenging in terms of the mathematics used. A reader less experienced with Fourier transforms is encouraged to start here. The line of the story loosely follows physics, programs, circuits. The story shows how to make precise measurements admit overpowering interference using a method called lock-in detection. Using Fourier series it analyzes the workings of the lock-in principle and some of its variants. Next, it discusses several ways of implementing the principle. Then the story applies the lock-in principle to signal conditioning for strain gauges—the sensor elements in force and torque sensors. The essay then turns to making simultaneous measurements with a single detector and states conditions so that the measurements do not disturb each other. Finally, it puts this principle to use when constructing the software and the circuit of an optical sensor for measuring moisture content in substrates such as paper.

Chapter 10 contains an essay on short-range radar for measuring speed and distance. Again the book's subtitle applies. Starting with the Doppler effect the story shows how to convert the information contained in the reflected signals at very high frequencies down into a much lower frequency range. It then unpacks the full Fourier apparatus in order to come up with an algorithm based on computing a discrete Fourier transform, for computing the speeds of the targets visible the radar systems. During this development the story shows us the relationship between the time we spend observing the reflected signal for a measurement and the speed resolution the algorithm produces. Next, it turns to measuring distances and introduces the frequency modulated continuous wave radar. After turning to a circuit, which uses a commercially available radar module, the story ends with some software issues and a sample measurement.

Chapter 11 is an essay on making nature compute Fourier transforms using two mirrors and a beam splitter. The story develops along the themes physics, programs, circuits again. The story discusses a spectrometer for infrared radiation. A spectrometer computes the intensity distribution with respect to wavelength in an incoming beam of light. The beam of light contains information about the chemical compositions of the substances it has interacted with. Therefore, these instruments are used for performing chemical analyses. After looking at light waves and their properties the story goes on to present several types of infrared spectrometers. It analyzes the trick with the two mirrors and the beam splitter—the Michelson interferometer—in detail. Next it presents the signal processing performed in the actual instrument. The essay finishes with covering auxiliary functions and the optical setup.

Acknowledgments

Many people supported me in writing this book. Werner Obermayr, Peter Porzer, and Josef Templ read early drafts of the complete book. Their numerous comments and corrections helped me improve the text tremendously. My former colleague Heimo Sirnik created all the schematic diagrams in this book. He incorporated the many changes I had patiently.

Friederike Kiefer and Johann Kiefer have been very supportive over the years. The many visits at their house have always been relaxing and productive at the same time. Johann also provided the initial funding for the work on the infrared spectrometer. My partner in the spectrometer work, Wolfgang Märzinger, taught me the physics necessary. He designed the optics and the mechanics of the instrument, and he is making the spectrometer an ongoing commercial success. Heinz Krenn provided guidance during the early phases. My project team at FH-Joanneum, Gernot Turner, Heimo Sirnik, Gernot Neißl, and Wolfgang Stocksreiter made Chap. 11 possible with their work. Gernot Kvas tuned the spectrometer's signal processing software.

My former colleagues Robert Okorn and Christian Netzberger and I had many inspiring discussions while we shared a lab.

Peter Meerwald provided thoughtful comments and valuable corrections for a number of chapters.

I have been member of Wolfgang Pree's group at the University of Salzburg for the last 6 years. He encouraged me to develop my ideas into this book. Andreas Naderlinger helped me with the Simulink® system, with the simulations in Chap. 8 in particular. He also gave me well-reasoned advice on early drafts of the manuscript. Stefan Resmerita took over the majority of the lab sections when we taught the material in this book together. His efforts gave me the time necessary for writing this book. Stefan Lukesch and Anton Pölzleitner commented on parts of the material.

Gerhard Bauer and his company Infolog helped me build some of the circuits in this book.

Stefan Huber and Roland Kwitt made helpful suggestions on earlier drafts of the material in Chap. 2. Martin Rieger reported several errors in an earlier draft of Chap. 10. Michael Kleber did some work on the signal conditioning circuit for strain gauges in Chap. 9.

Bruno Buchberger encouraged me when I was depressed over the slow progress I made in writing this book. My contacts at Springer UK, Charlotte Cross and Oliver Jackson cheered me up when I kept missing the deadlines I set myself.

My former students at FH-Joanneum and at the University of Salzburg inspired me to write this book. I hope they like it!

The writing of this book was partially supported by the Austrian Research Funding Association through the COMET program within the research network *Process Analytical Chemistry—Data Acquisition and Data-Processing (PAC)* under Contract 825340, by the Austrian Federal Ministry for Transport, Innovation and Technology, by the Austrian Federal Ministry of Economy, Family and Youth, and by the Federal State of Upper Austria.

Contents

Acronyms

AC	Alternating current
ACK	Acknowledge
ADC	Analog-to-digital converter
AM	Amplitude modulation
BJT	Bipolar junction transistor
CAN	Controller area network
CPLD	Complex programmable logic device
CW	Continuous wave
DAC	Digital-to-analog converter
DC	Direct current
DC	Signal content at 0 Hz
DDR SDRAM	Double data rate synchronous dynamic memory
DMA	Direct memory access
DRAM	Dynamic random access memory
DSP	Digital signal processor
DUT	Device under test
EDA	Electronic design automation
EEPROM	Electrically erasable programmable read only memory
EMC	Electromagnetic compliance
EMI	Electromagnetic interference
ESL	Equivalent series inductance (capacitor)
ESR	Equivalent series resistance (capacitor)
FET	Field effect transistor
FFT	Fast Fourier transform
Flash	Electrically erasable ROM, erased in pages
FM	Frequency modulation
FMCW	Frequency modulated continuous wave
FPGA	Field programmable gate array
FRED	Fast recovery epitaxial diode
FTIR	Fourier transform infrared

GAL	Generic array logic
HF	High frequency
I^{2}C	Inter-integrated circuit
IF	Intermediate frequency
IGBT	Isolated gate bipolar transistor
IR	Infrared
ISR	Interrupt service routine
LAN	Local area network
LASER	Light amplifcation by stimulated emission of radiation
LDO	Low dropout voltage regulator
LED	Light emitting diode
LIW	Long instruction word
LTI	Linear time-invariant
MCU	Microcontroller unit
MII	Media independent interface
MOS	Metal oxide semiconductor
MOSFET	Metal oxide semiconductor field effect transistor
NACK	Not acknowledge
NOOP	No operation
PAL	Programmable array logic
PCB	Printed circuit board
PFC	Power factor corrector
PFD	Phase frequency detector
PI	Proportional and integral (controller)
PID	Proportional integral differential (controller)
PLA	Programmable logic array
PLC	Programmable logic controller
PLL	Phase locked loop
PMSM	Permanent magnet synchronous motor
PV	Photovoltaic
PWM	Pulse width modulation
RAM	Random access memory
RF	Radio frequency
RMII	Reduced media independent interface
RMS	Root mean square
ROM	Read-only memory
SAR	Successive approximation analog-to-digital (converter)
SAW	Surface acoustic wave (filter)
SDRAM	Synchronous dynamic random access memory
SEPIC	Single-ended primary-inductance converter
SIMD	Single instruction multiple data
SNR	Signal-to-noise ratio
SPI	Serial peripheral interconnect
TAI	International atomic time
UART	Universal asynchronous receiver transmitter

VCO	Voltage controlled oscillator
VCXO	Voltage controlled quartz oscillator
VLIW	Very long instruction word
WCET	Worst-case execution time

Chapter 1
Models and Experiments

An embedded system is a computer system—software and hardware—that interacts with its physical environment—mainly without human intervention. It acquires its inputs from its physical environment via sensors; it impinges its outputs on this environment using actuators. When we set out to devise such a system we not only have to consider the software and the (digital) hardware, but also more importantly the properties of the sensors and actuators, and of the real world beyond this interface. The sensors and actuators and the apparatus they interface with might not be available. Either they may not exist at a certain stage of development or it may be too expensive or even impossible to furnish every single developer with a prototype of the system under development.

Instead of the real thing we use models as substitutes. A model is an approximate description of some reality—an abstraction from reality—covering what we believe at the time of model-creation to be the essentials. When developing an embedded system we have to model the behavior of the sensors, the actuators, and of the physical reality behind. We will abstract from the software and hardware by using appropriate models of computation. Many of the systems we are building have to satisfy precise timing constraints; they become real-time systems. The processors we use for building these systems often have very limited support for debugging software. Turning to a simulation system such as the Simulink® software instead, allows us to analyze the system under development and the interaction with its modeled environment at a level of detail impossible when developing on real hardware. When we are satisfied with the interactions between the models we are not done however. We have to quiz nature whether she agrees with our beliefs on how things work—we have to build a prototype and observe its behavior; we have to perform well thought-out experiments. In case some of the experiments produce unsatisfactory results we have to incorporate our new-found knowledge into our models by refining them and start all over.

© Springer International Publishing Switzerland 2015
P. Hintenaus, *Engineering Embedded Systems*, DOI 10.1007/978-3-319-10680-9_1

1.1 Hierarchy of Systems

The control unit of an industrial robot, for example, has to solve the equations of motion for the robot several times per second in order to make the robot move its hand along a given path while performing the desired change of the orientation of the piece it holds. The motor drive powering one of the robot's joints in turn has to move through a sequence of angular positions at precise points in time and with a given angular velocity at each position. The robot itself usually is part of a larger production cell. The cell controller coordinates the interaction of the cell's machinery by distributing the measurements taken by the cell's sensors and by providing coordinated commands to the cell's machines.

The controller for the motor drive has to deal with the physics of the motor, mainly electromagnetism and some mechanics. The robot controller is mostly concerned with the kinematics and the dynamics of the robot. It has to respect the limits imposed by the actual mechanism it controls. The kinematics deals with the geometry of the movements to be performed, while the dynamics considers the masses involved and the forces and torques necessary for the movements. The cell controller relies on the systems down the hierarchy to deal with the involved physics directly; for programming it we must respect the times necessary for the the required actions only.

This type of hierarchy is typical for large embedded systems. Each layer in such a hierarchy asks for the use of specialized platforms and specialized methods in the implementation of systems at that level.

1.2 The Bottom Layer—Control and Measurement

In this book, we focus exclusively on the systems at the bottom. Besides communication with the system-level above, a system at the bottom of the hierarchy typically implements a single function. In many cases, the type of function falls into one of the two broad categories, either controlling some process or providing measurements.

Often the process to be controlled converts energy from one form into another, for instance from chemical into mechanical as in combustion engines, from electrical into mechanical as in electric motors, or from one form of electrical energy into another form of electric energy as in photovoltaic inverters. Other processes requiring control are for example flows of substances, such as the flow of some liquid in a brewery.

A system populating the bottom layer typically must interact with its environment with a very high rate and at precise points in time. Sometimes the allowed error is less than a nanosecond. Such precision is not attainable in software, therefore specialized periphery, often in the form of application specific coprocessors operating with some degree of autonomy, is used to relieve the main processor from directly dealing with this interaction. Furthermore, power consumption should be low. Often such a system must operate off a small energy storage like a battery. Even when the system runs off the mains high power consumption produces large amounts of waste heat. When a number of systems have to operate in a confined space removal of the waste heat may be a challenge. Physical size and costs force us to build systems with a minimum number of hardware components. The amount of memory integrated together with the processors and the peripheral devices onto a single semiconductor chip determines to a large extent the size and the cost of the chip. An interface to memory devices external to the processor chip demands an increased number of connections, and therefore a more costly package for the chip. Unless the application absolutely needs large memories, we minimize the system's memory footprint and work with the memory resources provided on the processor chip. Lack of memory also keeps us from using complex operating systems. Quite often we work without an operating system at all. The little systems programming required, such as interrupt handling or scheduling, becomes part of the application.

1.3 Models of Physical Reality

Whenever feasible applying the laws of physics will give us very precise models. When these models become intractable or when we lack knowledge for applying the laws of physics, statistical inference may come to our rescue producing models from observations alone.

1.3.1 Models Based on First Principles

The laws of physics constitute mankind's beliefs about the behavior of inanimate objects. Before formulating these laws, physicists throughout the centuries have made quantitative observations of the behavior of nature; they have measured physical quantities. Next they have imagined conjectures fitting the measurements within the precision of the observations. Once a conjecture has been formulated, physicists can use the conjecture to make predictions about the behavior of nature. The more the predictions match with the associated measurements, the more accepted the conjecture becomes until it is established as a law of physics. A single contradictory observation, however, provided the measurement uncertainty is small enough, topples a law of physics, as nature is always right. At times the then accepted laws in different fields of physics have produced contradictory predictions; the current physical knowledge at that time has been inconsistent. Eventually, physicists have developed more refined theories replacing the older laws. At the end of the nineteenth century for example, electrodynamics, the theory describing electrical and magnetic phenomena and Newtonian mechanics were the accepted theories. Both theories were supported with vast numbers of experiments. However, these two theories provided contradictory predictions. In 1905, Einstein, based on the work by H.A. Lorentz and H. Poincaré introduced a new theory for mechanics, special relativity, resolving these contradictions (Einstein 1905). Newtonian mechanics survived as a very good approximation describing the behavior of macroscopic bodies moving at moderate speeds.

Table 1.1 The prefixes in the SI system

Prefix	Symbol	Factor	Prefix	Symbol	Factor
yotta	Y	10^{24}	yocto	y	10^{-24}
zetta	Z	10^{21}	zepto	z	10^{-21}
exa	E	10^{18}	atto	a	10^{-18}
peta	P	10^{15}	femto	f	10^{-15}
tera	T	10^{12}	pico	p	10^{-12}
giga	G	10^{9}	nano	n	10^{-9}
mega	M	10^{6}	micro	μ	10^{-6}
kilo	k	10^{3}	milli	m	10^{-3}
hecto	h	10^{2}	centi	c	10^{-2}
deca	da	10^{1}	deci	d	10^{-1}

Most laws of physics come in the form of equations between physical quantities. A physical quantity q is a product of powers of base quantities. In general, the exponents of the powers of the base quantities can be positive or negative rational numbers. A base quantity is the product of its numerical quantity value and its base unit. The seven types of base quantities in the *Système International d'Unités* (SI) (ISO 2009) are length, unit meter [m], mass, unit kilogram [kg], time, unit second [s], electric current, unit ampere [A], amount of substance, unit mol [mol], temperature, unit kelvin [K], and luminous intensity, unit candela [cd]. Throughout this book, whenever we talk of units alone we write unit expressions in square brackets. In the SI system, a unit can be preceded by a single prefix, indicating a multiplier. Table 1.1 lists the possible prefixes.

Besides scalars, vectors, matrices, and tensors composed of physical quantities all having the same unit, and the same prefix arise frequently. The usual laws of arithmetic apply. When we add physical quantities it is necessary that the summands all have the same unit and the same prefix. This condition is not sufficient: the magnitude of some torque for example and some work both have the unit N m, but their sum makes no sense; torque is a vector, while work is a scalar. Moreover, it is necessary for the two sides of an equation to have the same units. In any model we build, checking the units helps us to get the equations correct.[1]

1.3.1.1 Example—Mechanics

Let us consider a point-shaped body with mass m having position $\vec{r}(t)$ at time t. The unit of the position is meter [m]. Then the velocity $\vec{v}(t)$ of the body is the derivative of its position with respect to time,

---

[1] Unit mismatches can have dire consequences. In 1999, NASA lost the Mars Climate Orbiter due to mixing newton and pound as units of force.

$$\vec{v}(t) = \frac{d\vec{r}(t)}{dt},$$

and the unit of velocity is $[m\,s^{-1}]$. If the body is completely undisturbed it will either rest or move with constant velocity along a straight line. The acceleration $\vec{a}(t)$ of the body is the derivative of its velocity with respect to time,

$$\vec{a}(t) = \frac{d\vec{v}(t)}{dt} = \frac{d^2\vec{r}(t)}{dt^2}.$$

The unit of acceleration is $[m/s^2]$. Changing the velocity of the body requires the force $\vec{F}(t)$,

$$\vec{F}(t) = \frac{d(m\vec{v}(t))}{dt}.$$

The unit of force is newton, $[N = kgm/s^2]$. Using the Newtonian approximation that the mass m of the body is constant we may write $\vec{F}(t) = m\vec{a}(t)$. The momentum $\vec{p}(t)$ of the body is

$$\vec{p}(t) = m\vec{v}(t).$$

Let us look at two point-shaped bodies completely undisturbed from the outside. If the first body exerts a force $\vec{F}$ onto the second, then the second body exerts the force $-\vec{F}$ onto the first; the forces balance. This is Newton's third law. As a consequence the total momentum $\vec{p} = \sum_i^n m_i \vec{v}_i(t)$ of n point-shaped bodies with masses $m_1, \ldots, m_n$ moving with velocities $\vec{v}_1(t), \ldots, \vec{v}_n(t)$ and exerting forces onto each other, but being completely undisturbed from the outside is constant; we say that momentum is conserved.

A constant force $\vec{F}$ moving the point-shaped body a displacement $\vec{s}$ performs the work W,

$$W = \vec{F} \cdot \vec{s}.$$

The unit of work is joule, $[J = Nm = kgm^2/s^2]$. Power is the rate with which work is done,

$$P = \frac{dW(t)}{dt}.$$

The unit of power is watt $[W = Js^{-1}]$.

There is another physical quantity—energy with the unit joule—that cannot be produced nor destroyed but merely be converted from one form into another. A body with mass m moving with speed v has the kinetic energy $mv^2/2$ for example. The work $W = mv^2/2$ was necessary for accelerating the body; in turn the body performs the same work when it is brought to a standstill. We may understand energy as stored work or as the ability to do work. The forms of energy are kinetic energy, gravitational energy, elastic energy, heat energy, electrical energy, radiant energy, chemical energy, nuclear energy, and mass energy. Heat energy plays a special role. While we can convert any other form of energy into heat energy completely, the laws of thermodynamics tell us that the converse is not true. As engineers we, therefore, consider practically any heat energy our devices produce as loss, which we have to remove from the device, most often at the expenditure of more energy.

Conservation of energy allows us to check the validity of any model we build based on the laws of physics. For any time span the sum of the energy entering some apparatus during the time span, the energy leaving the apparatus during this time span, and the net change for the time span of the energy stored in the apparatus is constant. If a model we built for some apparatus does not follow this rule, the model is most certainly wrong and we have to find the errors in our reasoning.[2] In an electric motor, for example, energy enters in electrical form, leaves in the form of kinetic energy, and is stored as electric and as kinetic energy. The energy stored in a loss-less motor, for example, does not change while the motor operates with constant angular velocity and produces constant torque. Under these conditions the electric power consumed by this motor must equal the kinetic power produced by the motor.

Most of the preceding definitions and laws are rather specialized—for example, they deal with point-shaped bodies only. We may apply these definitions and laws approximatively to situations where the size of the bodies is small compared to the distances involved. Whether such a model predicts the real behavior well enough for our purposes can only be answered by performing carefully designed experiments. If the predictions turn out to be too coarse for the task at hand we have to refine the model. While being closer to the true behavior a more refined model typically is harder to understand, and will require more computational resources.

The conservation laws make sweeping statements about the behavior of the inanimate world. Their general applicability makes them so powerful, both when building models and when checking whether a given model is in accordance with the laws of physics.

1.3.2 Models Based on Statistical Inference

Engineers are lucky that physicists have discovered an enormous body of knowledge describing the behavior of the inanimate world. Often, however, we have to provide solutions in situations where the application of these laws to the situation at hand is either not possible due to limited knowledge or leads to models so complicated that these models are computationally intractable. In such a situation we can try to build a model using regression or machine-learning methods. When building a regression

---

[2] Alternatively, we may rush to the patent office and apply for protection of our new-found form of perpetual motion.

model for a system we must produce tuples of inputs to the system and measurements of the corresponding answers. A regression method then produces a model using certain assumptions about the relationship between the inputs and the outputs. Many methods assume this relation to be linear. In general, regression methods disrespect physics. The models they produce will not obey physics' conservation laws. Given no better alternatives these models can be very useful, provided they are applied with sufficient care.

Besides being employed during development, this type of models is routinely used to interpret measurement data. Many substances, for example, show a distinct wavelength dependent absorption pattern when subjected to infrared light; the substance leaves a "fingerprint" in the light it transmits. This allows us to do chemical analyses with light alone. Faced with a mixture of known substances with unknown concentrations we can use an infrared spectrometer for recording the absorption pattern of the mixture. Using a regression model built from mixtures of the same substances with known concentrations and the associated spectra, we can produce estimates for the unknown concentrations. The estimates can be very precise, provided the unknown concentrations are reasonably close to concentrations in the mixtures used for making the model.

1.4 Models of Computation

The signals a sensor produces when measuring some physical quantity typically vary continuously with time; they are analog signals. Some analog processing such as filtering is necessary before the signal can be quantized in time and value. Besides models for describing the digital computations performed by digital circuits and microprocessors, we need models for describing analog computation.

1.4.1 Finite State Machines

A finite state machine consists of the finite set of states S, the finite set of inputs I called the input alphabet, the finite set of outputs O called the output alphabet, the state transition function $t: (S \times I) \rightarrow S$, the output function, $o: (S \times I) \rightarrow O$ and the initial state s_0. A state machine computes the output sequence $y: \mathbb{N}_0 \rightarrow O$, $n \mapsto y_n$, from the input sequence $x: \mathbb{N}_0 \rightarrow I, n \mapsto x_n$ according to the rules

$$s_{n+1} = t(s_n, x_n)$$
$$y_{n+1} = o(s_n, x_n),$$

for all $n \in \mathbb{N}_0$. This type of state machine is due to Mealy (1955), while the type introduced by Moore (1956) does not take the input into account when computing the new output. We can tabulate both the state-transition function and the output function, but a state transition diagram communicates the purpose of a state machine better. In a state transition diagram every state is represented by a bubble and every transition is represented by an arc. Each arc is labeled with the input symbol required for the arc to be taken and with the output symbol produced. An arc labeled **default** takes care of all inputs not matching any other arc. An arc originating in thin air indicates the initial state.

State machines do not encode the passage of time explicitly. Usually, we identify the appearance of the input x_n with a point in time t_n. We assume the state machine to change state, and to produce a new output at these points in time instantly and sit idle in-between. This way a state machine becomes a discrete-time model.

Consider a microwave oven. The electromagnetic waves used for cooking can cause severe burns whenever a living being is subjected to them. Therefore, the oven may only generate these waves, while the door, which provides shielding, is closed and locked. The state machine in Fig. 1.1 provides controls for a lock preventing the door from being opened and for the microwave generator. It assumes that the sensor

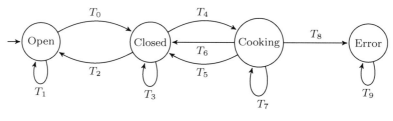

	Input		Input		Input
T_0	*door-closed*	T_4	*start-pressed*	T_8	*door-opened*
T_1	**default**	T_5	*time-expired*	T_9	**default**
T_2	*door-opened*	T_6	*stop-pressed*		
T_3	**default**	T_7	**default**		

State	Input	Output
Open	**default**	*disable-RF*
Close	*start-pressed*	*lock-door-enable-RF*
Close	**default**	*disable-RF*
Cooking	*time-expired*	*disable-RF-unlock-door*
Cooking	*stop-pressed*	*disable-RF-unlock-door*
Cooking	*door-opened*	*disable-RF-unlock-door*
Error	**default**	*disable-RF*

Fig. 1.1 State transition diagram, transitions, and output function of a state machine for a microwave oven

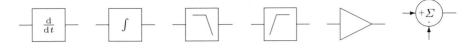

Fig. 1.2 Differentiator block, integrator block, lowpass filter block, highpass filter block, gain block, and summation block

for the position of the door produces a *door-closed* input provided the door is closed when the oven is switched on. The state Error is entered when the state machine receives a *door-opened* input for a door it believes to be locked. This might indicate a problem either with the lock or with the sensor for the position of the door. The microwave generator is disabled as the oven can no longer operate safely.

1.4.2 Signal and Data Flow—Block Diagrams

A block diagram consists of blocks connected by directed arcs. A block applies an operation, such as integration over time, to its inputs and provides the results as its outputs. The inputs and outputs of blocks usually are real-valued; other data types are possible. An arc represents a connection for passing an output of one block to the input of another; connections do not introduce delays. Typical blocks include differentiators, integrators, lowpass filters, highpass filters, gain blocks, and blocks for summation. A lowpass filter lets the slowly varying part at its input pass to its output while it attenuates the fast changing part; a highpass filter operates the other way round. Figure 1.2 lists some commonly occurring blocks, Appendix A contains a more comprehensive list.

Block diagrams are a natural way for describing measurement and control systems. The flexibility of block diagrams—one can always invent another block—makes them a good tool for informally communicating ideas during systems development. This flexibility allows us to also include some structural information into block diagrams, by mixing pure block diagrams with schematic descriptions of microprocessor systems for example.

When we use block diagrams as modeling tools we require a more rigorous semantics. Most often the passing of time in block diagram models is assumed to be continuous. Provided, we can describe every single block in a block diagram by a system of equations, we can convert this block diagram into another large system of equations. A numerical solution of this large system of equations then produces an approximation to the modeled behavior. Internally, the solvers for these systems of equations discretize time using either a fixed time-step or a variable one. Controlling this time-step allows us to trade simulation time for more precise results.

1.5 Structural Models

Often models describe both, structure and behavior. A circuit diagram, for example, Chap. 3, can be simulated in order to derive the behavior of the described circuit. The connectivity information contained in the circuit diagram is used as input for designing a printed circuit board for the circuit. Purely structural models such as module or class hierarchies are not used in this book.

1.6 Bibliographic Notes

The book by Lee and Seshia (2011) covers techniques for modeling, designing and analyzing embedded software and hardware. In Marwedel (2010) a broad overview over the knowledge and the techniques relevant when designing embedded systems is given. The book by Wolf (2014) presents high-performance architectures for embedded systems such as tablets and smart phones.

The book (Tipler and Llewellyn 2007) provides a gentle introduction to physics. Several introductory texts to physics at the undergraduate level are available, one is Meschede and Gerthsen (2010). The classic book by Feynman et al. (2011) is at the same time highly readable and very precise. For a take on classical mechanics by a computer scientist see the book by Sussman and Wisdom (2001). For a broad overview over the physics at the basis of today's technology see for example the book by Gershenfeld (2000). Gershenfeld's other book (1999) covers techniques for the mathematical modeling of physical systems.

For the use of statistical inference in chemistry see the book by Varmuza and Filzmoser (2009) and the one by Brereton (2003).

The book by Lee and Varaiya (2011) discusses several extensions to finite state machines. For modeling and simulation technology for the design of embedded systems see for example (Ptolemaeus 2014).

1.7 Exercises

Exercise 1.1 Consider a plane rigid body rotating around a fixed axis perpendicular to the body's plane. Let the point $\vec{r}$ be located on the body at a distance r from the axis of rotation. What arc length s does the point $\vec{r}$ travel when the body undergoes an angular displacement of θ?

Exercise 1.2 For the rigid body in Exercise 1.1, what is the body's angular velocity?

Exercise 1.3 For the rigid body in Exercise 1.1, what is the body's angular acceleration?

Exercise 1.4 Consider a force $\vec{F}$ in the plane of the body in Exercise 1.1 acting on the body. What is the torque the force $\vec{F}$ exerts on the body?

Exercise 1.5 Assume that the rigid body in Exercise 1.1 is point-shaped. What is the body's momentum of inertia around the axis of rotation?

References

Brereton R (2003) Chemometrics: data analysis for the laboratory and chemical plant, Wiley, New York

Einstein A (1905) Zur Elektrodynamik bewegter Körper. Annalender physik 17(10):891–921, doi:10.1002/andp.19053221004

Feynman R, Leighton R, Sands M, Gottlieb M (2011) The Feynman lectures on physics. Basic Books

Gershenfeld N (1999) The nature of mathematical modeling. Cambridge University Press, Cambridge

Gershenfeld N (2000) The physics of information technology. Cambridge series on information and the natural sciences. Cambridge University Press, Cambridge

ISO (2009) ISO 80000: Quantities and Units: Part 1 - Part 14. ISO

Lee EA, Seshia SA (2011) Introduction to embedded systems, a cyber-physical systems approach. http://LeeSeshia.org

Lee EA, Varaiya P (2011) Structure and interpretation of signals and systems, 2nd edn. http://LeeVaraiya.org

Marwedel P (2010) Embedded system design: embedded systems foundations of cyber-physical systems. Embedded systems. Springer, New York

Mealy GH (1955) A method for synthesizing sequential circuits. Bell Syst Tech J 34(5):1045–1079

Meschede D, Gerthsen C (2010) Gerthsen physik. Springer, New York

Moore EF (1956) Gedanken experiments on sequential machines. Automata studies, Princeton U, pp 129–153

Ptolemaeus C (ed) (2014) System design, modeling, and simulation using Ptolemy II. Ptolemy.org, http://ptolemy.org/systems

Sussman GJ, Wisdom J, Mayer ME (2001) Structure and interpretation of classical mechanics. MIT Press, Cambridge

Tipler P, Llewellyn R (2007) Modern physics. W. H. Freeman, California

Varmuza K, Filzmoser P (2009) Introduction to multivariate statistical analysis in chemometrics. CRC Press, Boca Raton

Wolf M (2014) High-performance embedded computing: architectures, applications, and methodologies, 2nd edn. Morgan Kaufmann Publishers Inc, San Francisco

Chapter 2
Tools—Mainly Mathematics

For building models of physical systems we need mathematical tools. In order to model physical phenomena we have to look at lines, surfaces, and vector fields in three-dimensional space. The notion of a signal allows us to concentrate on the information contained in some time-varying physical quantity without thinking about the physical representation. The notion of a system likewise describes the information processing performed by some apparatus without resorting to the physical properties of the apparatus. We assume the reader is familiar with the material. Therefore, we quickly list the definitions and properties we need in the sequel. For a pedagogical treatment of the material see Lee and Varaiya (2011), Kreyszig (2010), Franklin et al. (2010).

For learning about inanimate nature we have to probe her and measure her response. In order to do so we must familiarize ourselves with instrumentation for generating, capturing, and analyzing real-world signals.

2.1 Complex Numbers

Let i denotes the imaginary unit, $i^2 = -1$. A complex number $c \in \mathbb{C}$ is the sum $c = a + ib$ for $a, b \in \mathbb{R}$. We denote the real part a of c with $\Re(c)$ and the imaginary part b with $\Im(c)$. The complex number $c^* = a - ib$ denotes the complex conjugate of c. Euler's relation[1] (2.1) allows us to represent points on the unit-circle in a simple way,

$$e^{i\phi} = \cos\phi + i\sin\phi, \tag{2.1}$$

where $\phi \in \mathbb{R}$. What makes Euler's relation so remarkable is that it links in a very useful way a purely algebraic concept—complex numbers—to geometry. Identifying the complex number $c = a + ib = |c|e^{i\angle c}$ with the point $\vec{c} = (a, b)$ in the plane $|c|$ is

---

[1] Richard P. Feynman, Nobel laureate in Physics 1965, called Euler's relation the most remarkable formula in mathematics. We better remember this formula.

© Springer International Publishing Switzerland 2015
P. Hintenaus, *Engineering Embedded Systems*, DOI 10.1007/978-3-319-10680-9_2

the distance between the origin and the point $\vec{c}$. The angle $\angle c$ is the angle between the x-axis and the line through the origin and $\vec{c}$. Multiplying c and $e^{i\phi}$ amounts to rotating the point $\vec{c}$ counterclockwise by the angle ϕ around the origin, $ce^{i\phi} = |c|e^{i\angle c}e^{i\phi} = |c|e^{i(\angle c+\phi)} = (a+ib)(\cos\phi + i\sin\phi) = (a\cos\phi - b\sin\phi) + i(a\sin\phi + b\cos\phi)$.

Whenever you encounter some identity involving complex numbers, which is not immediately obvious, you can most probably derive it from Euler's relation.

2.2 Line and Surface Integrals

We consider a curve C in space with the endpoints $\vec{a}$ and $\vec{b}$. Let C have a parametric representation $\vec{r}$: $\{t \in \mathbb{R}: a \le t \le b\} \to \mathbb{R}^3$,

$$t \mapsto \vec{r}(t) = \begin{pmatrix} x(t) \\ y(t) \\ z(t) \end{pmatrix}$$

for $a, b \in \mathbb{R}$, such that $\vec{r}(a) = \vec{a}$ and $\vec{r}(b) = \vec{b}$. To each value $a \le t_0 \le b$ corresponds a point of C whose position vector is $\vec{r}(t_0)$. The parametric representation imparts a direction onto C, from the start point $\vec{r}(a)$ to the end point $\vec{r}(b)$. If the curve C has a parametric representation $\vec{r}$ such that $\vec{r}(t_0)$ is continuous and has a continuous derivative $\vec{r}'(t_0) = \frac{d\vec{r}}{dt}(t_0)$, which is not identical to the zero-vector for $a \le t_0 \le b$, then the curve C has a unique tangent direction at any point, which varies continuously along C; then we declare the curve to be smooth.

Vector fields are functions that assign a vector to every point in space. Let $\vec{F}: \mathbb{R}^3 \to \mathbb{R}^3$ be a continuous vector field. The line integral of $\vec{F}$ along the smooth curve C having a representation $\vec{r}$ with the properties above is

$$\int_C \vec{F} \cdot d\vec{l} = \int_a^b \vec{F}(\vec{r}(t)) \cdot \frac{d\vec{r}(t)}{dt}\, dt.$$

A piecewise smooth curve C consists of a finite number of smooth curves $C_1, \ldots, C_n$ such that the end point of C_i and the start point of C_{i+1} coincide for $1 \le i < n$. The line integral along C is the sum of the line integrals along the curves C_i. A curve is closed if its start point $\vec{a}$ and its endpoint $\vec{b}$ coincide. The line integral along a closed curve is denoted by $\oint_C \vec{F} \cdot d\vec{l}$.

Let us consider now a surface S in three-dimensional space. Let S have a parametric representation $\vec{r}: R \to \mathbb{R}^3$,

$$(u, v) \mapsto \vec{r}(u, v) = \begin{pmatrix} x(u, v) \\ y(u, v) \\ z(u, v) \end{pmatrix},$$

associating with every point (u_0, v_0) in some connected point set R of the uv-plane a point of S whose position vector is $\vec{r}(u_0, v_0)$. The point set R is connected if and only if any two points $\vec{p}_1, \vec{p}_2 \in R$ can be connected by finitely many line segments with each line segment contained entirely in R. The boundary B of the surface S is the set of all points p such that every neighborhood of p contains points in S as well as points not in S.

If the parametric representation $\vec{r}(u, v)$ is continuous and has continuous partial derivatives $\frac{\partial \vec{r}}{\partial u}(u_0, v_0)$ and $\frac{\partial \vec{r}}{\partial v}(u_0, v_0)$ with

$$\vec{N}(u_0, v_0) = \left(\frac{\partial \vec{r}}{\partial u} \times \frac{\partial \vec{r}}{\partial v} \right)(u_0, v_0) \neq \vec{0}$$

for all points $(u_0, v_0) \in R$ then the surface S has at any point $\vec{r}(u_0, v_0)$ a unique tangent plane and a unique normal whose direction is given by $\vec{N}(u_0, v_0)$, both varying continuously across S; the surface is smooth. We denote the normal vector with unit length as $\vec{n} = \vec{N}/|\vec{N}|$.

A smooth surface S is orientable if and only if the positive normal direction at an arbitrary point $\vec{p}$ of S can be continued in a unique and continuous way to the whole surface. Let $\vec{F} : \mathbb{R}^3 \to \mathbb{R}^3$ be a continuous vector field. Let $\vec{r}$ be a parametric representation with the properties stated above of the smooth orientable surface S. The integral of $\vec{F}$ through the surface S is

$$\iint_S \vec{F} \cdot \vec{n} \, \mathrm{d}A = \iint_R \vec{F}(\vec{r}(u, v)) \cdot \left(\frac{\partial \vec{r}(u, v)}{\partial u} \times \frac{\partial \vec{r}(u, v)}{\partial v} \right) \mathrm{d}u \, \mathrm{d}v.$$

The Möbius strip is an example for a smooth surface that cannot be oriented, see e.g. (Kreyszig 2010).

A piecewise smooth surface S consists of finitely many smooth surfaces $S_1, \ldots, S_n$ and their boundaries. A piecewise smooth surface is orientable if and only if each surface S_i is orientable and for each pair of surfaces, which have part of their boundaries in common, S_i and S_j with $i \neq j$, the directions at the common part of the boundaries run against each other. The direction of the boundary of the surface S_i is set so that the normal and the boundary are right handed.

2.3 Discrete-Time Signals and Systems

A discrete-time signal s is defined at equally spaced times $t_n = n\sigma$, $n \in \mathbb{Z}$, where $\sigma \in \mathbb{R}, \sigma > 0$ is the time-step. It maps the t_n to the elements of some set $\mathbb{A}$. Therefore, the discrete-time signal s can be identified with the function $s : \mathbb{Z} \to \mathbb{A}, n \mapsto s_n$. We use subscript notation for discrete time signals throughout this book. Sometimes, it is convenient to restrict the domain of a discrete-time signal to the nonnegative integers. The signal s is periodic with period $p \in \mathbb{N}$ if and only if $s_n = s_{n+p}$ for all n in the domain of s. Most often the image of the signal s comprises either real or complex numbers, but more complicated situations arise. A video, for example, maps time to pictures. A picture in turn may be represented by a tuple of three two-dimensional arrays of real numbers, one array each for the colors red, green, and blue. In this book we restrict ourselves to real-valued, $\mathbb{A} = \mathbb{R}$, or complex-valued, $\mathbb{A} = \mathbb{C}$, signals.

The signal $\delta \colon \mathbb{Z} \to \mathbb{R}$,

$$n \mapsto \delta_n = \begin{cases} 1 & \text{for } n = 0 \\ 0 & \text{for } n \neq 0, \end{cases}$$

is called the discrete impulse. For an arbitrary signal $x \colon \mathbb{Z} \to \mathbb{C}$ the sifting property of the discrete impulse,

$$\sum_{n'=-\infty}^{\infty} x_{n'} \delta_{n-n'} = x_n, \tag{2.2}$$

holds for all $n \in \mathbb{Z}$. This property is easy to argue, but its frequent occurrence earns it a name.

The unit-step signal, $u \colon \mathbb{Z} \to \mathbb{R}$ is the sum of the discrete impulse,

$$n \mapsto u_n = \sum_{n'=-\infty}^{n} \delta_{n'} = \begin{cases} 0 & \text{for } n < 0 \\ 1 & \text{for } n \geq 0. \end{cases}$$

A discrete-time system[2] S uniquely transforms the input signal $x \colon \mathbb{Z} \to \mathbb{R}, n \mapsto x_n$, into the output signal $y \colon \mathbb{Z} \to \mathbb{R}, n \mapsto y_n$. We denote this fact by writing

$$y_n = (S(x))_n.$$

While we have written the definition for real-valued inputs and outputs a definition with the obvious modifications will do for complex-values input and output signals.

Discrete-time systems model digital feedback-controllers, for example, quite well. Such a controller measures the response of the process it is supposed to control at equally spaced times. From each measurement it computes a control output, which acts on the process via an actuator shortly after the controller has taken the measurement.

2.3.1 Linear Time-Invariant Systems

The discrete-time system S is linear if and only if the system allows superposition, that is

---

[2] We restrict our discussion to single-input single-output systems.

$$(S(au + bv))_n = a\,(S(u))_n + b\,(S(v))_n$$

for all signals $u\colon \mathbb{Z} \to \mathbb{R}$ and $v\colon \mathbb{Z} \to \mathbb{R}$ and for all $a \in \mathbb{R}$ and $b \in \mathbb{R}$. The system S is time-invariant if and only if

$$y_{n-k} = (S(\tilde{x}))_n$$

for all signals $x\colon \mathbb{Z} \to \mathbb{R}$ and $y\colon \mathbb{Z} \to \mathbb{R}$ and for all $k \in \mathbb{Z}$ where $y_n = (S(x))_n$ and the signal $\tilde{x}$, $\tilde{x}_m = x_{m-k}$ is the time-shifted signal x. Obvious modifications cover the complex-valued case.

2.3.2 Impulse Response and Convolution

Let $u\colon \mathbb{Z} \to \mathbb{C}$ and $v\colon \mathbb{Z} \to \mathbb{C}$ be complex-valued discrete-time signals. The signal $u * v\colon \mathbb{Z} \to \mathbb{C}$,

$$n \mapsto (u * v)_n = \sum_{n'=-\infty}^{\infty} u_{n'} v_{n-n'},$$

is called the convolution of u with v. The convolution is commutative, $u * v = v * u$. The shifting property (2.2) becomes $x = x * \delta$.

The impulse response of the linear time-invariant discrete-time system S is $h\colon \mathbb{Z} \to \mathbb{R}$, $n \mapsto h_n = (S(\delta))_n$. The impulse response describes the system S fully; the response $y\colon \mathbb{Z} \to \mathbb{R}$ of S to an arbitrary input $x\colon \mathbb{Z} \to \mathbb{R}$ is the convolution of h with x,

$$n \mapsto y_n = (S(x))_n = (h * x)_n. \tag{2.3}$$

The obvious modifications cover complex-valued inputs and outputs. When the impulse response $h_{n'}$ is nonzero for some $n' < 0$, computing y_n in (2.3) requires knowledge of $x_{n'}$ for $n' > n$. Having to know the future in order to produce a valid answer in the present is beyond a humble engineer's abilities. We call the system S causal if and only if $h_{n'} = 0$ for all $n' < 0$. Noncausal systems are useful only for processing retrospectively signals that have been recorded in advance.

2.3.3 Circular Convolution

For periodic signals the infinite sum in the definition of the convolution is not only unnecessary, but it also introduces problems with convergence. The circular convolution circumvents these problems by summing over a single period only. Let $x\colon \mathbb{Z} \to \mathbb{C}$ and $y\colon \mathbb{Z} \to \mathbb{C}$ be complex-valued periodic discrete-time signals with period p. The signal

$$(x \circledast y)_n = \sum_{n'=0}^{p-1} x_{n'} y_{n-n'}$$

for $n \in \mathbb{Z}$ is called the circular convolution of x and y.

2.4 Continuous-Time Signals and Systems

A continuous-time signal s maps real numbers to the elements of some set $\mathbb{A}$. It usually is a function. The domain of s most often is understood as time but other interpretations are possible. In Chap. 11 we will encounter signals whose domain is one-dimensional space instead of time. The signal $s \colon \mathbb{R} \to \mathbb{A}$, $t \mapsto s(t)$ is periodic with period $p \in \mathbb{R}$, $p > 0$ if and only if $s(t + p) = s(t)$ for all $t \in \mathbb{R}$. We restrict ourselves again to real-valued or complex-valued signals. Although some continuous-time signals are not functions we will use function-notation throughout this book.

A signal, which is not a function, is the Dirac delta, $\delta(t)$. For all $t \in \mathbb{R}$, $t \neq 0$ the Dirac delta is zero, $\delta(t) = 0$ and for all real $\epsilon > 0$,

$$\int_{-\epsilon}^{\epsilon} \delta(t)\, dt = 1.$$

For getting an intuitive understanding[3] let us consider a strike to some object. We can describe such a strike by a function $P \colon \mathbb{R} \to \mathbb{R}$, $t \mapsto b(t)$. Let us assume the strike begins at time $-d/2$ and ends at time $d/2$ for some duration $d \in \mathbb{R}$, $d > 0$. Before the impact at time $-d/2$ the strike does not transfer energy to the object, $b(t) = 0$ for $t < -d/2$. After the impact, the power $P(t)$ delivered to the object at time t will rise sharply. We assume that the strike delivers maximum power at time 0. Between time 0 and time $d/2$ the power drops off to zero again. The total energy transferred by the strike is $\int_{-\epsilon}^{\epsilon} P(t) dt$ for $\epsilon > d/2$. When we consider a sequence of strikes all delivering the energy of 1 J with shorter and shorter duration the Dirac delta will be the limit of this sequence for $d \to 0$.

The Dirac delta has the sifting property; for a function $x \colon \mathbb{R} \to \mathbb{C}$, $t \mapsto x(t)$ being continuous at $t \in \mathbb{R}$

$$x(t) = \int_{-\infty}^{\infty} x(\tau)\delta(t - \tau)\, d\tau \tag{2.4}$$

holds.

---

[3] A precise mathematical treatment involves the theory of distributions and measure theory. This is well beyond the scope of this book.

The unit-step signal, $u : \mathbb{R} \to \mathbb{R}$ is the integral of the Dirac delta,

$$t \mapsto u(t) = \int_{-\infty}^{t} \delta(\tau) d\tau = \begin{cases} 0 & \text{for } t < 0 \\ 1 & \text{for } t \geq 0. \end{cases}$$

The instantaneous power $P(s(t))$ at time t of the complex valued signal $s : \mathbb{R} \to \mathbb{C}$, $t \mapsto s(t)$ is $P(s(t)) = s(t)s(t)^*$. The average power $\mathcal{P}(s)$ of s is

$$\mathcal{P}(s) = \lim_{T \to \infty} \frac{1}{T} \int_{-T/2}^{T/2} s(t)s(t)^* dt. \qquad (2.5)$$

A continuous-time system[4] S uniquely transforms the input signal $x : \mathbb{R} \to \mathbb{R}$, $t \mapsto x(t)$ into an output signal $y : \mathbb{R} \to \mathbb{R}$, $t \mapsto y(t)$. We denote this fact by writing

$$y(t) = (S(x))(t).$$

The definition covers real-valued inputs and real-valued outputs. For complex-valued inputs and outputs obvious modifications apply.

Continuous-time systems model electrical networks quite well, see Chap. 3. Such a network with one input and one output reacts to a voltage at its input with a voltage at its output.

2.4.1 Linear Time-Invariant Systems

The continuous-time system S is linear if and only if the system allows superposition, that is

$$(S(au + bv))(t) = a (S(u))(t) + b (S(v))(t)$$

for all signals $u : \mathbb{R} \to \mathbb{R}$ and $v : \mathbb{R} \to \mathbb{R}$ and for all $a \in \mathbb{R}$ and $b \in \mathbb{R}$. The system S is time-invariant if and only if

$$y(t - t') = (S(\tilde{x}))(t)$$

for all signals $x : \mathbb{R} \to \mathbb{R}$ and $y : \mathbb{R} \to \mathbb{R}$ and for all $t' \in \mathbb{R}$ where $y(t) = (S(x))(t)$ and $\tilde{x}(t) = x(t - t')$. Again the obvious modifications will cover the complex-valued case.

---

[4] We again restrict our discussion to single-input single-output systems.

2.4.2 Impulse Response and Convolution

Let $x: \mathbb{R} \to \mathbb{C}$ and $y: \mathbb{R} \to \mathbb{C}$ be complex-valued signals. The signal

$$(x * y)(t) = \int\limits_{-\infty}^{\infty} x(\tau) y(t - \tau) \, d\tau$$

is called the convolution of x with y. Equation (2.4) becomes $x = x * \delta$.

The impulse response of the linear time-invariant continuous-time system S is $h: \mathbb{R} \to \mathbb{R}, t \mapsto h(t) = (S(\delta))(t)$. The impulse response describes the system S fully; the response $y: \mathbb{R} \to \mathbb{R}$ of S to an arbitrary input $x: \mathbb{R} \to \mathbb{R}$ is the convolution of h with x,

$$t \mapsto y(t) = (S(x))(t) = (h * x)(t). \tag{2.6}$$

When the impulse response $h(t')$ is nonzero for some $t' < 0$ computing $y(t)$ in (2.6) requires knowledge of $x(t')$ for $t' > t$. The future's not ours to see; therefore we call the system S causal if and only if $h(t') = 0$ for all $t' < 0$. Noncausal systems serve as theoretical tools only. The obvious modifications cover systems with complex valued inputs and outputs.

2.4.3 Circular Convolution

For periodic signals the improper integral in the definition of the convolution is not only unnecessary but it also introduces problems with convergence. The circular convolution circumvents these problems by integrating over a single period only. Let $x: \mathbb{R} \to \mathbb{C}$ and $y: \mathbb{R} \to \mathbb{C}$ be complex-valued periodic signals with period p. The signal

$$(x \circledast y)(t) = \frac{1}{p} \int\limits_{0}^{p} x(\tau) y(t - \tau) \, d\tau$$

is called the circular convolution of x and y.

2.5 The Four Fourier Transforms

Under very general conditions we can describe a signal s as the weighted sum of phase-shifted sinusoids using the Fourier transform appropriate for the type of signal at hand. This representation is called the spectrum of s. The distribution of the amplitude with respect to frequency is the amplitude spectrum of s; the distribution of the phase with respect to frequency is the phase spectrum of s. Distinguishing between discrete-time and continuous-time signals, and between periodic and aperiodic signals gives rise to four possibilities. Discrete-time signals have periodic Fourier transforms, continuous-time signals have aperiodic Fourier transforms. Periodic signals have discrete Fourier transforms and aperiodic signals have continuous Fourier transforms.

2.5.1 Periodic Continuous-Time Signals—The Fourier Series

Let $s: \mathbb{R} \to \mathbb{C}, t \mapsto s(t)$ be a continuous complex-valued periodic signal with period $p \in \mathbb{R}$, that is $s(t + p) = s(t)$ for every real number t. The complex Fourier coefficients of s constitute the sequence $S: \mathbb{Z} \to \mathbb{C}$,

$$n \mapsto S_n = \frac{1}{p} \int_{-\frac{p}{2}}^{\frac{p}{2}} s(t) e^{-in\omega_0 t} \, dt. \tag{2.7}$$

where $\omega_0 = \frac{2\pi}{p}$. As the integrand is periodic with period p the bounds can be chosen freely, as long as the integration domain spans exactly one period. The signal s can be represented by its Fourier series

$$s(t) \sim \sum_{n=-\infty}^{\infty} S_n e^{in\omega_0 t}, \tag{2.8}$$

regardless of convergence. If the signal s is piecewise continuous in the interval $-\frac{p}{2} \leq t \leq \frac{p}{2}$, i.e. s has only finitely many finite jumps in the interval, and s is of bounded variation in the interval, then equality holds in (2.8) for all t where s is continuous. At a point t_0, where s is discontinuous, the sum in (2.8) is the average of the left-hand and the right-hand limit of s at t_0. The conditions above are called the Dirichlet conditions. The variation $V_a^b(s)$ of s in the interval $[a, b] = \{x \in \mathbb{R}: a \leq x \leq b\}$ is defined as

$$V_a^b(s) = \sup_{\substack{m \in \mathbb{N} \\ t_1 < \cdots < t_m \in [a,b]}} \sum_{i=1}^{m-1} |s(t_{i+1}) - s(t_i)|.$$

The variation of $f(t) = \sin \frac{1}{t}$ for $t \neq 0$ and $f(0) = 0$ for example is not bounded in the interval $[0, 2\pi]$, the variation of $g(t) = t^2 \sin \frac{1}{t}$ in the same interval, however, is bounded (Table 2.1).

Table 2.1 Some continuous-time periodic signals and their Fourier series coefficients

$s(t)$	$S_n = (\mathcal{F}(s))_n$	Comment
$x(t)h(t)$	$(X * H)_n$	
$(x \circledast h)(t)$	$X_n H_n$	
$x(t)$ where x is real-valued	X_n where $X_n = X^*_{-n}$	
$x(t)$ where $x(t) = x(-t)^*$	X_n where X is real-valued	
$s(t) = \begin{cases} 1 & \text{if } \frac{-p}{4} \le t < \frac{p}{4} \\ -1 & \text{if } \frac{p}{4} \le t < \frac{3p}{4} \\ s(t-p) & \text{otherwise} \end{cases}$	$S_n = \begin{cases} 0 & \text{if } n \text{ even} \\ 2\frac{(-1)^{n/2+1}}{n\pi} & \text{if } n \text{ odd} \end{cases}$	Square wave
$s(t) = \begin{cases} t + \frac{p}{4} & \text{if } \frac{-p}{2} \le t < 0 \\ -t + \frac{p}{4} & \text{if } 0 \le t < \frac{p}{2} \\ s(t-p) & \text{otherwise} \end{cases}$	$S_n = \begin{cases} 0 & \text{if } n \text{ even} \\ \frac{p}{\pi^2 n^2} & \text{if } n \text{ odd} \end{cases}$	Triangle
$s(t) = \begin{cases} t & \text{if } \frac{-p}{2} \le t < \frac{p}{2} \\ s(t-p) & \text{otherwise} \end{cases}$	$ip\dfrac{(-1)^n}{2n\pi}$	Sawtooth

The periodic signals with period p, $x: \mathbb{R} \to \mathbb{C}$ and $h: \mathbb{R} \to \mathbb{C}$, are arbitrary. Their Fourier coefficients constitute the sequences $X: \mathbb{Z} \to \mathbb{C}$ and $H: \mathbb{Z} \to \mathbb{C}$ respectively

For a periodic real-valued signal $s: \mathbb{R} \to \mathbb{R}$ the following notation often is more convenient. Using $\cos\phi = \frac{1}{2}(e^{i\phi} + e^{-i\phi})$ and $\sin\phi = \frac{i}{2}(e^{-i\phi} - e^{i\phi})$ we can rewrite (2.8) as

$$s(t) \sim a_0 + \sum_{n=1}^{\infty}(a_n \cos n\omega_0 t + b_n \sin n\omega_0 t), \qquad (2.9)$$

where

$$a_0 = \frac{1}{p}\int_{-\frac{p}{2}}^{\frac{p}{2}} s(t)\,dt,$$

$$a_n = \frac{2}{p}\int_{-\frac{p}{2}}^{\frac{p}{2}} s(t)\cos n\omega_0 t\,dt,$$

$$b_n = \frac{2}{p}\int_{-\frac{p}{2}}^{\frac{p}{2}} s(t)\sin n\omega_0 t\,dt.$$

The frequency $\frac{\omega_0}{2\pi}$ is called the fundamental frequency or the first harmonic, while the frequency $\frac{n\omega_0}{2\pi}$ is called the frequency of the $(n-1)$st overtone, or the nth harmonic. We can rewrite (2.9) further into a form containing amplitudes and phases,

$$s(t) \sim A_0 + \sum_{n=1}^{\infty} A_n \cos{(n\omega_0 t - \phi_n)},$$

where the offset A_0 is $A_0 = a_0$, the amplitude A_n of the nth harmonic is $A_n = \sqrt{a_n^2 + b_n^2}$ and the phase ϕ_n of the nth harmonic is $\phi_n = \arctan{\frac{b_n}{a_n}}$.

2.5.2 Periodic Discrete-Time Signals—the Discrete Fourier Transform

Let $s\colon \mathbb{Z} \to \mathbb{C}, n \mapsto s_n$ be a discrete-time complex-valued periodic signal with period p. The coefficients of the discrete Fourier transform $\mathcal{F}(s)$ of s constitute the periodic sequence $S\colon \mathbb{Z} \to \mathbb{C}$,

$$k \mapsto S_k = \sum_{n=0}^{p-1} s_n e^{-in\omega_0 k}, \qquad (2.10)$$

with period p, where $\omega_0 = \frac{2\pi}{p}$. The signal s can be represented in terms of the Fourier transform coefficients by

$$s_n = \frac{1}{p} \sum_{k=0}^{p-1} S_k e^{ik\omega_0 n}. \qquad (2.11)$$

As the summands in both sums are periodic with period p, the bounds in both sums can be chosen freely, as long as the summations cover exactly one period. For a real-valued signal s we can rewrite this equation for odd periods p as

$$s_n = \frac{1}{p} S_0 + \frac{2}{p} \sum_{k=1}^{(p-1)/2} \Re(S_k) \cos{k\omega_0 n} - \Im(S_k) \sin{k\omega_0 n}, \qquad (2.12)$$

using $e^{i\phi} = \cos\phi + i\sin\phi$ and $S_{-k} = S_k^*$.

When measuring a real-valued signal s' it is often a perturbed version of a true signal s. The perturbations, called noise, can be added for example by the inevitable imperfections of the measurement apparatus used for capturing s'. Let $s'_n = s_n + x_n$ where the x_n are normal distributed independent random variables all with zero mean and variance σ^2. The real part $\Re(S'_k)$ and the imaginary part $\Im(S'_k)$ of each Fourier transform coefficient S'_k then are random variables with mean $\Re(S_k)$ and

Table 2.2 Some discrete-time periodic signals and their discrete Fourier transforms

s_n	$S_k = (\mathcal{F}(s))_k$
$x_n h_n$	$\dfrac{1}{p} \displaystyle\sum_{k'=0}^{p-1} X_{k'} H_{k-k'} = \dfrac{1}{p}(X \circledast H)_k$
$(x \circledast h)_n$	$X_k H_k$
x_n where x is real-valued	X_k where $X_k = X^*_{-k}$
x_n where $x_n = x^*_{-n}$	X_k where X is real-valued
$e^{i2\pi f n}$ where $f = m/p$, $m \in \mathbb{Z}$, $m \neq 0$	$S_k = \begin{cases} 1 & \text{if } k \in \{m + ip : i \in \mathbb{Z}\} \\ 0 & \text{otherwise} \end{cases}$

The discrete-time periodic signals with period p, $x : \mathbb{Z} \to \mathbb{C}$ and $h : \mathbb{Z} \to \mathbb{C}$, are arbitrary, $X : \mathbb{Z} \to \mathbb{C}$ and $H : \mathbb{Z} \to \mathbb{C}$ are their discrete Fourier transforms

$\Im(S_k)$ respectively, and variance $\frac{p}{2}\sigma^2$, (Schoukens and Renneboog 1986). When reconstructing the signal s'_n according to (2.12) from the Fourier transform coefficients S'_k the real and imaginary parts of the Fourier transform coefficients are scaled with $2/p$. These scaled values then have variance $\frac{2}{p}\sigma^2$. The ratio $\frac{p}{2}$ of the variance of the s'_n and the variance of the real and imaginary parts of the scaled S'_k is called the processing gain of the discrete Fourier transform (Table 2.2).

2.5.3 Discrete-Time Signals—The Discrete-Time Fourier Transform

Let $s : \mathbb{Z} \to \mathbb{C}$, $n \mapsto s_n$ be a discrete complex-valued signal. The discrete-time Fourier transform $\mathcal{F}(s)$ of s is the function $S : \mathbb{R} \to \mathbb{C}$,

$$\omega \mapsto S(\omega) = \sum_{n=-\infty}^{\infty} s_n e^{-i\omega n}. \tag{2.13}$$

If s is absolutely summable, i.e., $\sum_{n=-\infty}^{\infty} |s_n|$ converges to a finite value, then the discrete-time Fourier transform exists and is finite for all ω. The transform is continuous and periodic with period 2π. The discrete signal s can be recovered by the inverse transform $\mathcal{F}^{-1}(S)$,

$$s_n = \frac{1}{2\pi} \int_{-\pi}^{\pi} S(\omega)e^{i\omega n}\, d\omega. \tag{2.14}$$

Table 2.3 Some discrete-time signals and their discrete-time Fourier transforms

s_n	$S(\omega) = (\mathcal{F}(s))(\omega)$
$x_n h_n$	$\dfrac{1}{2\pi} \displaystyle\int\limits_{0}^{2\pi} X(\Omega) H(\omega - \Omega)\, d\Omega = (X \circledast H)(\omega)$
$(x * h)_n$	$X(\omega) H(\omega)$
x_n where x is real-valued	$X(\omega)$ where $X(\omega) = X(-\omega)^*$
x_n where $x_n = x^*_{-n}$	$X(\omega)$ where X is real-valued
$\delta_{n-n'}$	$e^{-i\omega n'}$
$e^{i\omega_0 n}$	$2\pi \delta(\omega - \omega_0)$

The discrete-time signals $x: \mathbb{Z} \to \mathbb{C}$ and $h: \mathbb{Z} \to \mathbb{C}$ are arbitrary, $X: \mathbb{R} \to \mathbb{C}$ and $H: \mathbb{R} \to \mathbb{C}$ are their discrete-time Fourier transforms

As the integrand is periodic with period 2π, one can choose the bounds freely, as long as the integration spans one period.

A finite discrete signal $s: \{0, \ldots, p - 1\} \to \mathbb{C}$ can be extended to the signal $s': \mathbb{Z} \to \mathbb{C}$ by setting $s'_n = s_n$ for $n \in \{0, \ldots, p - 1\}$ and $s'_n = 0$ for $n \notin \{0, \ldots, p - 1\}$, or to the signal $s'': \mathbb{Z} \to \mathbb{C}$ by extending it periodically, i.e. $s''_n = s_{n \bmod p}$. Then the discrete Fourier transform coefficients S''_k of s'' are samples of the discrete-time Fourier transform S' of s',

$$S''_k = S'\left(\frac{2\pi}{p} k\right). \tag{2.15}$$

2.5.3.1 Frequency Response of Discrete-Time Linear Time-Invariant Systems

When we subject the discrete-time linear time-invariant system S to the input signal $x: \mathbb{Z} \to \mathbb{C}, n \mapsto x_n = e^{i\omega n}$ for $\omega \in \mathbb{R}$, then according to (2.3) the answer is $h * x$ where h is the impulse response of S. Expanding the convolution we get

$$(x * h)_n = \sum_{n'=-\infty}^{\infty} h_{n'} e^{i\omega(n-n')} = e^{i\omega n} \sum_{n'=-\infty}^{\infty} h_{n'} e^{-i\omega n'}.$$

Provided the sum exists we can rewrite this as $(x * h)_n = H(\omega) e^{i\omega n}$ where H, the frequency response of the system S, is the discrete-time Fourier transform of h (Table 2.3).

The response to $x': \mathbb{Z} \to \mathbb{R}$,

$$n \mapsto \cos \omega n = \frac{1}{2}\left(e^{i\omega n} + e^{-i\omega n}\right)$$

is $y': \mathbb{Z} \to \mathbb{C}$,

$$n \mapsto y'_n = \frac{1}{2} \left(H(\omega)e^{i\omega n} + H(-\omega)e^{-i\omega n} \right).$$

If the impulse response h is real-valued then $H(-\omega) = H(\omega)^*$ and

$$y'_n = |H(\omega)| \cos(\omega n + \angle H(\omega)),$$

where $\angle H(\omega)$ is the phase response of the system S at the angular frequency ω. The response of a discrete-time linear time-invariant system with real-valued impulse response to a sinusoidal input is a sinusoidal output with the same frequency as the input but with shifted phase.

Let us consider the discrete-time system S described by the equation

$$y_n = x_n + \alpha y_{n-1} \tag{2.16}$$

for $0 < \alpha < 1$. Let $x_n = 0$ for $n < n'$ and some $n' \in \mathbb{N}$. Provided the outputs y_n are all zero for $n < n'$ the system is linear and time invariant. Its impulse response is $h: \mathbb{Z} \to \mathbb{R}$,

$$n \mapsto S(\delta)_n = h_n = \begin{cases} 0 & \text{for } n < 0 \\ \alpha^n & \text{for } n \geq 0. \end{cases}$$

Therefore, the response to an arbitrary signal $x: \mathbb{Z} \to \mathbb{R}, n \mapsto x_n$ is $y: \mathbb{Z} \to \mathbb{R}$

$$n \mapsto y_n = \sum_{k=0}^{\infty} x_{n-k} \alpha^k.$$

The frequency response is $H: \mathbb{C} \to \mathbb{C}$,

$$\omega \mapsto H(\omega) = \frac{e^{i\omega}}{e^{i\omega} - \alpha}.$$

For plotting the frequency response for one period, Fig. 2.1, we compute gain and phase of the frequency response, $H(\omega) = |H(\omega)|e^{i\angle H(\omega)}$ and plot the gain $|H(\omega)|$ and the phase $\angle H(\omega)$ separately. We plot the gain on a logarithmic scale. In electrical engineering the decibel, [dB], denotes a ratio of powers. A ratio of 100 corresponds to 10dB, a ratio of 2 to about 3dB. More precisely, the ratio of two powers P_1 and P_2 is

$$10 \log_{10} \frac{P_1}{P_2} \text{ dB}.$$

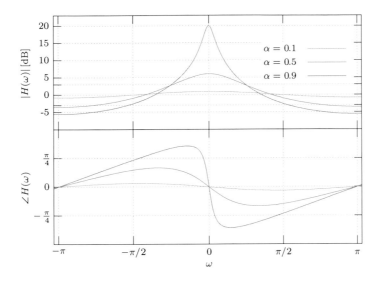

Fig. 2.1 Gain and phase of the discrete-time system described by Eq. (2.16)

Given two sinusoidal signals with voltage-amplitudes a_1 and a_2 we consider the ratio of the powers P_1 and P_2. The power P_1 is the power the first signal makes a resistor dissipate, while P_2 is the power the second signal makes the same resistor dissipate. A resistor is an electrical component with two terminals dissipating power P, which is proportional to V^2 when the voltage V is applied across it, see Sect. 3.3. Therefore, the power-ratio of the two sinusoidal signals is

$$10\log_{10}\frac{P_1}{P_2}\,\mathrm{dB} = 10\log_{10}\frac{a_1^2}{a_2^2}\,\mathrm{dB} = 20\log_{10}\frac{a_1}{a_2}\,\mathrm{dB}.$$

2.5.4 Continuous-Time Signals—The Continuous-Time Fourier Transform

Let $s\colon \mathbb{R} \to \mathbb{C}$, $t \mapsto s(t)$ be a continuous aperiodic complex-valued signal. The Fourier transform $\mathcal{F}(s)$ of s is the function $S\colon \mathbb{R} \to \mathbb{C}$,

$$\omega \mapsto S(\omega) = \lim_{T\to\infty} \int_{-T/2}^{T/2} s(t)\mathrm{e}^{-i\omega t}\,\mathrm{d}t. \tag{2.17}$$

The transform is continuous and aperiodic. The signal s can be recovered by the reverse transform $\mathcal{F}^{-1}(S)$

$$s(t) \sim \frac{1}{2\pi} \lim_{\Omega \to \infty} \int_{-\Omega/2}^{\Omega/2} S(\omega)e^{i\omega t}\, d\omega. \tag{2.18}$$

If the signal s is absolutely integrable, i. e., the integral $\int_{-\infty}^{\infty} |s(t)|\, dt$ converges to a finite value, and s is piecewise continuous in every finite interval, and s is of bounded variation in every finite interval, then the equality in (2.18) holds for all t where s is continuous. At a point t_0, where s is discontinuous, the integral in (2.18) is the average of the left-hand and the right-hand limit of s at t_0.

If the signal s is periodic with period p having Fourier series coefficients S_n, then the continuous-time Fourier transform $S(\omega)$ of s is

$$S(\omega) = 2\pi \sum_{n=-\infty}^{\infty} S_n \delta\left(\omega - \frac{2\pi}{p}n\right). \tag{2.19}$$

Let $s' : \mathbb{Z} \to \mathbb{C}$ be a sampled version of the signal s, $n \mapsto s'_n = s(\tau n)$, where τ is the sample period. The discrete-time Fourier transform S' of s' is related to the Fourier transform S of s by

$$S'(\omega) = \frac{1}{\tau} \sum_{k=-\infty}^{\infty} S\left(\frac{\omega - 2\pi k}{\tau}\right). \tag{2.20}$$

If the signal s is band-limited such that $S(\omega) = 0$ for $\omega \le \frac{-\pi}{\tau}$ or $\omega \ge \frac{\pi}{\tau}$ the sum reduces to

$$S'(\omega) = \frac{1}{\tau} S\left(\frac{\omega}{\tau}\right).$$

2.5.4.1 Frequency Response of Continuous-Time Linear Time-Invariant Systems

When we subject the continuous-time linear time-invariant system S to the input signal $x : \mathbb{R} \to \mathbb{C}, t \mapsto x(t) = e^{i\omega t}$ for $\omega \in \mathbb{R}$, then according to (2.6) the answer is $h * x$ where h is the impulse response of S. Expanding the convolution we get $(x * h)(t) = \int_{-\infty}^{\infty} h(\tau)e^{i\omega(t-\tau)}d\tau = e^{i\omega t} \int_{-\infty}^{\infty} h(\tau)e^{-i\omega\tau}d\tau$. Provided the integrals exist we can rewrite this as $(x * h)_n = H(\omega)e^{i\omega t}$ where H, the frequency response of the system S, is the continuous-time Fourier transform of h (Table 2.4).

The response to $x' : \mathbb{R} \to \mathbb{R}, t \mapsto \cos \omega t = \frac{1}{2}\left(e^{i\omega t} + e^{-i\omega t}\right)$ is $y' : \mathbb{R} \to \mathbb{C}$, $t \mapsto y'(t) = \frac{1}{2}\left(H(\omega)e^{i\omega t} + H(-\omega)e^{-i\omega t}\right)$. If the impulse response h is real-valued then $H(-\omega) = H(\omega)^*$ and

Table 2.4 Some continuous-time signals and their continuous-time Fourier transforms

$s(t)$	$S(\omega) = (\mathcal{F}(s))(\omega)$	Comment
$x(t)h(t)$	$\dfrac{1}{2\pi}\displaystyle\int_{-\infty}^{\infty} X(\Omega)H(\omega-\Omega)\,d\Omega$ $= \dfrac{1}{2\pi}(X * H)(\omega)$	
$(x * h)(t)$	$X(\omega)H(\omega)$	
$\dfrac{dx(t)}{dt}$	$i\omega X(\omega)$	
$x(t)$ where x is real-valued	$X(\omega)$ where $X(\omega) = X(-\omega)^*$	
$x(t)$ where $x(t) = x(-t)^*$	$X(\omega)$ where X is real-valued	
$\delta(t-T)$	$e^{-i\omega T}$	
$e^{i\Omega t}$	$2\pi\delta(\omega-\Omega)$	
$x(t-T)$	$e^{-i\omega T}X(\omega)$	
$e^{i\Omega t}x(t)$	$X(\omega-\Omega)$	
$s(t) = \begin{cases} 1 & \text{if } -p \le t \le p \\ 0 & \text{otherwise} \end{cases}$	$\dfrac{2\sin\omega p}{\omega}$	Boxcar
$s(t) = \begin{cases} t+p & \text{if } -p \le t < 0 \\ -t+p & \text{if } 0 \le t < p \\ 0 & \text{otherwise} \end{cases}$	$\dfrac{2-2\cos\omega p}{\omega^2}$	Triangle

The signals $x: \mathbb{R} \to \mathbb{C}$ and $h: \mathbb{R} \to \mathbb{C}$ are arbitrary, $X: \mathbb{R} \to \mathbb{C}$ and $H: \mathbb{R} \to \mathbb{C}$ are their Fourier transforms. The numbers $T \in \mathbb{R}$ and $\Omega \in \mathbb{R}$ are arbitrary, the number $p \in \mathbb{R}$, $p > 0$ is arbitrary

$$y'(t) = |H(\omega)|\cos(\omega t + \angle H(\omega)),$$

where $\angle H(\omega)$ is the phase response of the system S at the angular frequency ω. The response of a continuous-time linear time-invariant system with real-valued impulse response to a sinusoidal input is a sinusoidal output with the same frequency as the input but with shifted phase.

2.6 Noise

Any common signal processing apparatus introduces errors to the signals they process. A portion of these errors may be attributable to inadequacies of the apparatus; the aleatory behavior of nature at the microscopic level, however, introduces the rest. We call these errors noise. Any practical signal s will be degraded by noise. Most often the observed signal s will be the sum of the noiseless signal s' and some added noise n, $s(t) = s'(t) + n(t)$. The signal to noise ratio $\text{SNR}(s)$ then is the ratio of the average power of s' to the average power of n,

$$\text{SNR}(s) = \frac{P(s')}{P(n)}.$$

The signal to noise ratio is usually stated in decibel.

The spectral power density describes the frequency content of noise.[5] For each frequency f the spectral power density of a signal is the power the signal contains in a small frequency band around f. Band-limited white noise has a power spectral density being constant up to some band limit above which it drops to zero. Many electronic devices introduce noise with a spectral power density proportional to $1/f$ into the signals they process. These devices negatively affect low-frequency signals most, forcing us to make precision measurements at frequencies well away from 0Hz.

2.7 The Z-Transform

Let $x: \mathbb{Z} \to \mathbb{C}$, $n \mapsto x_n$ be a discrete-time signal. The Z-transform of x, $\hat{X}: \text{roc}(x) \to \mathbb{C}$ is defined as

$$z \mapsto \hat{X}(z) = (\mathcal{Z}(x))\,(z) = \sum_{n=-\infty}^{\infty} z^{-n} x_n. \tag{2.21}$$

The set $\text{roc}(x) \subseteq \mathbb{C}$ is the region of convergence of the Z-transform of x. When we substitute $z = re^{i\omega}$ in (2.21) we recognize that the Z-transform is actually the discrete-time Fourier transform of the signal $x': \mathbb{Z} \to \mathbb{C}$, $n \mapsto x_n' = r^{-n} x_n$. The sum in (2.21) converges exactly for those $r \in \mathbb{R}$ for which the sum $\sum_{n=-\infty}^{\infty} |r^{-n} x_n|$ converges to a finite value.

If x is right-sided, that is $x_n = 0$ for all $n < N$ and some $N \in \mathbb{Z}$, the region of convergence is the whole complex plane with the exception of a disc around the origin, $\text{roc}(x) = \{z \in \mathbb{C}: |z| > r'\}$ for some real r'; furthermore, if N is nonnegative the Z-transform of x converges also in the limit for $z \to \infty$. If x is left-sided, that is $x_n = 0$ for all $n > N$ and some $N \in \mathbb{Z}$, the region of convergence is a disc around the origin; for $N \leq 0$ the region of convergence includes the origin, $\text{roc}(x) = \{z \in \mathbb{C}: |z| < r'\}$ for some real r'; for $N > 0$ the origin is excluded, $\text{roc}(x) = \{z \in \mathbb{C}: z \neq 0, |z| < r'\}$. For any other signal x the region of convergence of x has annular shape, $\text{roc}(x) = \{z \in \mathbb{C}: r_1 < |z| < r_2\}$ for some real r_1, r_2; the

---

[5] A precise definition of the spectral power density of a noise signal requires the machinery of stochastic processes.

Table 2.5 Some discrete-time signals and their Z-transforms

w_n	$\hat{W}(z) = (\mathcal{Z}(w))(z)$	$\mathrm{roc}(w)$
$ax_n + by_n$	$a\hat{X}(z) + b\hat{Y}(z)$	$\mathrm{roc}(w) \supseteq \mathrm{roc}(x) \cap \mathrm{roc}(y)$
x_{n-M}	$z^{-M}\hat{X}(z)$	$\mathrm{roc}(x)$
$(x * y)_n$	$\hat{X}(z)\hat{Y}(z)$	$\mathrm{roc}(w) \supseteq \mathrm{roc}(x) \cap \mathrm{roc}(y)$
x_n^*	$\hat{X}(z^*)^*$	$\mathrm{roc}(x)$
x_{-n}	$\hat{X}(z^{-1})$	$\{z : z^{-1} \in \mathrm{roc}(x)\}$
nx_n	$-z\dfrac{\mathrm{d}\hat{X}(z)}{\mathrm{d}z}$	$\mathrm{roc}(x)$
$a^{-n}x_n$	$\hat{X}(az)$	$\{z : az \in \mathrm{roc}(x)\}$
δ_{n-M}	z^{-M}	$\mathbb{C}$
$u_n = \begin{cases} 0 & \text{if } n < 0 \\ 1 & \text{if } n \geq 0 \end{cases}$	$\dfrac{z}{z-1}$	$\{z \in \mathbb{C} : |z| > 1\}$
$a^n u_n$	$\dfrac{z}{z-a}$	$\{z \in \mathbb{C} : |z| > |a|\}$
$a^n u_{-n}$	$\dfrac{a}{a-z}$	$\{z \in \mathbb{C} : |z| < |a|\}$

The signals $x : \mathbb{Z} \to \mathbb{C}$ and $y : \mathbb{Z} \to \mathbb{C}$ are arbitrary; $\hat{X} : \mathrm{roc}(x) \to \mathbb{C}$ and $\hat{Y} : \mathrm{roc}(y) \to \mathbb{C}$ are their Z-transforms. The numbers $M \in \mathbb{Z}$ and $a, b \in \mathbb{C}$ are arbitrary

region of convergence may also be empty in this case. The region of convergence is an integral part of the Z-transform; the Z-transform of some entirely different signals differ in the respective regions of convergence only (Table 2.5).

Provided the unit circle is part of the region of convergence of $\hat{X}$ the Z-transform evaluated on the unit circle yields the discrete-time Fourier transform of x,

$$X(\omega) = (\mathcal{F}(x))(\omega) = \hat{X}(e^{i\omega}).$$

2.7.1 Stability of Linear Time-Invariant Discrete-Time Systems

A linear time-invariant discrete-time system S is bounded-input bounded-output stable if and only if for all bounded inputs $x : \mathbb{Z} \to \mathbb{R}$ the output $y : \mathbb{Z} \to \mathbb{R}$, $n \mapsto y_n = (S(x))_n$ is also bounded. A signal $x : \mathbb{Z} \to \mathbb{C}$ is bounded if and only if $|x_n| < A$ for all $n \in \mathbb{Z}$ and for some $A \in \mathbb{R}$. A necessary and sufficient condition for stability is that the impulse response h of S is absolutely summable, that is $\sum_{n=-\infty}^{\infty} |h_n|$ converges to a finite value. Moreover, the system S is stable if and only if the unit circle is part of the region of convergence of the transfer function $\hat{H}$ of S, which is the Z-transform of the impulse response h of S.

The transfer function of many linear time-invariant discrete-time systems is the quotient of two polynomials, $\hat{H}(z) = \frac{A(z)}{B(z)}$. The zeros of the numerator A are called the zeros of $\hat{H}$; the zeros of the denominator B are the poles of $\hat{H}$. A causal linear time-invariant discrete-time system S is stable if and only if all poles of the transfer function $\hat{H}$ of S lie within the unit circle.

Using the Z-transform we can argue about the stability and the frequency response of a composition of discrete-time systems in which some of the constituents are instable.

2.8 The Two-Sided Laplace Transform

Let $x: \mathbb{R} \to \mathbb{C}, t \mapsto x(t)$ be a continuous-time signal. The two-sided Laplace transform[6] of x, $\hat{X}: \mathrm{roc}(x) \to \mathbb{C}$ is defined as

$$s \mapsto \hat{X}(s) = (\mathcal{L}(x))(s) = \int_{-\infty}^{\infty} x(t)\mathrm{e}^{-st}\mathrm{d}t. \qquad (2.22)$$

The set $\mathrm{roc}(x)$ is the region of convergence of the Laplace transform of x. When we substitute $s = \sigma + i\omega$ with $\sigma, \omega \in \mathbb{R}$ in (2.22) we recognize that the Laplace transform is actually the continuous Fourier transform of the signal $x': \mathbb{R} \to \mathbb{C}$, $t \mapsto x'(t) = x(t)\mathrm{e}^{-\sigma t}$. If the signal x' satisfies the Dirichlet conditions the integral in (2.22) converges. Assuming the signal x is piecewise continuous and of bounded variation in every finite interval—assumptions satisfied by the signals appearing in practice—then the integral in (2.22) converges provided $\int_{-\infty}^{\infty} |x(t)\mathrm{e}^{-\sigma t}|\mathrm{d}t$ converges to a finite value.

If x is right-sided, that is $x(t) = 0$ for all $t < T$ and some $T \in \mathbb{R}$, the region of convergence is $\mathrm{roc}(x) = \{s \in \mathbb{C}: \Re(s) > \sigma'\}$ for some real σ'. If x is left-sided, that is $x(t) = 0$ for all $t > T$ and some $T \in \mathbb{R}$, the region of convergence is $\mathrm{roc}(x) = \{s \in \mathbb{C}: \Re(s) < \sigma'\}$ for some real σ'. For any other signal x the region of convergence of x is $\mathrm{roc}(x) = \{s \in \mathbb{C}: \sigma_1 < \Re(s) < \sigma_2\}$ for some real σ_1, σ_2; the region of convergence may also be empty in this case. The region of convergence is an integral part of the Laplace transform; the Laplace transform of some entirely different signals differ in the respective regions of convergence only (Table 2.6).

Let $x: \mathbb{R} \to \mathbb{C}, t \mapsto x(t)$ be a signal with $x(t) = 0$ for $t < 0$; let the signal $X: \mathbb{C} \to \mathbb{C}$ be the Laplace transform of x. Let the imaginary axis be part of $\mathrm{roc}(x)$.

---

[6] In the one-sided Laplace transform integration ranges from 0 to ∞. The one-sided transform is useful for solving initial value problems involving linear differential equations. See also Appendix C.

Table 2.6 Some continuous-time signals and their Laplace transforms

$w(t)$	$\hat{W}(s) = (\mathcal{L}(w))(s)$	roc(w)
$ax(t) + by(t)$	$a\hat{X}(s) + b\hat{Y}(s)$	roc(w) $\supseteq$ roc(x) $\cap$ roc(y)
$x(t - \tau)$	$e^{-s\tau}\hat{X}(S)$	roc(x)
$(x * y)(t)$	$\hat{X}(s)\hat{Y}(s)$	roc(w) $\supseteq$ roc(x) $\cap$ roc(y)
$x^*(t)$	$\hat{X}(s^*)^*$	roc(x)
$x(ct)$	$\dfrac{\hat{X}(s/c)}{\lvert c \rvert}$	$\{s \in \mathbb{C}: s/c \in \text{roc}(x)\}$
$tx(t)$	$-\dfrac{d\hat{X}(s)}{ds}$	roc(x)
$e^{at}x(t)$	$\hat{X}(s - a)$	$\{s \in \mathbb{C}: s - a \in \text{roc}(x)\}$
$\displaystyle\int_{-\infty}^{t} x(\tau)d\tau$	$\dfrac{\hat{X}(s)}{s}$	roc(w) $\supseteq \{s \in \text{roc}(x): \Re(s) > 0\}$
$\dfrac{dx(t)}{dt}$	$s\hat{X}(s)$	roc(w) $\supseteq$ roc(x)
$\delta(t - \tau)$	$e^{-s\tau}$	$\mathbb{C}$
$u(t) = \begin{cases} 0 & \text{if } t < 0 \\ 1 & \text{if } t \geq 0 \end{cases}$	$\dfrac{1}{s}$	$\{s \in \mathbb{C}: \Re(s) > 0\}$
$e^{-at}u(t)$	$\dfrac{1}{s+a}$	$\{s \in \mathbb{C}: \Re(s) > -\Re(a)\}$
$-e^{-at}u(-t)$	$\dfrac{1}{s+a}$	$\{s \in \mathbb{C}: \Re(s) < -\Re(a)\}$

The signals $x: \mathbb{R} \to \mathbb{C}$ and $y: \mathbb{R} \to \mathbb{C}$ are arbitrary; $\hat{X}: \text{roc}(x) \to \mathbb{C}$ and $\hat{Y}: \text{roc}(y) \to \mathbb{C}$ are their Laplace transforms. The numbers $a, b \in \mathbb{C}$ and $c, \tau \in \mathbb{R}$ are arbitrary

The final value theorem then allows us to compute the limit of $f(t)$ for $t \to \infty$ from the Laplace transform X,

$$\lim_{t\to\infty} x(t) = \lim_{s\to 0} sX(s). \tag{2.23}$$

Provided the imaginary axis is part of the region of convergence of x evaluating the Laplace transform along the imaginary axis yields the continuous-time Fourier transform of x,

$$X(\omega) = (\mathcal{F}(x))(\omega) = \hat{X}(i\omega).$$

2.8.1 Stability of Linear Time-Invariant Continuous-Time Systems

A linear time-invariant continuous-time system S is bounded-input bounded-output stable if and only if for all bounded inputs $x\colon \mathbb{R} \to \mathbb{R}$ the output $y\colon \mathbb{R} \to \mathbb{R}$, $n \mapsto y(t) = (S(x))(t)$ is also bounded. A signal $x\colon \mathbb{R} \to \mathbb{R}$ is bounded if and only if $|x(t)| < A$ for all $t \in \mathbb{R}$ and for some $A \in \mathbb{R}$. A necessary and sufficient condition for stability is that the impulse response h of S is absolutely integrable, that $\int_{-\infty}^{\infty} |h(t)|\mathrm{d}t$ converges to a finite value.

The transfer function $\hat{H}$ of the system S is the Laplace transform of the impulse response of S, $\hat{H}\colon \mathbb{C} \to \mathbb{C}$,

$$s \mapsto \hat{H}(s) = \frac{\hat{Y}(s)}{\hat{X}(s)} = \mathcal{L}(h)(s),$$

where $\hat{X}$ is the Laplace transform of an input x and $\hat{Y}$ is the Laplace transform of the corresponding output y. The system S is stable provided the imaginary axis is part of the region of convergence of the transfer function $\hat{H}$ of S.

The transfer function of many linear time-invariant continuous-time systems is the quotient of two polynomials, $\hat{H}(z) = \frac{A(z)}{B(z)}$. The zeros of the numerator A are called the zeros of $\hat{H}$; the zeros of the denominator B are the poles of $\hat{H}$. A causal linear time-invariant continuous-time system S is stable provided all poles of the transfer function $\hat{H}$ of S have a negative real part.

Using the Laplace transform, we can argue about the stability and the frequency response of a composition of continuous-time systems in which some of the constituents are instable.

Let us consider the continuous-time system S described by the equation

$$y(t) = x(t) - \frac{1}{\beta} \int\limits_{-\infty}^{t} y(\tau)\mathrm{d}\tau, \tag{2.24}$$

for $\beta > 0$. Applying the Laplace transform to both sides and rearranging we get

$$\frac{\hat{Y}(s)}{\hat{X}(s)} = \hat{H}(s) = \frac{s\beta}{1 + s\beta}.$$

There are two possibilities for the region of convergence of the impulse response h of S, $\mathrm{roc}(h) = \{s \in \mathbb{C}\colon \Re(s) < -1/\beta\}$ describing a noncausal system and $\mathrm{roc}(h) = \{s \in \mathbb{C}\colon \Re(s) > -1/\beta\}$ describing a stable causal system. The frequency response of S for the second choice is

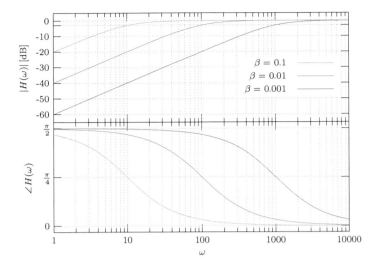

Fig. 2.2 Bode plot of the continuous-time system described by Eq. (2.24)

$$H(\omega) = \hat{H}(i\omega) = \frac{\omega\beta}{\omega\beta - i}.$$

We plot the magnitude and the phase of the frequency response separately, Fig. 2.2. For the magnitude we use the decibel scale. For the angular frequency ω we choose a logarithmic scale. Inspecting this so-called Bode plot we recognize the system S as a highpass filter.

2.9 Differential Equations, State-Space Models

The system of n linear differential equations with constant coefficients

$$\frac{dv_1(t)}{dt} = a_{1,1}v_1(t) + a_{1,2}v_2(t) + \cdots + a_{1,n}v_n(t) + b_1x(t)$$

$$\frac{dv_2(t)}{dt} = a_{2,1}v_1(t) + a_{2,2}v_2(t) + \cdots + a_{2,n}v_n(t) + b_2x(t)$$

$$\vdots$$

$$\frac{dv_n(t)}{dt} = a_{n,1}v_1(t) + a_{n,2}v_2(t) + \cdots + a_{n,n}v_n(t) + b_nx(t)$$

$$y(t) = c_1v_1(t) + c_2v_2(t) + \cdots + c_nv_n(t) + dx(t),$$

where $a_{i,j} \in \mathbb{R}$, $b_i \in \mathbb{R}$, $c_i \in \mathbb{R}$ and $d \in \mathbb{R}$ for $1 \leq i, j \leq n$, constitute the state-space model of a linear time-invariant continuous-time system S with one input $x(t)$ and one output $y(t)$. The signals $v_1(t), \ldots, v_n(t)$ are the states. For simulating the system S with tools like the Simulink® system we may either use the state space model directly or translate it into a block diagram. In order to construct such a block diagram we introduce for each state $v_i(t)$ in the state-space model an integrator into the block diagram. The output of the integrator produces the state $v_i(t)$, while the input of the integrator is wired to a block diagram computing the right-hand side of the equation for $\frac{dv_i(t)}{dt}$ in the state-space model. The outputs of the integrators are fed back into the right-hand sides. Another block diagram computes the output from the input and the states. For an example see Fig. 2.6.

The state space representation is not limited to linear systems. As long as we manage to transform the system of differential equations into equations defining the n states of the form

$$\frac{dv_i(t)}{dt} = f_i(v_1(t), v_2(t), \ldots, v_n(t), x(t))$$

and an equation defining the output of the form

$$y(t) = g(v_1(t), v_2(t), \ldots, v_n(t), x(t))$$

we have a state space model suitable for simulation.

2.9.1 The Harmonic Oscillator

The apparatus of differential equations was invented for modeling physical systems. These systems can be mechanical, electrical, optical or any other domain imaginable. Let us, for example, consider a body with mass m suspended from the ceiling via a spring, Fig. 2.3. We assume the spring has no mass and that the spring follows Hook's law. More precisely, we assume that the force $F_S(t)$ the spring exerts on the body is proportional to the displacement $y(t)$ of the body from rest, $F_S(t) = -ky(t)$ where k is the stiffness of the spring. Measurements show that Hook's law describes the action of a spring made from steel reasonably well. The body is attached to a damper, which is fixed to the floor. We assume the damper exerts a force on the body proportional to the body's velocity, $F_D(t) = -c\frac{dy(t)}{dt}$. The constant of proportionality c is called

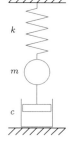

Fig. 2.3 A body is suspended from the ceiling with a spring. The body is connected to a damper, which is fixed to the floor

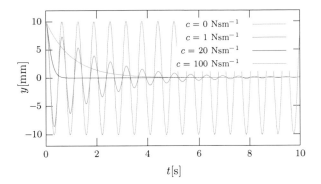

Fig. 2.4 Behavior of a mass-spring-damper system for different damping coefficients. The body has a mass of 1 kg, the spring a stiffness of 100 Nm^{-1}. The damping coefficient $c = 20$ Nsm^{-1} results in the system being critically damped

the damping constant. No external forces shall act on our contraption. According to Newton's second law of motion, the sum of these two forces have to balance the force F_m required for accelerating the body, $F_m(t) = m\frac{d^2 y(t)}{dt^2}$. Equating $F_m(t) = F_S(t) + F_D(t)$ and rearranging gives us the second order differential equation

$$m\frac{d^2 y(t)}{dt^2} + c\frac{dy(t)}{dt} + ky(t) = 0, \tag{2.25}$$

the equation of a harmonic oscillation.

Depending on the magnitude of the damping constant c, we can identify four different types of solutions for (2.25), Fig. 2.4. Let $\alpha = \frac{c}{2m}$. If $c = 0$ the general solution has the form

$$y(t) = A \cos \omega_0 t + B \sin \omega_0 t,$$

where $\omega_0 = \sqrt{\frac{k}{m}}$, which describes an undamped harmonic oscillation. For $c^2 < 4mk$ the general solution is

$$y(t) = e^{-\alpha t} \left(A \cos \omega' t + B \sin \omega' t \right),$$

where $\omega' = \sqrt{\frac{k}{m} - \frac{c^2}{4m^2}}$, which describes a damped harmonic oscillation. For $c^2 = 4mk$ the general solution is

$$y(t) = (A + Bt)e^{-\alpha t},$$

which describes a critically damped system. A critically damped system will return to rest without oscillation as fast as possible. At last, for $c^2 > 4mk$ the general solution is

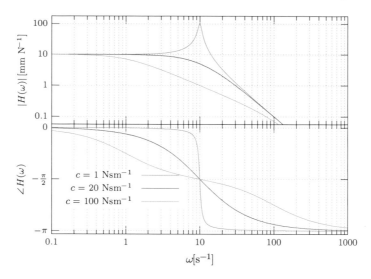

Fig. 2.5 Frequency response of the driven harmonic oscillator for different damping coefficients. The damping coefficient $c = 1$ Nsm^{-1} results in a peak in the amplitude response indicating resonance at that frequency

$$y(t) = Ae^{-(\alpha-\beta)t} + Be^{-(\alpha+\beta)t},$$

where $\beta = \sqrt{\frac{c^2}{4m^2} - \frac{k}{m}}$, which describes an overdamped system.

In order to handle an external driving force $x(t)$ acting on the body we have to change (2.25) into

$$m\frac{d^2 y(t)}{dt^2} + c\frac{dy(t)}{dt} + ky(t) = x(t), \qquad (2.26)$$

the equation of a driven harmonic oscillator. For analyzing the behavior of this system, we compute the system's transfer function $\hat{H}(s)$ by applying the Laplace transform to both sides of (2.26) and rearranging,

$$\hat{H}(s) = \frac{\hat{Y}(s)}{\hat{X}(s)} = \frac{1}{ms^2 + cs + k}.$$

The frequency response $H(\omega) = \hat{H}(i\omega)$, Fig. 2.5, shows a peak of the amplitude at

$$\omega_r = \sqrt{\frac{k}{m} - \frac{c^2}{2m^2}}$$

for $c^2 < 2mk$. When the external force excites the system close to that frequency, the amplitude of the response will be much larger than for external forces at other frequencies. This behavior is called resonance; $f_r = \frac{\omega_r}{2\pi}$ is the resonant frequency

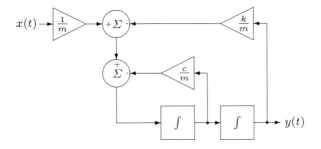

Fig. 2.6 Block diagram for Eq. (2.26)

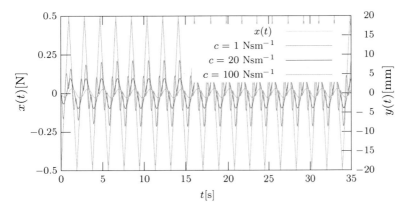

Fig. 2.7 Numerical simulation of the mass-spring-damper system. The underdamped system shows resonance at the third harmonic of the driving force, while the critically damped and the overdamped system show no such reaction

of the driven mass-spring-damper system. Unexpected resonance can be disastrous; the large excursions associated with it can overload structures.

Alternatively, we can bring (2.26) into state-space form,

$$\frac{dv_1(t)}{dt} = v_2(t)$$

$$\frac{dv_2(t)}{dt} = -\frac{c}{m}v_2(t) - \frac{k}{m}v_1(t) + \frac{1}{m}x(t)$$

$$y(t) = v_1(t),$$

and transform it into a block diagram for numerical simulation, Fig. 2.6. When we drive the system with a periodic, triangular force with about one third of the frequency f_r, then the content of the driving force at the third harmonic drives the underdamped system into resonance, Fig. 2.7. The driving force sets in at time 0, while it was zero before. The critically damped system and the overdamped system show no recognizable reaction to the excitation at the third harmonic. The simulations in Fig. 2.7 start out with the systems at rest. Each of the three systems takes some time for reaching a steady response. The output of a linear time-invariant system S to a sinusoidal input $x(t)$, which sets at time 0, $x(t) = u(t)a\cos(\omega t + \phi)$, is the sum of the so-called transient response, which for a stable system S tapers off after some time, and the steady state response. The steady state response has the same frequency as the input. A frequency response-based analysis of the system S provides us with the steady state information only.

2.9.2 The Pendulum

An ideal pendulum, Fig. 2.8, consists of a massless stiff rod of length L and a point-shaped body with mass m. One end of the rod attaches to the mass, while the other connects to a joint. The joint allows the rod together with the body to rotate freely on a horizontal axis. Gravity pulls with a force of mg at the mass. When the rod is at an angle $\phi(t)$ toward the vertical the force $F_g = -mg\sin\phi(t)$ tries to restore the pendulum to the vertical position. The constant $g = 9.81\,\mathrm{ms}^{-2}$ is the Earth's gravitational pull. The force F_g has to balance the force $F_m = mL\frac{d^2\phi(t)}{dt^2}$ for the tangential acceleration of the pendulum, $F_m = F_g$. Rearranging yields

$$mL\frac{d^2\phi(t)}{dt^2} + mg\sin(\phi(t)) = 0.$$

This differential equation is nonlinear. We introduce a damping force F_D, which is proportional to the velocity of the mass, $F_D = -Lc\frac{d\phi(t)}{dt}$, then $F_m = F_D + F_g$ and

$$\frac{d^2\phi(t)}{dt^2} + \frac{c}{m}\frac{d\phi(t)}{dt} + \frac{g}{L}\sin(\phi(t)) = 0.$$

Fig. 2.8 A body with mass m connected to a rod, a pendulum, is allowed to pivot on a horizontal axle. Whenever the pendulum deviates from the vertical the Earth's gravitation provides a restoring force

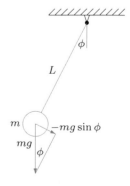

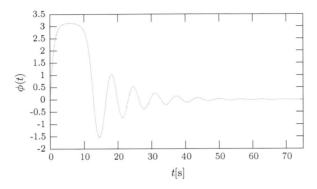

Fig. 2.9 Behavior of a pendulum starting with an initial angular velocity. The length of the rod is $L = 10\,\mathrm{m}$, the mass is $1\,\mathrm{kg}$ and the damping constant is $c = 0.2\,\mathrm{Nsm^{-1}}$

We can convert this equation into state-space form and simulate the pendulum's behavior numerically, Fig. 2.9. Note that the time for one oscillation depends on the amplitude of the oscillation. The nonlinear nature of the driven pendulum puts its analysis beyond the means of the mathematical tools we discuss.

2.10 Feedback Control

Often we want the output $y(t)$ of a causal system S_1 to track a reference signal $r(t)$. More precisely, we want to find an input $u(t)$ for the system S_1 such that $(S_1(u))(t)$ is as close to $r(t)$ as possible. Control engineers study this problem extensively. The system S_1, called the plant, consists of an actuator for converting the input $u(t)$ into some physical quantity—such as the position of a valve or the speed of a motor—and the process producing the output.

Provided we know the plant S_1 exactly we may try to find a causal system S_2, a controller, such that the series composition of the controller S_2 with the plant S_1 tracks the reference $r(t)$ reasonably well, in other words that $(S_1(S_2(r)))(t)$ is reasonably close to $r(t)$. The transfer function of this series composition is $\hat{G}_2(s)\hat{G}_1(s)$, where $\hat{G}_1$ is the transfer function of the plant S_1 and $\hat{G}_2$ is the transfer function of the controller S_2. In this open-loop operation the controller cannot react to disturbances in the plant as it lacks information about the output $y(t)$. More severely, open-loop operation does not work when the plant S_1 is unstable. While we may think that we can cancel unwanted poles p with $\Re(p) \geq 0$ in the transfer function $\hat{G}_1(s)$ of the plant S_1 with corresponding zeros of $\hat{G}_2(s)$, deteriorating conditions over the lifetime of the plant and the controller will move poles and zeros rendering such an attempt futile.

The physics of the plant S_1 determines whether the plant is stable or not. For controlling a possibly instable plant, which is subjected to disturbances, we have to feed back information about the output $y(t)$ to the controller S_2. In the unity feedback composition, Fig. 2.10 left, the output $y(t)$ is subtracted from the reference $r(t)$ producing the error signal $e(t)$. The controller S_2 tries to minimize the error signal by producing the input $u(t)$ for the plant S_1 according to an appropriate control law. Feeding back the output to the controller creates a closed control loop; subtracting the output from the reference makes the feedback negative. The unity feedback composition has the transfer function

$$\hat{H}(s) = \frac{\hat{G}_1(s)\hat{G}_2(s)}{1 + \hat{G}_1(s)\hat{G}_2(s)},$$

where $\hat{G}_1(s)$ is the transfer function of the plant and $\hat{G}_2(s)$ is the transfer function of the controller, while the feedback composition in Fig. 2.10 right, has the transfer function

$$\hat{H}'(s) = \frac{\hat{G}_1(s)}{1 + \hat{G}_1(s)\hat{G}_2(s)}.$$

The latter composition appears in the analysis of circuits involving operational amplifiers. An operational amplifier is an electronic amplifier, which amplifies the difference between its two inputs, ideally with infinite gain. The operational amplifier then acts as plant, while the so-called feedback network acts as controller.

In both feedback compositions the poles of the plant's transfer function $\hat{G}_1(s)$ cancel. By designing an appropriate controller S_2 we can place the poles and zeros of the chosen feedback composition structure as we like.

Sometimes we do not know the transfer function of the plant, but we still can measure the frequency response $H(\omega)$ of the plant S_1, provided the plant is stable. We consider a proportional controller with the transfer function $\hat{G}_2(s) = K_P$ for some gain $K_P \in \mathbb{R}$, $K_P > 0$ and assume that the unity feedback system built from S_1 and S_2 is stable for small gains K_P and becomes unstable for large gains. The phase margin then is $\angle H(\omega') + \pi$ where $|K_P H(\omega')| = 1$ for $\omega' \in \mathbb{R}$, $\omega' > 0$. If this phase margin is positive then the unity feedback composition of S_1 and S_2 is stable. A controller introducing delay, due to a software implementation for example, eats up some of the phase margin designed into the control system, in that way

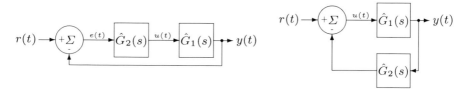

Fig. 2.10 Two possible feedback compositions of one system, the plant, with transfer function $\hat{G}_1(s)$ and one, the controller, with transfer function $\hat{G}_2(s)$. The unity feedback composition to the *left* is used for building control systems, while the composition to the *right* appears in the analysis of circuits involving operational amplifiers

diminishing the quality of the control. Therefore, unnecessary delays in a controller's implementation shall be avoided.

A proportional controller employed in the unity feedback configuration feeds a scaled version of the error signal to the plant, $u(t) = K_{\mathrm{P}}e(t)$. We look at the response of the control loop to a unit-step input. The closed-loop transfer function is

$$\hat{H}(s) = \frac{K_{\mathrm{P}}\hat{G}_1(s)}{1 + K_{\mathrm{P}}\hat{G}_1(s)}.$$

The Laplace transform of the step response is $\hat{H}(s)/s$. Assuming stability of the closed-loop system and using the final value theorem (2.23) we compute the steady state,

$$\lim_{t\to\infty} y(t) = \lim_{s\to 0} s\frac{\hat{H}(s)}{s} = \hat{H}(0) = \lim_{s\to 0}\frac{K_{\mathrm{P}}\hat{G}_1(s)}{1 + K_{\mathrm{P}}\hat{G}_1(s)}.$$

The steady state response is unequal to one whenever $\lim_{s\to 0}\hat{G}_1(s)$ is finite. For making the steady state error $|1 - \hat{H}(0)|$ disappear the transfer function $\hat{G}_1(s)$ of the plant must have a pole at 0. Increasing the controller's gain K_{P} will reduce the steady state error at the cost of increasing the overshoot of the output $y(t)$ whenever the reference $r(t)$ changes.

In order to make the steady state error vanish for plants having no pole at 0 we augment the proportional controller by an integrator resulting in a proportional plus integral controller, its control equation is

$$u(t) = K_{\mathrm{P}}\left(e(t) + \frac{1}{T_1}\int_{t_0}^{t} e(\tau)d\tau\right), \tag{2.27}$$

where t_0 is the time the control loop starts to operate. All signals associated with the control loop are assumed to be zero before t_0. The controller's transfer function $\hat{G}_2(s)$ is

$$\hat{G}_2(s) = K_{\mathrm{P}}\left(1 + \frac{1}{T_1 s}\right).$$

As rule of thumb, increasing the proportional gain K_{P} makes the control loop react faster to changes of the reference at the cost of increased overshoot. It also reduces the steady state error. Decreasing the integral time T_1 speeds up the control loop too, again at the cost of increased overshoot.

Real plants have physical limits beyond which they cannot operate. A valve for example cannot operate beyond fully opened or fully closed. When such operational limits prevent the plant's output to reach the reference, the integrator's output will either grow or shrink without bounds. When the reference later returns to a reachable value it will take the integrator considerable time for winding down to sensible values.

Fig. 2.11 Basic structure of
the software realization of a
proportional and integral
controller

```
1: procedure PICONTROLLER(yₖ,rₖ)
2:     eₖ ← rₖ − yₖ
3:     uₖ ← uₖ₋₁ + Kₚ (eₖ + T/T₁ eₖ − eₖ₋₁)
4:     output uₖ
5:     eₖ₋₁ ← eₖ
6:     uₖ₋₁ ← uₖ
7: end procedure
```

During this time the input to the plant will still exceed the plant's limits. A crude anti-windup strategy limits the value of the integrator's output by temporarily switching off the integrator.

The procedure PICONTROLLER in Fig. 2.11 is the implementation of a proportional and integral controller in software. It must be called periodically with period T. We arrive at this discrete-time realization by approximating the integral in (2.27) with a sum. The procedure accepts the plant's output measured for the kth invocation y_k and the reference r_k as parameters. The error e_{k-1} and the output u_{k-1}, both from the previous invocation, constitute the state of the controller. It is good practice to pass the plant's new input before updating the controller's state, in order to avoid unnecessary delays.

For reducing the overshoot we may add a differential term to the proportional plus integral controller resulting in a proportional integral differential controller,

$$u(t) = K_{\mathrm{P}} \left(e(t) + \frac{1}{T_1} \int_{t_0}^{t} e(\tau)\mathrm{d}\tau + T_{\mathrm{D}} \frac{\mathrm{d}e(t)}{\mathrm{d}t} \right).$$

As a rule of thumb increasing the derivative time T_{D} reduces the overshoot. The differential term, however, is sensitive to noise added when measuring the plant's output $y(t)$.

In many control systems several control loops are nested into each other. The outermost control loop in such a cascaded control scheme processes the reference $r(t)$. The output of the controller in the outer loop is the reference for the next inner control loop. When controlling an electric motor, for example, the innermost loop controls the torque the motor produces. The next outer loop controls the velocity by passing the torque required for achieving the desired velocity as reference to the innermost loop. In drives used for positioning, the outermost loop controls the position by passing the velocity required for arriving and holding the desired position to the middle control loop.

Fig. 2.12 A function generator produces periodic signals within a frequency range from 1 μHz
up to several megahertz

2.11 Instrumentation for Producing and Measuring Signals

When we engineer an embedded system our ideas eventually must prove themselves
in the physical reality. In order to quantify the performance of our system we must
stimulate it with well-defined inputs, and measure and analyze its responses. We
limit our discussion to basic instrumentation for general purpose use.

2.11.1 The Function Generator

A function generator[7] produces periodic signals with frequencies ranging from zero
to several megahertz, Fig. 2.12. It offers the choice between several predefined waveforms; sine, rectangle, ramp, triangle, pulse, and noise are typical. A freely programmable waveform may also be available. Frequency, amplitude, and a constant offset
from zero can be set for the selected waveform. For the rectangle waveform we can
select the so-called duty cycle, that is the ratio between the time the signal is on and
the time the signal is off. For the pulse waveform we can set the width of the pulses.
The noise waveform typically is band-limited white noise.

In addition, a function generator offers basic modulation features. Modulation
means that the output signal results from the modification of a higher frequency
carrier signal with a lower frequency signal. Amplitude modulation multiplies the

---

[7] The ancestor of today's function generators is the HP-200A, an audio oscillator, which grew out
of Bill Hewlett's master's thesis at Stanford. The HP-200A is famous for the ingenious use of a light
bulb for providing negative feedback. Bill Hewlett and Dave Packard assembled these oscillators
starting 1939 in the garage behind Packard's house at 367 Addison Avenue in Palo Alto. One of the
first customers, buying eight HP-200B, were the Walt Disney Studios.

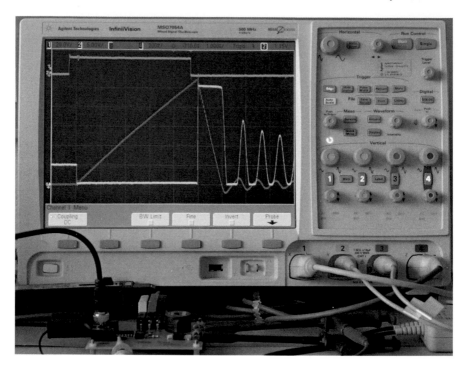

Fig. 2.13 Still life with oscilloscope. An oscilloscope simultaneously records and displays several signals. The length of a record can vary between less than a nanosecond and several hundred seconds. The range of possible amplitudes can span more than three decades

carrier with the lower frequency signal for producing the output. Frequency modulation shifts the frequency of the carrier according to the lower frequency signal.

A frequency sweep modulates the frequency of the output signal with a slow ramp. By analyzing the answer of some device to such a frequency sweep we can derive the device's frequency response. A linear sweep uses a sawtooth-shaped ramp. A logarithmic sweep uses an exponential ramp as modulation. For measurement instruments analyzing signals derived from the function generator's output the function generator provides a synchronization signal indicating, for example, the beginning of a sweep.

2.11.2 The Oscilloscope

The oscilloscope, scope for short, is the most important instrument for capturing and displaying signals, Fig. 2.13. A scope consists of a horizontal section, a vertical section, a trigger section[8] and a display for the waveforms. Each section has its own set of controls on the front panel. Modern scopes usually have several channels, two or four are common, for simultaneously capturing and displaying several signals.

The vertical section consists of the amplifiers for the channels. The controls for each channel are the gain, the channel's vertical position on the scope's screen, the coupling, and a switch for disabling the channel. A channel's gain is usually shown in a line above the screen's graticule. The unit the gain is stated in is the unit of the measured physical quantity per vertical division of the graticule. The control for the vertical position allows us to arrange the channels on the screen. A small triangle to the left of the graticule indicates zero for the channel. The coupling can be alternating current (AC) or direct current (DC). With the coupling set to AC a highpass filter, which blocks constant signal content, is introduced right after the channel's input connector. With DC coupling constant signal content is passed to the channel's amplifier.

When the trigger section recognizes a trigger event it commands the scope to start a new recording. The most basic type of trigger event is when the signal fed into one of the scope's channels crosses an adjustable trigger level, either from above or from below. A small triangle marked with the letter T to the left of the graticule indicates the trigger level. The operator can select between AC and DC coupling for the trigger source. Sophisticated trigger sections support triggers, for example, on pulse width, or on the appearance of some pattern on optional digital inputs. Practically, all scopes support an external trigger source.

The horizontal section provides the timing for the recordings. In analog scopes the horizontal section provided a ramp to the horizontal deflection coils of a cathode ray tube whenever a trigger section commanded a new recording. The horizontal coils deflected the electron beam inside the tube from left to right. Where the electron beam hit the phosphorus covering the tube's screen the phosphorus produced a spot of light. The output of the vertical amplifier drove the vertical coils, which deflected the beam in the vertical direction. The afterglow of the beam's trace provided a visual image of the signal's waveform. In modern digital scopes memory replaces the short-term storage action of the cathode ray tube. The outputs of the vertical amplifiers are digitized at a rate controlled by the horizontal section and written into the scope's acquisition memory. The size of the acquisition memory determines how long the scope can record after a trigger event. After the scope has displayed the content of the acquisition memory on its screen it is ready to make a new recording. The scope can operate either in single trigger mode or in continuous trigger mode. In single trigger mode, the first trigger event commands a recording and all subsequent trigger events are ignored. Single trigger mode helps analyzing nonrecurring events. In continuous trigger mode, the scope takes a record whenever the trigger section commands it, and the scope is ready. The scope displays subsequent recordings on top of each other. In the process it lets old recordings fade away.

---

[8] The basic concept of a scope has changed very little since Howard Vollum and Melvin Murdock, the founders of Tektronix, introduced the first practical oscilloscope, the Tektronix type 511, in 1948.

Digital scopes allow us to apply mathematical operations to the recordings in real time. The absolute value of the recording's Fourier transform is particularly useful for estimating the frequency content of a measured signal.

Scopes are not perfect either. High frequency signals are not reproduced faithfully. A scope's bandwidth specifies the frequency at which the displayed amplitude is down by 3dB from the true value. For observing digital waveforms with some accuracy the scope's bandwidth should be at least five times higher than the frequency of the fastest digital clock signal. For observing analog waveforms the scope's bandwidth shall be three times higher than the highest harmonic in the observed signals.

2.12 Bibliographical Notes

The book (Lee and Varaiya 2011) stands out for treating discrete-time and continuous-time equally. The book by Kreyszig (2010) provides a broad reference to many mathematical topics of use when doing engineering work. The book (Oppenheim and Schafer 1975) is still a very useful reference for discrete-time signals. Jänich's book (2001) covers advanced material such as partial differential equations in a very readable manner. The book by Gershenfeld (1999) contains an ample selection of mathematical methods for modeling physical systems.

The book (Middleton 1996) contains an in-depth treatise on signals influenced by noise. For the effects of noise on the discrete Fourier transform of a signal see (Schoukens and Renneboog 1986; Kester and Analog Devices 2003). The representation of a periodic signal by its Fourier series is proven for example in the book by Zygmund (2002a, Chap. 2, Theorem 8.1). The representation of a signal by its Fourier integral is proven for example in (Zygmund 2002b, Chap. 16, Theorem 1.3 and the text following it). For both continuous-time and discrete-time control systems see for example (Franklin et al. 2010, 1997). Both books contain many exemplary models of physical systems.

2.13 Exercises

Exercise 2.1 Several functions are used to suppress artifacts in the Fourier transform of functions which we have observed for a finite amount of time only. In this context these functions are called windows. The idea is to observe the signal $s : \mathbb{R} \to \mathbb{C}$, $t \mapsto s(t)$ only between times $-p$ and p for some positive real p and pad the observation $o(t)$ with zeros before and after. Let us consider the signal $s(t) = 1$, whose transform is the Dirac delta $\delta(\omega)$. The observation $o(t)$ is the appropriate boxcar function. The transform of the boxcar function shows us the errors we introduce into the Fourier transform by replacing the signal $s(t)$ by the observation $o(t)$. Plot the Fourier transform of a boxcar window.

Exercise 2.2 We can modify the error in the Fourier transform by transforming the signal $w(t)o(t)$ instead of the signal $o(t)$ for a window function $w(t)$. The triangle function is another popular window. Plot its Fourier transform. Compare it to the transform of the boxcar.

Exercise 2.3 The Welch window has the definition

$$w(t) = 1 - \frac{t^2}{p^2},$$

for $-p \le t \le p$, zero otherwise. Compute and plot its Fourier transform.

Exercise 2.4 The von-Hann window has the definition

$$w(t) = \frac{1}{2}\left(1 + \cos\frac{\pi t}{p}\right),$$

for $-p \le t \le p$, zero otherwise. Compute and plot its Fourier transform.

Exercise 2.5 The Blackman–Nuttall window has the definition

$$w(t) = 0.3635819 + 0.4891775\cos\frac{\pi t}{p} + 0.1365995\cos\frac{2\pi t}{p} + 0.0106411\cos\frac{3\pi t}{p},$$

for $-p \le t \le p$, zero otherwise. Compute and plot its Fourier transform.

Exercise 2.6 Consider the spring-mass-damper system in Fig. 2.3. Play with the damping constant c and observe the height and shape of the resonant peak.

Exercise 2.7 Derive the range of damping constants for which the spring-mass-damper system in Fig. 2.3 will exhibit resonance.

2.14 Lab Exercise

Exercise 2.8 Use a function generator to generate signals with different waveforms and an oscilloscope to visualize these signals. Familiarize yourself with the operation of these instruments. While exploring the modulation options of your signal generator observe both the resulting waveforms and the amplitudes spectrum of the waveform on your scope.

References

Franklin GF, Powell DJ, Emami-Naeini A (2010) Feedback control of dynamic systems, 6th edn. Pearson, Upper Saddle River
Franklin GF, Workman ML, Powell D (1997) Digital control of dynamic systems, 3rd edn. Addison-Wesley Longman Publishing Co., Inc., Boston

Gershenfeld N (1999) The nature of mathematical modeling. Cambridge University Press, Cambridge

Jänich K (2001) Analysis für Physiker und Ingenieure Funktionentheorie, Differentialgleichungen Spezielle Funktionen, Springer-Lehrbuch. Springer, Berlin

Kester W (2003) Mixed-signal and DSP design techniques, Analog Devices series. Analog Devices and Newnes, Amsterdam

Kreyszig E (2010) Advanced engineering mathematics. Wiley, New York

Lee EA, Varaiya P (2011) Structure and interpretation of signals and systems, 2nd edn. http://LeeVaraiya.org/

Middleton D (1996) An introduction to statistical communication theory: an IEEE press classic reissue. Wiley, New York

Oppenheim AV, Schafer R (1975) Digital signal processing. Prentice-Hall, Englewood Cliffs

Schoukens J, Renneboog J (1986) Modeling the noise influence on the Fourier coefficients after a discrete Fourier transform. IEEE Trans Instrum Meas 35(3):278–286

Zygmund A (2002a) Trigonometric series, vol I. Cambridge University Press, Cambridge

Zygmund A (2002b) Trigonometric series, vol II. Cambridge University Press, Cambridge

Chapter 3
Voltage, Current, Basic Components

Unavoidably an engineer will use basic electronic components when designing an embedded system for a specific task. A solid understanding of the mathematical modeling of the basic components and circuits will later enable us to design optimized systems. Measuring the behavior of such circuits is an integral part of the engineer's tool set.

3.1 Electric Charge, Voltage, and Current

In terms of electric charge, there are two kinds around, the positive and the negative ones. A positive and a negative charge attract each other, two positive charges repel each other, as do two negative ones. Idealizing we consider charges without spatial extent and in vacuum. The force $\vec{F}_{12}$ that a point charge Q_1 exerts on another point charge Q_2 is governed by Coulomb's law,

$$\vec{F}_{12} = \frac{Q_1 Q_2}{4\pi \epsilon_0 r^3} \vec{r},$$

where $\vec{r}$ is the vector from Q_1 to Q_2, and r is its length. The constant $\epsilon_0 = 8.8542 \times 10^{-12}\,\mathrm{A\,s\,V^{-1}\,m^{-1}}$ is the permittivity of vacuum. The unit of charge is coulomb, [C]. Charge is preserved; in any contraption with no connection to the outside the total charge is constant.

Voltage, usually denoted by the letter V, is measured between two points in a circuit. The unit is volt [$V = JC^{-1}$]. If work of one joule [J] is released when moving a charge of one coulomb from point a to point b, then there is a voltage of 1 V between a and b. Given an electrically conducting connection between the two points, current will flow from the more positive point a to the less positive point b.

Moving charges constitute electric currents. In technical documents, like circuit diagrams, equations and data sheets, currents are denoted by the letter I together with some appropriate subscripts. Electrical currents are measured in ampere [A].

© Springer International Publishing Switzerland 2015
P. Hintenaus, *Engineering Embedded Systems*, DOI 10.1007/978-3-319-10680-9_3

If two parallel conductors of infinite length and negligible circular cross section placed one meter apart in vacuum and carrying both the same constant current I are attracted by 2×10^{-7} N per meter of length, then the current I conducted by each of the conductors is one ampere. The force that makes the two conductors attract each other is the Lorentz force (3.2).

What at the macroscopic level appears as current at the microscopic level is the transport of electrically charged particles. These particles can be electrons drifting through a metallic conductor, ions drifting in a fluid or even charged particles shooting through vacuum. A current $I(t)$ at time t flowing into or out of a reservoir of charge changes the charge $Q(t)$ stored, $I(t) = \frac{dQ(t)}{dt}$; a constant electrical current of one ampere flowing for one second transports a charge of one coulomb, $[\text{C} = \text{A s}]$. The electrical current is one of the seven fundamental quantities in the *Système International d'Unités (SI)*.

3.1.1 The Electric and the Magnetic Field

Any distribution of electric charge in space creates an electric field $\vec{E}$. The positive charges are the sources of the electric field; the negative charges are the sinks. A force $\vec{F}(\vec{r})$ will act on a test charge Q at a point $\vec{r}$. A test charge is so small that it does not contribute to the overall electric field in an appreciable way. The electric field at point $\vec{r}$ is $\vec{E}(\vec{r}) = \vec{F}(\vec{r})/Q$, its unit is $[\text{N C}^{-1}]$.

Any current in space creates a magnetic field $\vec{B}$. The unit of $\vec{B}$ is tesla, $[\text{T} = \text{N A}^{-1}\text{m}^{-1}]$. More precisely, a current I flowing in the direction of the infinitesimal vector $d\vec{l}$ at point $\vec{r}_1$ produces the infinitesimal magnetic field (Law of Biot–Savart)

$$\vec{B}(\vec{r}_2) = \frac{\mu_0}{4\pi} \frac{I d\vec{l} \times \vec{r}}{r^3} \tag{3.1}$$

at point $\vec{r}_2$, where $\vec{r} = \vec{r}_2 - \vec{r}_1$ is the vector from $\vec{r}_1$ to $\vec{r}_2$, r is its length and $\mu_0 = 4\pi \times 10^{-7}\,\text{V s A}^{-1}\,\text{m}^{-1}$ is the magnetic constant. By integrating the right-hand side over the whole space the field $\vec{B}$ can be computed for each point $\vec{r}_2$ in space. Unlike the electric field, the magnetic field has neither sources nor sinks. The magnetic field exerts a force $\vec{F}$, the Lorentz force, on a point charge Q moving with velocity $\vec{v}$ with respect to $\vec{B}$,

$$\vec{F} = Q\vec{v} \times \vec{B}. \tag{3.2}$$

A changing magnetic field creates an electric field $\vec{E}$. The change of $\vec{B}$ through a closed conducting loop C induces a voltage across the terminals of the loop (Faraday's Law),

$$V_C = \oint_C \vec{E} \cdot d\vec{l} = -\frac{d\Phi_m}{dt}, \tag{3.3}$$

where

$$\Phi_m = \iint_S \vec{B} \cdot \vec{n} \, dA$$

is the magnetic flux of $\vec{B}$ through the loop C and S is a smooth orientable surface that is bounded by C. The magnetic flux does not depend on the chosen surface.

Any change of an electric field $\vec{E}$ creates a magnetic field too. The full interaction of electric and magnetic fields with each other and with charges and currents is described by a set of partial differential equations, Maxwell's equations, which cover electromagnetic phenomena including electric motors, radio, and optics. The full treatment is beyond the scope of this book.

3.1.2 Electric Circuits

An electric circuit consists of wires and components. Metals in general, aluminum, copper, and silver in particular, are good conductors. Electric circuits for electronics are usually built on printed circuit boards. The wires on such boards are realized from thin layers of copper deposited on sheets of non-conducting material.

In circuit diagrams, straight lines denote wires, Fig. 3.1. An arrow together with a name attached to a wire names the wire. Arrows with the same name connect the wires they are attached to. Depending on the situation, the direction of an arrow either indicates the flow of current or more commonly the flow of information. Using these arrows allows us to connect wires on different pages of a large circuit diagram. It also allows us to avoid a mess of lines within a page. Note that crossing lines do not denote an electrical connection unless the intersection is marked with a dot.

In direct current circuits, the voltages between the pairs of points of the circuit are constant, as are the currents through the wires of the circuit. In alternating current circuits, these voltages and currents vary with time.

In many circuits, a single point in the circuit is designated as ground. Then all voltages are measured with respect to this ground point. The voltage between a point in a circuit and ground is also called the electrical potential at that point. The symbols in Fig. 3.2 are commonly used for ground. Several occurrences of the same ground symbol in a circuit diagram denote the same electrical conductor, they are

Fig. 3.1 *Leftmost*, arrow for connecting wires, indicating information or current to flow out of the wire it is attached to. Next, arrow for connecting wires. Information or current flows into the wire. Next, two connected wires. *Rightmost*, two independent wires without a connection

Fig. 3.2 Symbols for ground.
Wires attached to the same
ground symbol are connected

GND

short circuited by convention. For drawing circuit diagrams, it is good practice to
let currents flow from the top of the page to the bottom, thus the top is at a higher
electrical potential than the bottom.

Two parts of a circuit which are not connected by a conductor are called galvani-
cally isolated. It is still possible to pass both energy and information across such an
isolation barrier by using either alternating electric or alternating magnetic fields, or
light. Galvanic isolation is commonly used for interfacing electronics to the outside
world where dangerously high voltages might prevail. For safety reasons, electronic
devices for medical use that can be connected to the mains during normal operation
must incorporate two galvanic isolation barriers between the mains and the patient.

3.1.3 The Kirchhoff Voltage Law and the Kirchhoff Current Law

If a number of wires and a number of components form a loop in a direct current
circuit then the sum of the voltages across the components around the loop is zero
(Kirchhoff's voltage law). For the direct current circuit in Fig. 3.3, the voltage V_1
between point A and B across U1 equals the voltage between A and B across U2.

If the wires 1 to n connect to a single point in a direct current circuit then
$\sum_{k=1}^{n} I_k = 0$, where I_1 is the current flowing through wire 1 into the connection,
I_2 is the current through wire 2 and I_n is the current through wire n (Kirchhoff's
current law), Fig. 3.4.

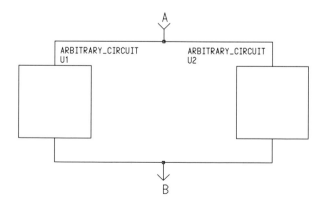

Fig. 3.3 In this direct current circuit, the voltage V_1 across the circuit U1 from A to B and the voltage
V_2 across the circuit U2 from A to B are equal because of Kirchhoff's voltage law

Fig. 3.4 According to Kirchhoff's current law in this direct current circuit, the sum of all currents flowing into the node is zero

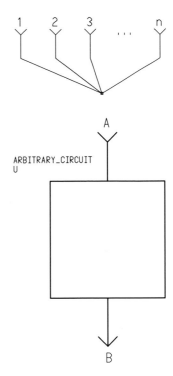

Fig. 3.5 The circuit U consumes the electrical power $P(t) = V(t)I(t)$ at time t, where $V(t)$ is the voltage between A and B, and $I(t)$ is the current through U both at time t

3.2 Electric Power

Let $V(t)$ be the voltage between point A and point B at time t in the circuit in Fig. 3.5; further let $I(t)$ be the current through the circuit U at t. The instantaneous electrical power consumed by a circuit at time t is

$$P(t) = V(t)I(t).$$

The energy consumed between the time t_1 and the time t_2 is

$$W = \int_{t_1}^{t_2} V(t)I(t)\,dt.$$

The circuit converts some of the energy W into heat, which typically is lost and has to be removed from the circuit; it stores another part and converts the last part into mechanical energy or light or any other useful form of energy.

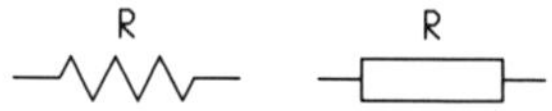

Fig. 3.6 Symbols for resistors, American-style (*left*), IEC-style (*right*)

Fig. 3.7 Ohm's Law. The voltage between A and B across the resistor R is proportional to the current through the resistor R

3.3 Resistors

A resistor is a component with two terminals. It turns electrical energy into heat. In circuit diagrams, resistors are denoted by the symbols in Fig. 3.6. According to Ohm's law (3.4), the voltage V between the points A and B across the resistor R in Fig. 3.7 is proportional to the current I from A to B through R,

$$V = RI. \tag{3.4}$$

The constant of proportionality, R, is called the resistance of the resistor R; its unit is ohm $[\Omega = \mathrm{V\,A^{-1}}]$.

The power P the resistor R dissipates is

$$P = VI = I^2 R = \frac{V^2}{R}.$$

The maximum power a resistor is able to dissipate without damage depends on its type and the ambient temperature.

Resistors are not perfect. The value of a resistor's resistance lies within a tolerance band of the resistance's nominal value. Tolerances of $\pm 10\,\%$, $\pm 5\,\%$, and $\pm 1\,\%$ are common. Besides the power rating, a resistor has a maximum voltage limit too. Exceeding this limit will permanently change the resistance. Moreover, a resistor's resistance depends on the resistor's temperature. Standard types show a temperature coefficient in the range of $\pm 10 \times 10^{-6}\,\mathrm{K^{-1}}$ to $\pm 100 \times 10^{-6}\,\mathrm{K^{-1}}$. For precise measurement circuits, resistors with a temperature coefficient in a range smaller than $\pm 1 \times 10^{-6}\,\mathrm{K^{-1}}$ can be obtained.

Engineers usually represent the temperature dependence of some physical quantity Q with a temperature coefficient α. The value Q_{T_0} of Q is given at some appropriate temperature T_0. Often $25°$ C is chosen for T_0. For estimating the value Q_T at some other temperature T, one computes

$$Q_T \approx Q_{T_0} \left(1 + \alpha(T - T_0)\right).$$

Fig. 3.8 In a series connection of resistors, *left* the total resistance is the sum of the individual resistances, in a parallel connection, *right* the total conductance is the sum of the conductances of the individual resistors

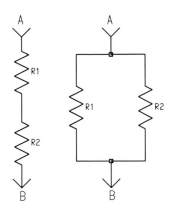

Data sheets often state worst-case estimates for the temperature coefficient α only.

The total resistance R between point A and B of the series connection of the resistors R1 and R2 in the circuit in Fig. 3.8, left side, is the sum of the individual resistors,

$$R = R_1 + R_2,$$

where R_1 is the resistance of R1 and R_2 is the resistance of R2. Sometimes the conductance of a resistor, the reciprocal of the resistance, is used. Its unit is siemens $[\mathrm{S} = \Omega^{-1}]$. The total conductance $\frac{1}{R}$ of a parallel connection of resistors, Fig. 3.8 right side, is the sum of the conductances of the individual resistors, $\frac{1}{R} = \frac{1}{R_1} + \frac{1}{R_2}$. Therefore, the resistance R of the parallel connection is

$$R = R_1 \parallel R_2 = \frac{R_1 R_2}{R_1 + R_2}.$$

3.4 Current and Voltage Sources

An ideal current source is a hypothetical component that adjusts the voltage between its two terminals to any value necessary for delivering its specified current, Fig. 3.9. Practical current sources can be built using semiconductors. Such a circuit exhibits a limited voltage range, the so-called range of compliance of the current source, for which it is able to drive its current.

An ideal voltage source is a hypothetical device with two terminals that sources or sinks any current necessary to maintain the specified voltage between the terminal

Fig. 3.9 Symbol for current sources. The *arrow* points in the direction of the current

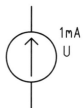

labeled + and the terminal labeled − or to produce the specified waveform, Fig. 3.10. Practical voltage sources can source or sink a limited range of current only; they exhibit a limited range of compliance.

When describing the operational principles of a circuit, using ideal current and voltage sources allows us to concentrate on these principles without cluttering our schematics with unnecessary detail.

A lab power supply typically has two adjustments for each of its outputs, one to set the voltage and one to set the current limit. As long as the current drawn by a load from an output stays within the set current limit the power supply delivers the voltage the output is set to. Once the load tries to draw more current, the supply reduces the voltage at the output, to keep the current within limits. It is good lab practice to set the current limit to a value just a little higher than required. In case of an inadvertent short, a prudently set current limit most often saves the device under test from damage.[1]

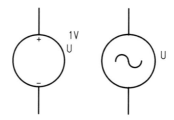

Fig. 3.10 Symbols for voltage sources. The source on the *left* generates a voltage of 1 V between the + terminal and the − terminal, the one on the *right* generates a sinusoidal voltage

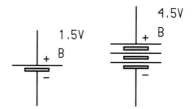

Fig. 3.11 Circuit symbol for batteries. The battery on the *left* consists of a single cell with an open-loop voltage, the voltage between the + terminal and the − terminal without a connected load, of 1.5 V, the one on the *right* consists of three cells

[1] It also saves the experimenter's reputation!

A battery is an electrochemical voltage source combined with an internal series resistor, Fig. 3.11. The value of the internal resistor depends on the type and condition of the battery. It gets larger when the battery is discharged. Be aware that the series resistance of a charged battery can be extremely small. Short circuiting the terminals of a battery might result in the explosive release of energy, possibly damaging the battery and its surroundings!

3.4.1 Equivalent Circuits

Any circuit with two terminals consisting of current and voltage sources and resistors can be replaced by either its Norton equivalent or its Thévenin equivalent.

The Norton equivalent of the circuit is a parallel combination of a current source and a resistor. To compute the current of the source and the resistance R of the resistor one considers two situations: First one computes or measures the open-loop voltage V_O across the two terminals of the circuit, that is the voltage between the two terminals without any further connection between them. Second one computes or measures the short-circuit current I_S, that flows through a conductor with resistance $0\,\Omega$ connecting the two terminals. The current of the source is I_S, the resistance R is $R = V_O/I_S$.

The Thévenin equivalent of the circuit is a series combination of a voltage source and a resistor. The voltage of the source is the open-loop voltage V_O of the circuit, the resistance R is $R = V_O/I_S$.

3.4.2 Voltage Dividers

A voltage divider, Fig. 3.12, consists of two resistors. Its output voltage V_o, the voltage across the load, depends on the input voltage V_i between point A and point B, the resistances R_1 and R_2 of the two resistors R1 and R2, and the current I_o drawn by the load U1. For an unloaded voltage divider, that is $I_o = 0$, the input current I_i is $I_i = \frac{V_i}{R_1+R_2}$, and therefore $V_o = V_i \frac{R_2}{R_1+R_2}$. The short-circuit current, $V_o = 0$, is $\frac{V_i}{R_1}$.

We assume a voltage source provides the voltage V_i. We consider the voltage divider as a circuit with two terminals to which the load connects. We can then replace the voltage source providing V_i and the two resistors R1 and R2 with their Thévenin equivalent, a series combination of a voltage source with a voltage V_S of

$$V_S = V_i \frac{R_2}{R_1 + R_2},\qquad(3.5)$$

and a series resistor with a resistance R_S of

$$R_S = R_1 \parallel R_2.$$

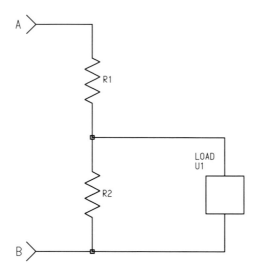

Fig. 3.12 Voltage Divider. The voltage across the load depends on the voltage between A and B, on the ratio of the resistance of R1 and of the resistance of R2, and on the current through the load

Fig. 3.13 Circuit symbol for a potentiometer. By adjusting the position of the middle terminal, the wiper, one can change the ratio of the resistances

Adjustable resistors, Fig. 3.13, usually come in the form of a voltage divider. This form is called a potentiometer. The position of the middle terminal, the wiper, determines the ratio of the resistances of the two resistors.

3.5 Voltage, Current, and Resistance Measurements

The voltage across a circuit is measured by placing a voltmeter in parallel with the circuit, Fig. 3.14. The resistance between the two terminals of an ideal voltmeter is infinite, so that it does not disturb the voltage it is measuring by drawing current. Modern instruments have a resistance in the range of $10\,G\Omega$ or higher. These instruments compare the voltage to be measured with a precise reference voltage in an analog-to-digital converter.

Currents are measured by routing the current through an ammeter, Fig. 3.15. In order to not disturb the measurement, the resistance between the two terminals

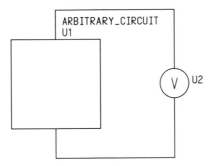

Fig. 3.14 Voltage measurement. For measuring the voltage across the arbitrary circuit U1, the voltmeter U2 is wired in parallel with the circuit

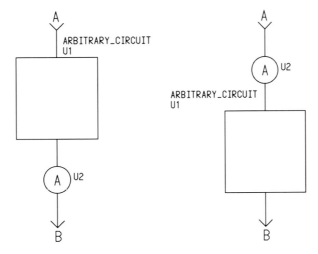

Fig. 3.15 Current measurement. For measuring the current through the arbitrary circuit U1, the ammeter U2 is wired in series with the circuit. Low-side current measurement (*left*), high-side measurement (*right*)

of an ideal ammeter is zero. One commonly used measurement principle senses the magnetic field produced by the current flowing in a wire. A current probe picks up this magnetic field using a ferromagnetic ring around the conductor and senses it with a combination of a coil and a hall-effect sensor, Fig. 3.16. The coil can only pick up alternating current, while the hall-effect sensor senses direct current too. Current probes capable of measuring up to 20 kA are available. The other common measurement principle routes the current through a resistor with small resistance and measures the voltage drop across this resistor. Moderate currents, up to several 10's of amperes, can be handled. The resistor, called shunt, typically has a resistance of 1–100 mΩ. The resistance is chosen so that the power dissipated by the resistor is manageable while still producing a useful voltage drop. The resistance of the wires that conduct the current to be measured is in the same order of magnitude as the

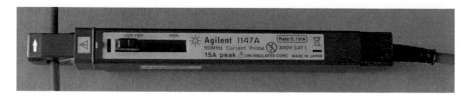

Fig. 3.16 Probe for measuring dynamic currents. The probe picks up and measures the magnetic field produced by the current in the conductor that is threaded through the probe's core. The *arrow* indicates the direction of positive currents

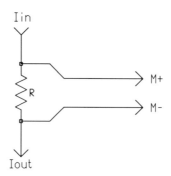

Fig. 3.17 Kelvin connection for current measurements. A separate measurement path avoids picking up the voltage drops along the current path

resistance of the shunt itself. In order to avoid erroneous results, one has to pick up the voltage drop across the shunt directly at the shunt. A kelvin connection, Fig. 3.17, provides two paths to and from the shunt R. The one path from Iin to Iout carries the current to be measured, while the other path provides the voltage drop between M+ and M− to the sensing circuit. This sensing circuit has high resistance, therefore only negligible current flows in the sensing path, making the path's resistance irrelevant.

Very often the voltages within a circuit are referenced to the voltage at the low side of the circuit, most often ground. When using a shunt for measuring the current through the circuit, placing the shunt in a high-side configuration, Fig. 3.15 right side, mandates the evaluation of a small voltage difference between two wires, which both might be at a high voltage. Designing such a measurement requires special attention. Placing the shunt at the low side of the circuit, Fig. 3.15 left side, circumvents the high voltages, but the circuit looses its reference connection; the circuit floats.

The resistance of a device is measured by observing the voltage drop across the device while forcing a small constant current through it. For convenience, many instruments measure the voltage drop across their terminals, Fig. 3.18, therefore the measured resistance is the resistance of the device plus the resistance of the measurement leads. For allowing the precise measurement of small resistances, some instruments offer four-wire setups where one pair of leads conducts the current and the other picks up the voltage drop, forming a kelvin connection. As the setups for

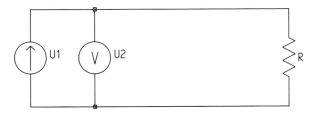

Fig. 3.18 Two-wire resistance measurement. The voltage drop is picked up at the terminals of the ohmmeter. For small resistances, voltage drop introduced by long measurement leads causes severe error

Fig. 3.19 Benchtop multimeter configured for voltage measurements. In order to support four-wire resistance measurements, this instrument has five terminals for connecting the device under test

resistance measurements are circuits by themselves, resistance cannot be measured while the device under test is part of another circuit.

The functions of a voltmeter, an ammeter, and an ohmmeter are often combined into a multimeter, Fig 3.19. Typically these instruments have at least three terminals, one for the positive lead in voltage and resistance measurements, one for the positive lead in current measurements, and a common terminal for the negative lead. The most dangerous mistake is to use a multimeter set up for current measurement for measuring voltages. This will result in a short circuit and high currents being conducted by the multimeter. If the current is high enough the fuse internal to the multimeter will not just blow but explode, destroying the instrument in the process! Well designed devices will at least contain the shrapnel to prevent further damage to the unhappy experimenter.

A resolution and accuracy of four decimal digits can be expected from a good handheld multimeter. Benchtop instruments routinely deliver 6 digits or more, but trade this resolution with a longer measurement time.

$$
\begin{array}{c}
\boxed{+\;+\;+\;+\;+\;+\;+\;+\;+\;+\;+\;+\;+} \\[4pt]
\vec{E} \quad
\begin{matrix}
\curlyvee & \curlyvee & \curlyvee & \curlyvee & \curlyvee & \curlyvee & \curlyvee & \curlyvee & \curlyvee & \curlyvee \\
\curlyvee & \curlyvee & \curlyvee & \curlyvee & \curlyvee & \curlyvee & \curlyvee & \curlyvee & \curlyvee & \curlyvee \\
\curlyvee & \curlyvee & \curlyvee & \curlyvee & \curlyvee & \curlyvee & \curlyvee & \curlyvee & \curlyvee & \curlyvee
\end{matrix} \\[4pt]
\boxed{-\;-\;-\;-\;-\;-\;-\;-\;-\;-\;-\;-\;-}
\end{array}
$$

Fig. 3.20 The electric field $\vec{E}$ inside a capacitor with parallel plates is homogeneous. What little field leaks into the surroundings of the capacitor can safely be ignored. *Positive* charges are indicated by $+$, *negative* ones by $-$

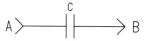

Fig. 3.21 The voltage across a capacitor is proportional to the time-integral of the current through it

3.6 Capacitors

A capacitor, Fig. 3.20, consists of two conducting plates which are placed in close proximity to each other with an isolator, the dielectric, between them. Each plate is connected to one of the capacitor's terminals. A capacitor stores energy in the electric field that forms between its plates once a voltage is applied to its terminals. Until equilibrium is reached this voltage drives a current, which deposits positive charge on one plate and negative charge on the other. The electric field is mostly confined to the space between the plates. We can safely ignore what little electric field leaks into the surroundings of a capacitor. Capacitors do not conduct direct current. They do, however, conduct alternating current. Therefore, they are used in alternating current or AC circuits. In Fig. 3.21, the voltage $V_C(t)$ between point A and point B across the capacitor C is proportional to the electric charge stored in the capacitor. The capacitance C of the capacitor C is the reciprocal of the constant of proportionality,

$$
V_C(t) = \frac{Q_C(t)}{C} = \frac{1}{C}\int_0^t I_C(\tau)\,d\tau, \tag{3.6}
$$

where t denotes time and $V_C(0) = 0$. When a constant current is forced through a capacitor, the voltage across the capacitor ramps up (or down) linearly. The unit of capacitance is farad, $[\mathrm{F} = \mathrm{A\,s\,V^{-1}} = \mathrm{C\,V^{-1}}]$. Typical values range from less than 1 pF for ceramic capacitors to several thousand μF for electrolytic capacitors. As the electric field in a capacitor is concentrated between the plates, capacitors mounted in close proximity to each other do not influence each other. In a parallel connection of capacitors, Fig. 3.22 left, the total capacitance C is the sum of the capacitances C_1 of the capacitor C1 and the capacitance C_2 of the capacitor C2,

Fig. 3.22 Parallel (*left*) and series (*right*) connection of capacitors

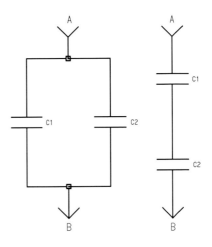

$$C = C_1 + C_2.$$

In a series connection, Fig. 3.22 right, the reciprocal of the total capacitance C is the sum of the reciprocal of the capacitance C_1 and the reciprocal of C_2,

$$\frac{1}{C} = \frac{1}{C_1} + \frac{1}{C_2}.$$

Capacitors have a voltage limit which may depend on the frequency of the applied voltage. Aluminum electrolytic capacitors and tantalum capacitors are polarized. Applying a voltage of reversed polarity destroys them.

Capacitors are used to conduct the alternating content of a signal from one part of a circuit to another. If the receiving side does not draw any current, there will be a constant voltage across the capacitor, (3.6). The capacitor blocks the constant signal content only. The two parts of the circuit are said to be AC coupled.

3.6.1 Energy Stored in a Capacitor

In order to compute the energy stored in a capacitor, we assume that the voltage across the capacitor is initially zero, $V_C(0) = 0$. We assume some continuous voltage $V_C(t)$ is applied across the capacitor, then the energy stored in the capacitor at time t is $W_C(t) = \int_0^t P_C(\tau)d\tau = \int_0^t V_C(\tau)I_C(\tau)d\tau = C\int_0^t V_C(\tau)\frac{dV_C(\tau)}{d\tau}d\tau$. The electric energy stored in a capacitor with voltage V_C across its terminals is, therefore,

$$W_C = \frac{C}{2}V_C^2.$$

3.7 Inductors

An inductor, Fig. 3.23, consists of a length of insulated wire wound into a winding. The two ends of the wire form the terminals of the inductor. An inductor stores energy in the magnetic field that forms as soon as an electric current flows through the wire. The voltage $V_L(t)$ between the point A and the point B across the inductor L in Fig. 3.24 is proportional to the time-derivative of the current $I_L(t)$ through the inductor,

$$V_L = L \frac{dI_L(t)}{dt}. \tag{3.7}$$

The constant of proportionality is called the inductance L of the inductor L, its unit is henry, $[\mathrm{H} = \mathrm{V\,s\,A}^{-1} = \Omega\mathrm{s}]$. The inductance of an inductor depends on the number of turns in its winding, on the geometry of the winding and on the material and geometry of the core onto which the winding is wound. Often ferromagnetic materials are used as cores to boost the inductance. When the current through an inductor with a ferromagnetic core gets too high, the core will saturate and the inductance will break down. Therefore, when designing with such inductors consult the data sheets for the saturation current of the inductor, and make sure that the current through the inductor in your circuit does not exceed this value. There are no general rules for the parallel connection and for the serial connection of inductors, because one has

Fig. 3.23 Cut through a toroidal inductor. A wire decorated with a *dot*, ⊙, carries a current emerging from the drawing plane, one decorated with a *cross*, ⊗, carries a current dipping into the drawing plane. The magnetic field $\vec{B}$ is confined to the space inside the winding. Outside the winding it is zero. Other geometries, for example solenoids, produce fields reaching out into the surrounding space

A ⟩───ⴰⴰⴰⴰⴰ───⟩ B

Fig. 3.24 The voltage across the inductor L is proportional to the time-derivative of the current through L to consider how the magnetic fields of the individual inductors interact in order to compute the inductance of the combined circuit. According to (3.7), the current I_L through an inductor is constant if and only if the voltage V_L is zero. Such behavior is useless in direct current circuits, therefore inductors are used in alternating current circuits only.

3.7.1 Energy Stored in an Inductor

In order to compute the energy stored in an inductor, we assume that the current $I_L(t)$ through the inductor is initially zero, $I_L(0) = 0$. We furthermore assume some continuous current $I_L(t)$ is forced through the inductor. Then the energy stored in the inductor at time t is $W_L(t) = \int_0^t P_L(\tau)\,d\tau = \int_0^t V_L(\tau)I_L(\tau)\,d\tau = L \int_0^t \frac{dI_L(\tau)}{d\tau} I_L(\tau)\,d\tau$. The magnetic energy stored in an inductor with current I_L through it, therefore, is

$$W_L = \frac{L}{2} I_L^2.$$

3.8 RLC Circuits

While the circuit Fig. 3.25 is not a direct current circuit, we still may use Kirchhoff's laws as long as the circuit is lumped. The length of the longest wire in a lumped circuit has to be very small when compared to the wavelength of the highest frequency appearing in such a circuit. Moreover, the electric and magnetic fields generated in the circuit's components have to be confined to the generating components.

For analyzing the behavior of the circuit, we let positive currents flow from top to bottom and from left to right. The current through the resistor $I_R(t)$ and through the inductor $I_L(t)$ has the same magnitude but opposite direction as the

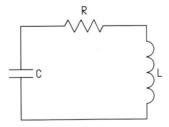

Fig. 3.25 An oscillating circuit. Energy oscillates between being stored in the magnetic field inside the inductor L and being stored in the electric field inside the capacitor C. The resistor R converts a portion of the energy being transferred back and forth into heat

current through the capacitor $I_C(t)$ because of Kirchhoff's current law, $I_C(t) = -I_R(t) = -I_L(t)$. The voltage drop across the capacitor $V_C(t)$ is the sum of the voltage drops across the resistor $V_R(t)$ and the inductor $V_L(t)$ because of Kirchhoff's voltage law, $\frac{1}{C}\int_0^t I_C(\tau)d\tau = V_C(t) = V_R(t) + V_L(t) = -RI_C(t) - L\frac{dI_C(t)}{dt}$. Taking the derivative $\frac{d}{dt}$ on both sides results in

$$L\frac{d^2 I_C(t)}{dt^2} + R\frac{dI_C(t)}{dt} + \frac{1}{C}I_C(t) = 0, \tag{3.8}$$

which is the equation of a damped harmonic—i.e. sinusoidal—oscillation.

We verify conservation of energy. According to Sect. 2.9.1 and assuming that the voltage across the capacitor is V_0 at time $t = 0$ and that the current through the capacitor is I_0 at $t = 0$ the solution is

$$I_C(t) = e^{-\frac{Rt}{2L}}\left(-\frac{I_0 CR + 2CV_0}{\sqrt{4LC - C^2 R^2}}\sin\omega^* t + I_0\cos\omega^* t\right)$$

$$V_C(t) = e^{-\frac{Rt}{2L}}\left(\frac{V_0 CR + 2LI_0}{\sqrt{4LC - C^2 R^2}}\sin\omega^* t + V_0\cos\omega^* t\right)$$

for $R^2 < \frac{4L}{C}$, where $\omega^* = \sqrt{\frac{1}{LC} - \frac{R^2}{4L^2}}$. The energy stored in the capacitor at time t is $W_C(t) = \frac{C}{2}(V_C(t))^2$, the energy stored in the inductor at time t is $W_L(t) = \frac{L}{2}(I_C(t))^2$ and the energy dissipated by the resistor between time 0 and t is $W_R(t) = R\int_0^t (I_C(\tau))^2 d\tau$. The total energy W is

$$W = W_C(t) + W_L(t) + W_R(t) = \frac{C}{2}V_0^2 + \frac{L}{2}I_0^2,$$

which is constant. The plots in Fig. 3.26 show how the energies $W_R(t)$, $W_L(t)$ and $W_C(t)$ evolve.

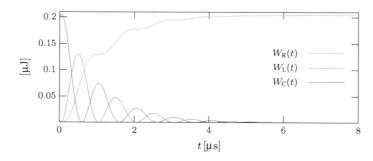

Fig. 3.26 The energies W_C, W_L and W_R for $L = 1\,\mu H$, $C = 100\,nF$, $R = 2\,\Omega$, $V_0 = 2\,V$ and $I_0 = 100\,mA$. Energy in the circuit in Fig. 3.25 oscillates between being stored in the capacitor and being stored in the inductor while the resistor converts more and more energy into heat

3.9 Alternating Current Circuits

For describing the response of a circuit composed of resistors, capacitors, and inductors when driven with a sinusoidal voltage or current of frequency $f = \frac{\omega}{2\pi}$, one uses complex voltages and complex currents. The real part of a complex voltage or of a complex current is the in-phase component, while the imaginary part is the quadrature (out-of-phase) component. The quotient of the voltage across a circuit and the current through it is called the impedance of the circuit. The impedance is usually denoted by the letter Z together with an appropriate subscript. The imaginary part of a circuit's impedance is called the circuit's reactance, the real part its resistance. The reactance of a resistor is zero.

When a sinusoidal current $I(t) = I_0 \cos \omega t$ flows through a capacitor with capacitance C, the voltage across this capacitor is $V_C(t) = \frac{1}{C} \int_0^t I_0 \cos \omega \tau d\tau = \frac{1}{\omega C} I_0 \sin \omega t$. The voltage V_C across the capacitor lags the current I_C through it; V_C is out of phase with I_C by $\frac{-\pi}{2}$. Therefore, the impedance of a capacitor is

$$Z_C = \frac{1}{i\omega C}.$$

When the current $I(t)$ flows through an inductor L with inductance L, the voltage across this inductor is $V_L(t) = L \frac{dI_0 \cos \omega t}{dt} = -\omega L I_0 \sin \omega t$. The voltage V_L across the inductor leads the current I_L through it, the phase difference is $\frac{\pi}{2}$. The impedance of an inductor is

$$Z_L = i\omega L.$$

Using Kirchhoff's current law and Kirchhoff's voltage law together with Ohm's law generalized to complex voltages and currents, $V = ZI$, the response of any circuit to a sinusoidal excitation can be analyzed. If the excitation is not sinusoidal, one has to resort to the capacitor and inductor laws.

The circuit in Fig. 3.27 can be understood as a voltage divider composed of complex impedances. When excited by a sinusoidal voltage $V_0(t) = V_0 \cos \omega t$ at point A amplitude and phase of the output voltage V_0 at point B, the frequency response $H(\omega)$, can be computed by substituting the resistance R of the resistor R for R_1 and

Fig. 3.27 A first-order lowpass filter

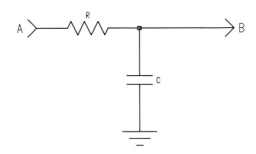

the impedance $\frac{1}{i\omega C}$ of the capacitor C for R_2 in (3.5),

$$V_o = V_0 \frac{\frac{1}{i\omega C}}{R + \frac{1}{i\omega C}} = V_0 \frac{1 - i\omega RC}{\omega^2 R^2 C^2 + 1} = V_0 H(\omega). \tag{3.9}$$

By substituting 0 for ω in (3.9), we see that the circuit passes a constant voltage undamped. The voltage-gain $G(\omega)$ of the circuit is

$$G(\omega) = \left| \frac{1 - i\omega RC}{\omega^2 R^2 C^2 + 1} \right| = \frac{1}{\sqrt{\omega^2 R^2 C^2 + 1}}. \tag{3.10}$$

For computing the angular frequency ω_c, at which $|V_o| = \frac{V_0}{\sqrt{2}}$ we solve $G(\omega_c) = \frac{1}{\sqrt{2}}$ for ω_c. The positive solution is $\omega_c = \frac{1}{RC}$. Therefore the cutoff frequency f_c is

$$f_c = \frac{\omega_c}{2\pi} = \frac{1}{2\pi RC}.$$

When we substitute $\frac{1}{RC}$ for ω in (3.9), we get $V_o = V_0 \frac{1-i}{2}$. Therefore, the response leads the excitation by $\frac{\pi}{4}$ at the cutoff frequency. For $\omega \gg \omega_c$, the gain approximately is $\frac{1}{\omega RC}$. Thus, for a large frequency f, well away from the cutoff frequency f_c, doubling the frequency cuts the gain of the circuit in half,

$$G(4\pi f) \approx \frac{G(2\pi f)}{2}.$$

The circuit is a lowpass filter. It contains a single component with frequency-dependent impedance, the capacitor C, therefore it is first order. The filter's Bode plot, Fig. 3.28, represents the gain and the phase of the filter in compact form. The step response, Fig. 3.29 shows the answer of the filter to a unit step.

We can analyze the circuit in Fig. 3.30 using a similar argument. We excite the circuit at point A with the sinusoidal voltage $V_0(t) = V_0 \cos \omega t$ and observe the circuit's response V_o at point B. The impedance Z of the inductor L paralleled with the capacitor C is $Z = L \parallel C = \frac{i\omega L}{1-\omega^2 LC}$, where L is the inductance of L and C is the capacitance of C. Substituting in (3.5) gives us the amplitude and phase of the response,

$$V_o = V_0 \frac{i\omega L}{i\omega L - R\left(\omega^2 LC - 1\right)} = V_0 H(\omega),$$

where R is the resistance of the resistor R. The amplitude of the response has a maximum at $\omega_c = \frac{1}{\sqrt{LC}}$, its center frequency f_c is $f_c = \frac{1}{2\pi\sqrt{LC}}$. At this frequency, the parallel combination of L and C has infinite impedance, the circuit resonates. The circuit's Bode plot, Fig. 3.31 sketches the response. The resonant peak becomes more pronounced the larger the resistance R is.

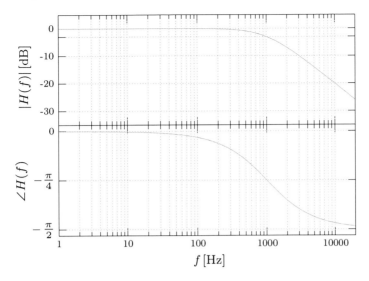

Fig. 3.28 Bode plot of a first-order lowpass filter with a cutoff frequency of 1 kHz

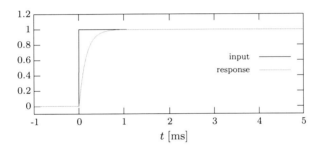

Fig. 3.29 Step response of a first-order lowpass filter with a cutoff frequency of 1 kHz

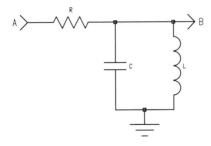

Fig. 3.30 A resonant filter circuit. The inductor shortens low-frequency content to ground, while the capacitor shortens high-frequency content

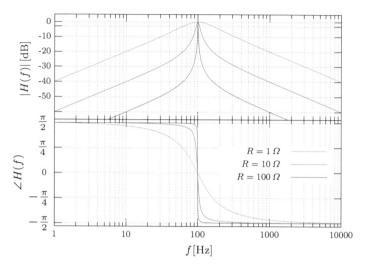

Fig. 3.31 Bode plot of the circuit in Fig. 3.30. The center frequency f_c is 100 Hz

The average magnitude of a time-dependent voltage $V(t)$ and a time-dependent current $I(t)$ is often given by the root-mean-square voltage V_{RMS} or the root-mean-square current I_{RMS},

$$V_{RMS} = \sqrt{\frac{\int_{t_0}^{t_1} V^2(t)\,dt}{t_1 - t_0}},$$

$$I_{RMS} = \sqrt{\frac{\int_{t_0}^{t_1} I^2(t)\,dt}{t_1 - t_0}}.$$

The voltage V_{RMS} is the constant positive voltage which when applied across some resistor makes the resistor dissipate the same power as does the voltage $V(t)$ on average when applied across the resistor; the current I_{RMS} is the positive constant current which when sent through some resistor makes the resistor dissipate the same power as does the current $I(t)$ on average when sent through the resistor. The time between t_0 and t_1 has to be chosen long enough so that the fluctuations of $V(t)$ or $I(t)$ do not influence the value of V_{RMS} or I_{RMS} significantly.

Let us assume we apply a sinusoidal voltage $V(t) = V \cos \omega t$ across some circuit with two terminals. Let us assume furthermore that $V(t)$ causes a current $I(t) = I \cos(\omega t + \phi)$ to flow. The average power P consumed by the circuit depends on the power factor $\cos \phi$,

$$P = V_{RMS} I_{RMS} \cos \phi.$$

Fig. 3.32 Three-phase
alternating current supply.
The three voltage sources
produce three sinusoidal
voltages, all with the same
frequency and the same
amplitude. The phases of any
two pairs are $2\pi/3$ apart

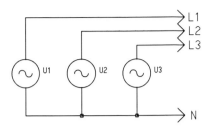

3.10 Three-Phase Circuits

A three-phase supply, Fig. 3.32, consists of three voltage sources. The voltage source
U1 produces the voltage $V_1(t) = V\cos(\omega t)$, the source U2 the voltage $V_2(t) = V\cos(\omega t - 2\pi/3)$, and the source U3 the voltage $V_3(t) = V\cos(\omega t + 2\pi/3)$. The
wires L1, L2, and L3 are the phases of the supply, the wire N is the neutral. A balanced
load draws from the three phases of a three-phase supply three currents having the
same waveforms and the same phase relations with the respective voltages. When
a three-phase supply is loaded with a balanced load, the neutral wire carries no
current. The phase difference between the phase currents and their respective phase
voltages in a three-phase system with balanced load is called ϕ; the power factor
$\cos\phi$ determines how much power such a system transmits.

The electrical grid for distributing electrical power is mostly a three-phase system.
In Europe, the mains frequency is 50 Hz in North America it is 60 Hz. Light loads
may operate from a single phase only, while heavy loads must operate from all three
phases and be balanced. In order to keep the phase currents to a minimum for a given
power consumption, loads are required to show a power factor $\cos\phi$ close to one.

3.11 Bibliographical Notes

Physics texts such as (Meschede and Gerthsen 2010; Tipler and Llewellyn 2007)
cover electrostatics and electrodynamics in detail. For a more technical approach
see introductory texts to electrical engineering, e.g. (Harriehausen and Schwarzenau
2013; Rizzoni and Kearns 2014). The classic book by Horowitz and Hill (1989)
contains an easy introduction to electrical engineering. Many application notes by
the late Jim Williams in (Dobkin and Williams 2011, 2012) contain sections on
measurement techniques.

3.12 Exercises

Exercise 3.1 Sketch the voltage at point A and the current through the inductor L
in the circuit in Fig. 3.33. The switch SW conducts for 3 μs. The component D is a
transient voltage suppressor. As long as the absolute value of the voltage across it is

Fig. 3.33 Circuit for
Exercise 3.1. The switch SW
conducts for 3 μs

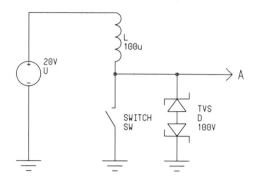

less than 100 V it is an open circuit. With larger voltages across it D behaves like a
short circuit.

Exercise 3.2 Show that energy is preserved in the circuit in Fig. 3.25 with $R = 0$.

Exercise 3.3 Analyze the circuit in Fig. 3.34.

Exercise 3.4 Analyze the circuit in Fig. 3.35.

Exercise 3.5 Compute the root-mean-square voltage V_{RMS} for a sinusoidal voltage
with amplitude V.

Exercise 3.6 Compute the voltages between the phases of a three-phase supply.

Fig. 3.34 Circuit for
Exercise 3.3

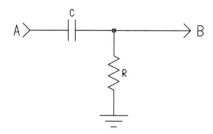

Fig. 3.35 Circuit for
Exercise 3.4

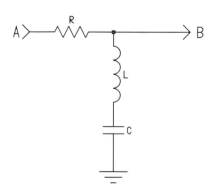

3.13 Lab Exercises

Exercise 3.7 Build the circuit in Fig. 3.33 and verify its behavior.

Exercise 3.8 Build the circuit in Fig. 3.27 and verify its behavior. Use a function generator as source and set it up to logarithmically sweep the frequency band of interest.

Exercise 3.9 Build the circuit in Fig. 3.30 and verify its behavior. Use a function generator as source.

Exercise 3.10 Build the circuit in Fig. 3.34 and verify its behavior.

Exercise 3.11 Build the circuit in Fig. 3.35 and verify its behavior.

References

Dobkin B, Williams J (eds) (2011) Analog circuit design: a tutorial guide to applications and solutions. Elsevier, Amsterdam

Dobkin B, Williams J (eds) (2012) Analog circuit design volume 2: immersion in the black art of analog design. Elsevier, Amsterdam

Harriehausen T, Schwarzenau D (2013) Moeller Grundlagen der Elektrotechnik. Springer, Heidelberg

Horowitz P, Hill W (1989) The art of electronics. Cambridge University Press, Cambridge

Meschede D, Gerthsen C (2010) Gerthsen Physik. Springer, Heidelberg

Rizzoni G, Kearns JA (2014) Principles and applications of electrical engineering. Mc Graw Hill, New York

Tipler P, Llewellyn R (2007) Modern physics. W. H Freeman, New York

Chapter 4
Digital Electronics

Digital electronics is the basic technology for today's computers and for most of today's communications equipment. Mastering digital electronics allows us to move some tasks an embedded system has to perform into dedicated digital hardware. Such hardware implementation usually exhibits much higher performance than a software implementation of the same task executing on a processor. The temporal behavior of bistable storage elements forces us to embrace a synchronous design discipline. Following this discipline produces circuits with predictable timing properties. Tasks demanding highly predictable timing, therefore, benefit most from being implemented directly in digital hardware.

4.1 Inputs

A digital circuit consists of components and of wires connecting the components' inputs and outputs. It drives wires carrying a logic One to a voltage close to the voltage of its positive supply rail and wires carrying a logic Zero to a voltage close to the voltage of its negative supply rail. Between the voltage band representing a logic One and the voltage band representing a logic Zero sits the forbidden zone. The larger the forbidden zone the more immune the circuit is to noise.

A logic input expects the connecting wire to either sit at a voltage level representing a logic One or at a voltage level representing a logic Zero. When being driven into the forbidden zone for a time longer than is required for switching between the logic states it penalizes such unruly behavior with elevated power consumption. In extreme cases oscillation and destruction are possible. The time it takes a wire to swing from logic Zero to logic One is the rise time t_r; the time it takes to swing from logic One to logic Zero is the fall time t_f. A logic input requires rise and fall times in the range of 30 ns. An analog signal, which may sit in the forbidden zone for an indefinite amount of time, is not suitable for driving a logic input.

A Schmitt trigger input can be driven with an analog signal. When the signal at the input transitions slowly from 0 V to the supply voltage the input initially

© Springer International Publishing Switzerland 2015
P. Hintenaus, *Engineering Embedded Systems*, DOI 10.1007/978-3-319-10680-9_4

interprets the signal as Zero. As soon as the signal reaches the threshold V_{T+} the input interprets the signal as One. When the signal ramps down from the supply voltage toward $0\,V$ the input initially interprets the signal as One. As soon as the signal reaches the threshold V_{T-} the input changes its interpretation of the signal to Zero. The thresholds obey $V_{T+} > V_{T-}$, the Schmitt trigger has a hysteresis of $V_H = V_{T+} - V_{T-}$. A large hysteresis improves immunity to noise. A Schmitt trigger input circuit shows elevated power consumption while its input signal sits in the middle between the supply rails.

4.2 Outputs

A push-pull output stage establishes a low-impedance connection between the wire it connects to and the positive supply rail in order to drive the wire to a One. In order to drive the wire to a Zero it establishes a low-impedance connection between the wire and the negative supply rail. Connecting two push-pull outputs to the same wire is an error. The first output may try to drive the wire to a One while simultaneously the second tries to drive the same wire to a Zero. This results in a low-impedance connection between the supply rails and excessive power consumption.

A tri-state output buffer has an additional enable input besides its signal input. As long as the output is disabled the buffer effectively disconnects itself from the wire connected to it. When it is enabled it behaves like a push-pull output. As long as at most a single output is enabled several tri-state outputs can connect to a single wire. This allows building bidirectional buses. Typically a bus master, a microprocessor, for example, is responsible for providing the enable signals for the tri-state drivers in such a way that no two drivers are enabled simultaneously. In order to establish a valid logic level when the wire is not driven by any output a pull-up or pull-down device is added to the wire. A resistor between the wire and the positive rail acts as a pull-up, while a resistor between the wire and the negative rail acts as a pull-down. Pull-up and pull-down devices typically have resistances between $1\,k\Omega$ and $10\,k\Omega$. An alternative consuming less power is a bus hold device, also known as keeper latch, which pulls the wire with a high impedance to the logic level the wire was last driven to. Bus hold devices sometimes are built into logic inputs.

An open-drain output pulls the wire connected to it to Zero by establishing a low-impedance connection between the wire and the negative supply rail. It does not pull the wire to One; it disconnects itself from the wire instead. Therefore, a pull-up device has to provide the logic One. Several open-drain outputs may connect to the same wire. If at least one of them drives Zero onto the wire, overriding the pull-up, the wire will carry the Zero. Otherwise the pull-up will drive the wire to One.

4.2.1 The Nine Logic Values

Several outputs may drive a single wire. For representing the impedance of the different outputs we need more than two values for describing a wire's state. An idealized

Table 4.1 Function table of the wire operation

	U	X	0	1	Z	W	L	H	–
U	U	U	U	U	U	U	U	U	U
X	U	X	X	X	X	X	X	X	X
0	U	X	0	X	0	0	0	0	X
1	U	X	X	1	1	1	1	1	X
Z	U	X	0	1	Z	W	L	H	X
W	U	X	0	1	W	W	W	W	X
L	U	X	0	1	L	W	L	W	X
H	U	X	0	1	H	W	W	H	X
–	U	X	X	X	X	X	X	X	X

wire is a commutative and associative operation, Table 4.1, combining the logic values driven onto it. The idealization does not take into account the time for signal propagation.

When simulating a logic circuit the uninitialized value, U, serves as initial value for all wires in a circuit. Persisting uninitialized values in a simulation hint at errors in the simulated design. The Forcing Unknown, X, indicates error conditions such as bus conflicts. The Forcing Zero, 0, represents a low-impedance connection to the negative supply rail, while the Forcing One, 1, represents a low-impedance connection to the positive supply rail. A push-pull output imprints Forcing Zeros and Forcing Ones onto the wire it connects to. A disabled tri-state output drives the high-impedance value, Z, onto the wire it connects to. Termination structures, necessary when dealing with long wires, drive the Weak Unknown, W, onto the wires they connect to. A pull-down device drives a Weak Zero, L onto the wire it connects to, while a pull-up device drives a Weak One, H. The don't care value, –, is used in truth tables for indicating that the output value is irrelevant for a combination of input values or that in input is irrelevant for a certain combination of the other inputs. Don't Cares allow logic synthesis tools to generate simpler circuits. Inputs have very high impedance. Therefore, they do not distinguish between Forcing and Weak Zeros and between Forcing and Weak Ones.

4.3 Combinatoric Logic

Combinatoric logic is built from logic gates and wires. A logic gate has a number of inputs and a single push-pull output. A circuit is combinatoric if we can arrange its diagram with the circuit's inputs to the left and the circuit's outputs to the right. Furthermore, the gate whose output drives a wire must appear left of any gate which connects to the wire with one of its input and left of any output connected to the wire. Therefore, combinatoric circuits have no loops. They are stateless, and realize logic formulas. For building these circuits we have and-gates, Fig. 4.1, or-gates, Fig. 4.2, and inverters, Fig. 4.3, at our disposal. In circuit diagrams the supply rails

Fig. 4.1 And-gate with two
inputs and its truth table

A	B	Y
0	0	0
0	1	0
1	0	0
1	1	1

Fig. 4.2 Or-gate with two
inputs and its truth table

A	B	Y
0	0	0
0	1	1
1	0	1
1	1	1

Fig. 4.3 Inverter and its truth
table

A	Y
0	1
1	0

are shown only when necessary for understanding the circuit. Instead of drawing separate inverters we indicate inverted inputs and inverted outputs with small circles. When drawing a circuit we strive for presenting the circuit's function clearly. We delegate the minimization the number of gates used by the circuit to an electronic design automation (EDA) tool.

A multiplexer, Fig. 4.4, switches its output between two or more inputs under the control of select inputs.

A decoder, Fig. 4.5, decodes binary input. It has an output for any possible combination. The outputs of a decoder often drive inverted inputs. Therefore, they usually are inverted themselves.

A gate cannot react instantaneously to changes at its inputs. It changes its output to the new value within its propagation delay time t_{pd}. If several inputs change their value almost instantaneously, the output of the gate is Forcing Unknown until the time t_{pd} has elapsed.

The temporal behavior of logic circuits is described by timing diagrams. Such a diagram shows how the states of the circuit's wires evolve over time. Both a Weak and a Forcing One are represented by ⊞; a Weak and a Forcing Zero are represented by ⊞. A Weak and a Forcing Unknown shows as ■; a Don't Care shows as ⊞. High Impedance shows as ⊞ .

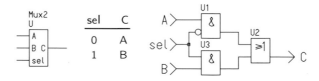

Fig. 4.4 Multiplexer with two inputs, its truth table and its circuit

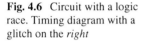

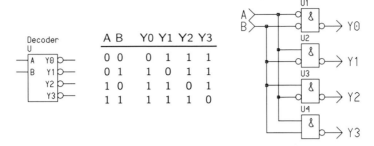

A B	Y0 Y1 Y2 Y3
0 0	0 1 1 1
0 1	1 0 1 1
1 0	1 1 0 1
1 1	1 1 1 0

Fig. 4.5 Two to four decoder, its truth table and its circuit

Fig. 4.6 Circuit with a logic race. Timing diagram with a glitch on the *right*

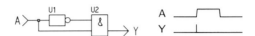

Consider the circuit in Fig. 4.6. Naively we may think that the output Y of the and-gate U2 is constant Zero. When the input A switches from Zero to One the two inputs of the and-gate may be both One momentarily because of the propagation delay of the inverter U1. Whether the output Y reflects this depends on the physical properties of the circuit. The output may become One momentarily, a so-called glitch; it may only make it to halfway between the supply rails before returning to Zero, a runt pulse; or it may not move much at all. A logic hazard occurs when two inputs to a circuit change their states (almost) simultaneously, or when two signals derived from the same inputs reconvene at the same gate.

4.4 Sequential Logic

Sequential circuits have state. Flip-flops store this state. The simple flip-flop, Fig. 4.7, is the building block for static random access memories. A Zero at the clear input C clears the flip-flop unconditionally. A Zero at the preset input P sets the flip-flop provided the input C is One.

A rising-edge triggered flip-flop changes its state at points in time determined by the rising edges, the transitions from Zero to One, of a signal fed into the flip-flop's clk input. A falling-edge triggered flip-flop reacts to the One to Zero transitions at its clk input. A rising-edge triggered D flip-flop, Fig. 4.8, copies at the points in time defined by the rising edges at its clk input the values at these times at its D input to

Fig. 4.7 Simple flip-flop and its truth table

C P	Q
0 –	0
1 0	1
1 1	Q

clk	D	Q
0	–	Q
1	–	Q
↑	D	D

en	clk	D	Q
0	–	–	Q
0	↑	–	Q
1	0	–	Q
1	1	–	Q
1	↑	D	D

Fig. 4.8 *Left side* D flip-flop and its truth table. *Right side* D flip-flop with clock enable and its truth table. The symbol ↑ indicates a rising edge

its Q output. The output Q will change within the flip-flop's propagation delay. The clock enable input en, when available, allows skipping clock edges.

4.4.1 Flip-Flops and Metastability

For ensuring correct operation a D flip-flop requires its D input to be stable before a rising edge on its clk input for its setup time t_{su}. Furthermore, it requires D to be still stable after the edge for its hold time t_h. The outputs of any gate or flip-flop react to changes at the inputs not earlier than specified by the minimum value for the propagation delay time t_{pd}. Within a logic family this time is longer than the hold time t_h of a flip-flop. The signal driving the clock enable input of a flip-flop must satisfy similar requirements. Violating the flip-flop's setup or hold time results in unpredictable behavior. The flip-flop's output may swing to a voltage in the middle of the forbidden zone, indicated with 〰 in timing diagrams, and stay at this voltage for an arbitrary time; the flip-flop has become metastable. Eventually, the output will fall back randomly either to Zero or to One, see Fig. 4.9. The times the flip-flop stays in the metastable state are distributed exponentially.

The signals driving the preset and clear inputs of a simple flip-flop must have a minimum pulse width. If this pulse width requirement is violated the flip-flop may become metastable. Metastability is not a deficiency of flip-flops; any bistable mechanism has a metastable state.

4.4.2 Synchronous Logic

Glitches and metastability seem to turn the design of digital logic into an endeavor with unpredictable outcome. Luckily we can identify a design discipline, the

Fig. 4.9 Metastable behavior resulting from a setup time violation

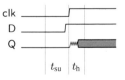

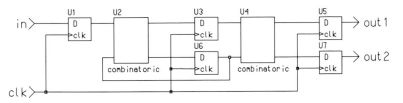

Fig. 4.10 Synchronous design style. All flip-flops receive their clocks from the same source. The inputs to combinatoric parts are provided by the outputs of flip-flops or by input signals synchronized to the global clock. The outputs of combinatoric blocks are caught by flip-flops

synchronous discipline, avoiding both problems. A synchronous circuit uses either rising-edge triggered or falling-edge triggered flip-flops only. For our discussion we choose the rising edge. The signals driving the clock inputs of the flip-flops, the clock signals, are periodic and exhibit constant frequency and phase relationships. Moreover, each input signal of the circuit is synchronous to one of the clock signals. More precisely, each edge of the input signal must fall into a short time interval after a rising edge of the associated clock signal. An input must be delayed for at least a flip-flop's hold time t_h. Outputs may still exhibit glitches. If a glitch-free output is required it must be the direct output of a flip-flop. The write-enable line of asynchronous memory for example must be glitch-free.

In the simplest case, Fig. 4.10, a single clock signal clk drives the clock inputs of all flip-flops. In timing diagrams arrows indicate the clock edges at which the flip-flops change state. Input signals, synchronous with the clock, or the outputs of flip-flops drive the inputs of combinatoric circuits. After each rising edge of the clock signal the outputs of the combinatoric parts may exhibit glitches and runt pulses. The flip-flops at the outputs of a combinatoric part hide this behavior from the rest of the system. The outputs must have settled down to the correct values at least a setup time before the next rising edge of the clock signal. The times required for the outputs of the combinatoric circuits to settle together with the setup time of the flip-flop determine the maximum clock frequency the circuit is able to operate at. Such a synchronous circuit is active right after each rising edge of the clock. For the rest of each clock cycle it sits idle.

More elaborate synchronous systems use not only the single clock signal clk but also signals derived by clock dividers from clk for triggering flip-flops. The frequencies of the derived signals are integer fractions of the frequency of the clock signal.

4.4.3 State Machines

When we implement a state machine in synchronous logic we may assume that the inputs are synchronous with the common clock signal. First we represent each state with a separate flip-flop. One flip-flop—representing the state the machine is in—is set; all others are reset. This scheme is called one-hot encoding. For establishing the state machine's initial state at startup we use D flip-flops with set or reset input,

clk	D	S	Q
0	–	–	Q
1	–	–	Q
↑	–	0	1
↑	D	1	D

clk	D	R	Q
0	–	–	Q
1	–	–	Q
↑	–	0	0
↑	D	1	D

Fig. 4.11 *Left side* D flip-flop with synchronous set and its truth table. *Right side* D flip-flop with synchronous reset and its truth table. The symbol ↑ indicates a rising edge

Fig. 4.11. If the set input S of a settable D flip-flop is Zero during a rising edge on the flip-flop's clock input the flip-flop is set regardless of the D input. If the reset input R of a resettable D flip-flop is Zero during a rising edge the flip-flop is reset. We represent the initial state of the state machine with a settable flip-flop and all other states with resettable flip-flops. A common reset signal, synchronous to clk, drives the single set and all reset inputs. When the reset signal becomes One the state machine starts to work.

Next we realize the state transitions. All gates we use in the sequel shall have the required number of inputs. To each state S we associate an or-gate. We wire the gate's output to the D-input of the flip-flop representing S. We consider each transition T terminating in S. Let S' be the state in which T originates. An and-gate represents T; its output is wired to an input of the or-gate associated with S. The output of the flip-flop representing S' is wired to an input of the and-gate. The other inputs of the and-gate are wired to the inputs of the state machine, directly for the inputs that must be One for the transition to be taken, via an inverter for those that must be Zero.

The circuit in Fig. 4.14 realizes the state machine in Fig. 4.12; Fig. 4.13 shows a timing diagram for this circuit. The flip-flops U0–U3 represent states 0–3; the flip-flop U4 represents the state Start. The gates U5 and U6 for example, realize the transition T_{12}, the gate U8 realizes T_{11}, U9 realizes T_7 and U10 realizes T_0.

Last we have to realize the output function. We consider each output O of the state machine. We wire the output of an or-gate to the output O. For every combination of a state S with the input values for which O is one we add an and-gate. We wire the output of the and-gate to an input of the or-gate associated with O. We wire the output of the flip-flop representing S to an input of the and-gate. The other inputs of the and-gate are wired to the inputs of the state machine, directly for the inputs that must be One for the output O to become One, via an inverter for the inputs that must be Zero for O to become One.

In Fig. 4.14 the gates U29–U33 realize the output function for up. For the output function of the cnt output we reuse these gates and add gates U34–U39.

4.4.4 Transparent Latch

Another storage element is the transparent latch, Fig. 4.15. While its latch enable input LE is One the latch copies the value of its D input to its output Q; the latch is transparent. The moment LE becomes Zero the latch freezes its output Q.

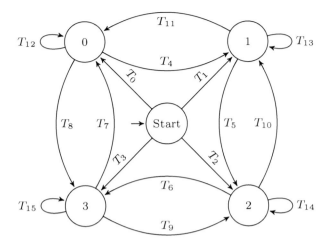

	Condition		Condition		Condition		Condition
T_0	$(A, B) = (0, 0)$	T_4	$(A, B) = (1, 0)$	T_8	$(A, B) = (0, 1)$	T_{12}	**default**
T_1	$(A, B) = (1, 0)$	T_5	$(A, B) = (1, 1)$	T_9	$(A, B) = (1, 1)$	T_{13}	**default**
T_2	$(A, B) = (1, 1)$	T_6	$(A, B) = (0, 1)$	T_{10}	$(A, B) = (1, 0)$	T_{14}	**default**
T_3	$(A, B) = (0, 1)$	T_7	$(A, B) = (0, 0)$	T_{11}	$(A, B) = (0, 0)$	T_{15}	**default**

State	A	B	cnt	up	State	A	B	cnt	up	State	A	B	cnt	up	State	A	B	cnt	up
0	0	0	0	0	1	0	0	1	0	2	0	0	0	0	3	0	0	1	1
0	1	0	1	1	1	1	0	0	0	2	1	0	1	0	3	1	0	0	0
0	1	1	0	0	1	1	1	1	1	2	1	1	0	0	3	1	1	1	0
0	0	1	1	0	1	0	1	0	0	2	0	1	1	1	3	0	1	0	0
Start	–	–	0	0															

Fig. 4.12 State transition diagram, transitions, and output function of a sample state machine. The state machine has two inputs A and B and two outputs cnt and up

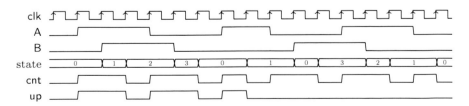

Fig. 4.13 Timing diagram of the circuit in Fig. 4.14 realizing the state machine in Fig. 4.12. The state is not visible outside the state machine

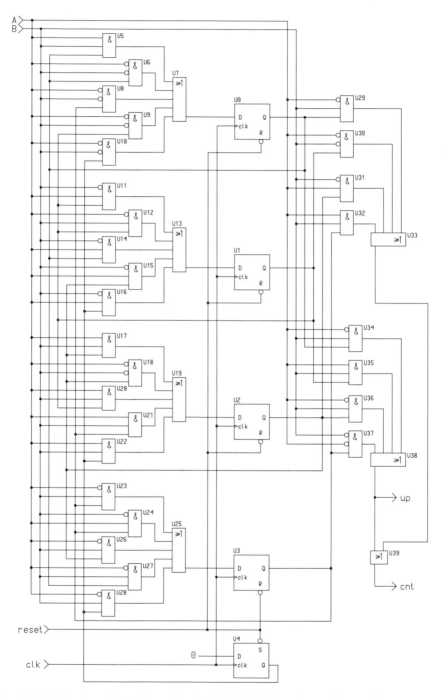

Fig. 4.14 Realization of the state machine in Fig. 4.12 in logic using one-hot encoding

LE D Q

0 –   Q

1 D   D

Fig. 4.15 A transparent latch and its truth table

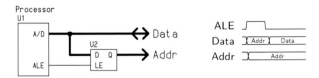

Fig. 4.16 Circuit and timing of a latch catching addresses off a multiplexed bus. Thick lines indicate parallel wires

Some processor chips use the same lines for transferring addresses and data. During a memory access the processor first sends the address and then switches the lines to sending or receiving data. It provides an address latch enable signal ALE to indicate when it outputs the address. A transparent latch catches the address and provides it to the memories for the rest of the access. Edge triggered flip-flops can be used, but they make the address available to the rest of the system only after the falling edge of ALE, while a transparent latch makes the address available shortly after the rising edge of ALE, Fig. 4.16.

In order to work perfectly a transparent latch requires its D input to be stable before a falling edge on its LE input for at least its setup time t_{su}. Furthermore, it requires its D input to be stable after a falling edge for at least its hold time t_h. Violating the setup time or the hold time of a transparent latch may cause metastability.

4.5 Clock Domains

The synchronous design discipline solves the problems associated with glitches and metastability. Several problems remain, though. Building a fully synchronous system forces us to reliably distribute a common clock signal to all parts of the system. Transmitting a high-frequency clock over long distances is technically challenging. Moreover, the clock source becomes a single point of failure. For these reasons practically all spatially dispersed systems have several clock sources. The physical world follows its own rhythms; signals originating there will be asynchronous with the clocks of any system processing these signals.

The systems we build consist of several clock domains. A clock domain has a single clock generator. All edges within a clock domain have fixed temporal relation to the domain's clock signal. A signal entering a clock domain must be synchronized

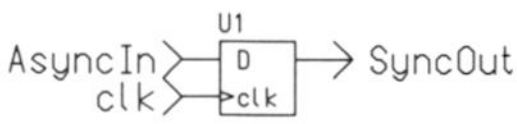

AsyncIn ▷──[U1 : D / clk]──▷ SyncOut

Fig. 4.17 A single stage synchronizer. The probability for it to become metastable is too high for this synchronizer to be useful

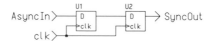

Fig. 4.18 A two stage synchronizer. In order to further reduce the probability of failure additional flip-flops can be added

to the domain's clock. The circuit in Fig. 4.17 will do that, but there is a hitch. At some point the signal to be synchronized, AsyncIn, will exhibit an edge violating the flip-flop's setup or hold time. The flip-flop may become metastable long enough to cause havoc in the downstream logic. The probability of malfunction is too high for this circuit to be useful.

In principal, we cannot build a perfect synchronizer. What we can do, however, is to reduce the probability of malfunction to very low levels by giving the synchronizing flip-flops time to drop out of the metastable state. The circuit in Fig. 4.18 does just that. Adding a third flip-flop will reduce the probability of failure further. Using metastability-hardened flip-flops helps too, but might come at the cost of increased power consumption.

4.6 Digital Building Blocks

A register consists of a number of flip-flops controlled by a common clock signal and, if available, by a common clock enable. Usually a register stores a word of data.

An accumulator is a register paired with an adder-subtractor circuit. In a single clock cycle an accumulator can initialize the register with the data at its inputs, clear the register, add or subtract the data at its inputs from the data stored in its register and store the result in its register.

Counters consist of flip-flops combined with combinatoric logic so that for each clock cycle the outputs take a step in the code the counter counts in. A simple binary counter counts up to a predetermined value, not necessarily a power of two, and wraps around to zero. An up-down counter takes a step up or a step down in each clock cycle depending on a control input.

A shift register converts a serial input into a parallel output or a parallel input into a serial output. It consists of a linear arrangement of flip-flops having a common clock. The output of one flip-flop is wired to the input of the next. In shift registers allowing parallel loading multiplexers are introduced between the flip-flops for switching between the load and the shift operation.

4.7 Realization of Logic Circuits

So-called random logic is composed of digital building blocks where each building block resides in a separate integrated circuit. A circuit board, designed for the circuit, realizes the connections between the building blocks. The building blocks usually are chosen from a single logic family. A logic family consists of integrated circuits having the same electrical specifications and compatible timing. In particular the minimum propagation delay times of the gates and the hold times of the flip-flops match. The individual integrated circuits implement a single building block, such as a counter, a decoder, or a shift register. Although the number of the building blocks available within a family has dwindled, we still can build small digital systems by hardwiring a handful of these chips together. Prototyping is rather painful, every bug forces us to rewire parts of the circuit.

Programmable logic chips allow us to configure the chips' functions freely using electronic design automation tools. A programmable logic chip consists of a number of identical cells and a configurable interconnection network. A cell consists of a configurable building block for combinatoric logic and a flip-flop. The actual structure of a programmable logic chip is of secondary importance to the designer. Design tools translate a circuit into the resources available on the chip automatically.

A complex logic device, CPLD for short, uses logic gates with configurable switches as building block for the combinatoric logic in each cell. Nonvolatile memory cells control the switches in the cells and in the interconnection network. Complex logic devices are available with up to 300 cells in packages with up to 200 connections usable for logic signals. They lend themselves well to the implementation of autonomous periphery in microprocessor-based systems. Most complex logic devices can be reconfigured. Fixing a bug requires us to erase the memory cells and reprogram them.

A field programmable gate array, FPGA for short, uses random access memory storing truth tables as building block for the combinatoric logic in each cell. Some of this memory can be combined for providing on-chip storage capabilities. The content of random access memory cells configures the interconnection network. An FPGA must be configured by initializing the memories before it can operate. Besides the cells and the interconnection network an FPGA may contain complex functions such as multipliers or even complete microprocessors. The largest FPGAs available contain more than two million cells in packages with up to 1,200 connections usable for logic. They lend themselves for the implementation of complete systems on a single chip. Changing the configuration entails rebooting the chip.

Custom logic chips offer the highest performance and flexibility. Developing such a chip, however, is costly. Fixing a bug usually requires a new production run.

Small designs can be accomplished by drawing circuit diagrams. Anything larger than a peripheral is better designed using a text-based hardware description language such as VHDL or Verilog (IEEE 2006, 2009). A synthesis tool translates a design into the primitives provided by the target technology. The result is a circuit diagram consisting of the target's primitives only. A so-called net list represents this circuit diagram. Another tool places the primitives on the chip and routes the connections. For finding bugs one simulates first at the design level without considering propagation delays, setup times, and hold times. After synthesis the design is simulated again, this time considering the propagation delays, setup times, and hold times of the primitives used. After placement and routing the design is simulated a third time also considering the delays introduced by the interconnections. Small designs can be completed using free tools provided by the silicon vendors. The design of a large digital system, however, is a complex project requiring an elaborate and costly toolchain.

4.8 Support Circuits

Clock generators, often integrated into logic chips, either use a quartz crystal or a ceramic resonator as timing element. Quartz crystals provide higher precision than ceramic resonators but are more costly and require more space. The maximum frequency of such a clock generator's output signal is limited to about 20 MHz. A frequency multiplier produces the high-frequency signal required by the logic circuit from the output of the clock generator.

A reset generator monitors the supply voltage. It keeps the system in the reset state until the supply voltage is within specifications. The voltage at which the reset generator releases the system should be halfway between the minimum voltage the system requires for operation and the minimum voltage the power supply delivers.

4.9 Analog Aspects of Fast Logic

When we realize a high-speed digital system on a circuit board, analog and high-frequency effects deserve our attention. Disregarding these effects will at best result in a system exhibiting excessive power consumption and excessive emission of radio-frequency interference. At worst the system will not work at all.

Logic inputs and logic outputs exhibit a parasitic capacitance to the negative supply rail of about 10 pF. When we design a bidirectional bus driven with tri-state outputs we have to ensure each bus line is at a valid voltage when no output drives the bus. When using a pull-up resistor we also have to obey the required rise time t_r of the inputs. The pull-up resistor has to provide enough current to load the paralleled parasitic capacitances from a logic Zero to a logic One within t_r. Let the supply voltage be 3.3 V, the minimum voltage for a logic One be 2.15 V, the rise time t_r be 33 ns, and the combined parasitic capacitance be 50 pF. Choosing a pull-up with a resistance of 560 Ω results in a voltage of about 2.3 V after the rise time. Whenever some output drives the bus line to a logic Zero a rather high current of about 6 mA flows through the pull-up. A bus hold device does not consume such a high current. Unlike a pull-up it does not impose a logic One onto the bus line. It merely keeps the bus line at the state to which a low-impedance output has pulled it previously. Holding the bus line requires enough current for the combined leakage currents of all inputs and outputs, about 50 μA in our example.

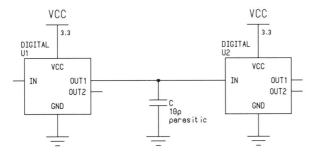

Fig. 4.19 A digital output driving an input via a wire with parasitic capacitance

The outputs of fast logic chips produce signals with rise and fall times of about 1 ns. These signals have an appreciable frequency content well above 3 GHz. A copper trace on a circuit board delays electrical signals by about 55 ps per centimeter of length. Therefore, a rising edge has a spatial length l of 18 cm. For a wire shorter than about $l/6$ we can assume that the wire distributes its state instantaneously to all inputs connected to it. Circuits smaller than $l/6$ are called lumped, larger ones are called distributed. In distributed circuits we have to take into account wave propagation effects on the wires, such as reflections, see, Johnson and Graham (1993).

A digital output has to charge or discharge the combined parasitic capacitance of the wire and the inputs it drives, Fig. 4.19. When the output is driven to One, the parasitic capacitance will store an energy of 50 pJ. When the output switches from One to Zero the parasitic capacitance has to be discharged by a current that flows into the output and further into ground. Thus, a 100 MHz square-wave signal transmitted across such a wire will make the output dissipate a power of 50 pJ 100 MHz = 5 mW. If the output produces rise and fall times of 1 ns, an average current of 33 mA will flow out of the output during a rising edge and into the output during a falling edge. The charge for these currents must be stored locally at the output. Digital chips have several connections for the positive and the negative supply rail. Each such pair of connections must be decoupled with a capacitor with excellent high-frequency properties. A capacitance of about 100 nF will do. These decoupling capacitors store the charge for the currents flowing during the fast edges. In order to close the electrical circuit the currents flowing from the output to the inputs over the wire have to flow back to the output in the ground connection. Given a choice the return current will take the path of the wire. Any detour from the ideal results in the emission of radio frequency interference. A ground plane, a solid layer of copper connected to the negative supply rail, creates ideal paths for all return currents. Together with a plane connected to the positive supply rail it forms a capacitor with excellent high-frequency properties.

In the circuit in Fig. 4.20 the digital component U1 drives the input of another digital component U2. The ground connection of U1 has an inductance of 20 nH which is typical for a trace with a length of about 3 cm on a circuit board combined with the inductance of the ground pin itself. When the output OUT2 of U1 swings from high to low within one nanosecond, an average current of 33 mA has to flow into OUT2 and out of the GND pin of U1 for the duration of the edge. For simplicity we

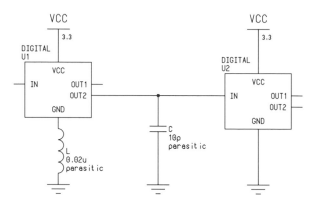

Fig. 4.20 A digital output driving an input via a wire with parasitic capacitance. The ground connection of the driving chip has parasitic inductance

assume that the current ramps up linearly to 66 mA during the first 500 ps and down linearly to 0 mA during the second 500 ps. The voltage at the GND pad of the silicon die inside of U1 will jump to $20\,\text{nH}\ 132\,\text{mA}\,\text{ns}^{-1} = 2.64\,\text{V}$ during the first half and to $-2.64\,\text{V}$ during the second, before returning to ground. Such an exertion, known as ground bounce, definitely interferes with the function of the circuit. The shortest possible connections of the GND pins to the ground plane minimizes the parasitic inductance and fixes ground bounce.

4.10 Bibliographic Notes

The construction of complex digital circuits is covered in great depth by many books on very large scale integrated circuit design. The book by Kaeslin is both comprehensive and readable (Kaeslin 2008). Before one embarks on the construction of a circuit board for high-speed logic the book by Johnson and Graham is required reading (Johnson and Graham 1993). Their second book provides even more in-depth information on distributing high-speed signals (Johnson and Graham 2003). The two most widely used hardware description languages are defined in IEEE (2009) and IEEE (2006). The discussion in 4.2.1 follows IEEE (1993). The book by Abelson and Sussman contains a beautiful program for simulating digital circuits written in Scheme (Abelson and Sussman 1996).

4.11 Exercises

Exercise 4.1 Design an edge detector. The edge detector has a clock input clk and a signal input A. The signal feeding A is synchronous to the clock. It has two outputs A↑ for indicating a rising edge at the beginning of the current clock cycle and A↓ for indicating a falling edge.

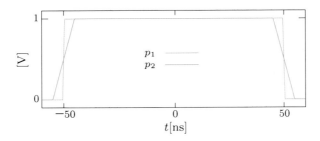

Fig. 4.21 The pulse p_1 has a rise and a fall time of 1 ns, the pulse p_2 of 10 ns

Exercise 4.2 Implement the state machine described in Fig. D.13.

Exercise 4.3 Compare the frequency content of the two pulses p_1 and p_2 in Fig. 4.21. The pulse p_1 has a rise and a fall time of 1 ns, while p_2 has a rise and a fall time of 10 ns.

4.12 Lab Exercises

Exercise 4.4 Implement the edge detector you designed in Exercise 4.1.

Exercise 4.5 Implement the circuit in Fig. 4.14.

Exercise 4.6 Implement the circuit you designed in Exercise 4.2.

References

Abelson H, Sussman GJ (1996) Structure and interpretation of computer programs, 2nd edn. MIT Press, Cambridge
IEEE (1993) IEEE Standard Multivalue Logic System for VHDL Model Interoperability (Std-logic1164). IEEE Std 1164–1993.doi:10.1109/IEEESTD.1993.115571
IEEE (2006) IEEE Standard for Verilog Hardware Description Language. IEEE Std 1364–2005 (Revision of IEEE Std 1364–2001).doi:10.1109/IEEESTD.2006.99495
IEEE (2009) IEEE Standard VHDL Language Reference Manual. IEEE Std 1076–2008 (Revision of IEEE Std 1076–2002).doi:10.1109/IEEESTD.2009.4772740
Johnson H, Graham M (1993) High speed digital design: a handbook of black magic. Prentice Hall Modern Semiconductor Design Series, Upper Saddle River
Johnson H, Graham M (2003) High-speed signal propagation: advanced black magic, 1st edn. Prentice Hall Press, Upper Saddle River
Kaeslin H (2008) Digital integrated circuit design: from VLSI architectures to CMOS fabrication, 1st edn. Cambridge University Press, New York

Chapter 5
Programmable Devices—Software and Hardware

Embedded systems usually are delivered as a combination of software and hardware. Unlike general purpose systems, where the hardware developer does not know which software the system will execute, we are here responsible for both—hardware and software, when developing an embedded system. This gives us the unique chance, and at the same time the daunting task to build hardware and software, together orchestrating a finely tuned solution to satisfy the customer's requirements in functionality, safety, and economical feasibility.

Embedded systems monitor and influence the environment they are working in. For successfully tackling such tasks, embedded systems must have a precise sense of time. Despite the speed of todays computers, providing precise timing with any certainty and in an economical way turns out to be a hard problem without a single, all-encompassing solution. Besides the processor, memories and peripherals determine the temporal behavior of our systems. In order to arrive at a reasonable compromise for achieving precise timing in a given design situation, we need an arsenal of lines of attack, both hardware and software based.

5.1 Challenges Particular to Embedded Systems Development

Unlike a conventional computer, an embedded system has to directly deal with the physics of its application. Programs typically abstract time away into just another variable if treating it at all. For embedded programs, however, time becomes very important. Nature marches on relentlessly, we cannot make it wait however hard we try. Therefore, our systems must bend to nature's will. For serving their purpose, they must acquire inputs from sensors and deliver results to actuators at precise instances in time. Think of controlling a car's air bags in case of a crash, firing them too early or too late by only a few milliseconds will make them completely ineffective. In a frontal impact against a fixed barrier with $50\,\mathrm{km\,h^{-1}}$ the driver's air bag must be fired 30 ms after the start of deceleration. The head of a driver restrained by a three-point harness touches the air bag after 54 ms. After 84 ms the driver's motion has been

© Springer International Publishing Switzerland 2015   95
P. Hintenaus, *Engineering Embedded Systems*, DOI 10.1007/978-3-319-10680-9_5

stopped; and after 150 ms the crash is over (Braess and Seiffert 2005). For the same situation, a passenger air bag is fired 40 ms after impact.

5.1.1 Time

Systems whose correctness or quality depends fundamentally on their temporal behavior are real-time systems. The instances in time at which such a system must interact with its environment, by acquiring inputs or delivering results, are called deadlines. Missing a soft deadline produces degraded but still useful results. Missing a firm deadline produces useless results but the consequences are still bearable. Missing a hard deadline has catastrophic results (Kopetz 2011). Systems having to obey hard deadlines are much harder and more costly to design than those having to deal with soft or firm deadlines only. The time for firing an air bag, for example, is a hard deadline.

5.1.2 Reliability

Individually and as society, we depend on many technical systems to function reliably. We expect a car's engine to start every time we turn the key. We expect electrical power to be delivered all year round without interruption. We expect production plants to operate uninterrupted but for scheduled maintenance periods. At the heart of all these systems sit embedded computers monitoring and controlling operations. These embedded systems have to operate flawlessly for very long timespans; human intervention for resetting the computers occasionally is not acceptable.

5.1.3 Safety

Embedded systems often control environments where their failure can result in massive loss of property, injury, and loss of life. Firing an air bag is a safety function. Reliable systems are not necessarily safe and safe systems are not necessarily reliable. A car, for example, can be perfectly reliable but its handling qualities may be so bad making it unsafe to drive. Methods for building safe systems not only have to include technical aspects but also societal and political ones. They deserve independent treatment (Leveson 1995, 2011).

5.1.4 Power

Many embedded systems are battery powered. While the performance of available batteries improves at a slow pace (Reddy 2010), much more endurance can be gained from making our devices more energy efficient. Every single instruction executed by some processor pumps some charge through the processor. Therefore, reducing the number of instructions a processor executes in a single second will help improve energy efficiency. Some systems even do away with the battery and rely on the little energy they can harvest from their environment. Such systems can be run indefinitely without requiring service. In order not to exhaust these systems' energy stores prematurely, very low power computation and communication is called for.

5.2 The Event Triggered Architecture

An event is a point in time where something interesting happens. So-called event generators are sources of events. A processor in an event triggered system wakes up at each event it is interested in, reacts to the event, and goes to sleep again. The systems we are considering consist of several clock domains. Even the simplest system has to deal with at least two clock domains; the world ticks differently than our systems' clock generators. When a processor reacts to an event which originates in a clock domain other than the processor's one, the event must be synchronized at the hardware level to the local clock of the processor's clock domain in order to curtail metastability. For each event generator g a processor listens to, the processor executes an event handler thread whose basic structure is sketched in Fig. 5.1.

Let us consider the implementation of state machines. We describe state machines destined for event-triggered systems slightly differently. Instead of stating the output function explicitly, we associate an action with each state transition. An action is executed whenever its associated transition is taken. In an event triggered system, the input sequence of a state machine will be implemented as a stream of events. Therefore, the state machine's transitions will be performed by event handlers. When the events of such a stream originate in different event generators in different clock domains, the order of closely spaced events is not defined anymore. Moreover, while one event handler performs a state transition for a state machine, another event handler might be woken up and may try to perform a different state transition for the same machine. In order to keep the implementation of the state machine correct, we have to synchronize the executions of the involved threads. In order to facilitate this analysis, we state for each transition the thread performing the transition.

Fig. 5.1 Basic structure of an event handler thread. The function NEXTEVENT returns the event, generator g has pending

```
1: thread EVENTHANDLER(g)
2:    loop
3:       wait for g
4:       e ← NEXTEVENT(g)
5:       handle e
6:    end loop
7: end thread
```

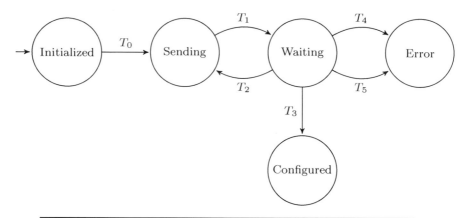

	Condition	Action	Thread
T_0		send first, start timer	APPLICATION
T_1	record sent	none	COMHANDLER
T_2	ACK & records left	send next, restart timer	COMHANDLER
T_3	ACK & no records left	stop timer	COMHANDLER
T_4	NACK	none	COMHANDLER
T_5	timeout	none	TIMER

Fig. 5.2 State transition diagram and transitions with actions of the machine of a configuration service for remote clients

Consider, for example, a network-based configuration service for some small remote subsystem. In order to configure the subsystem, a configuration master delivers a sequence of configuration records to the subsystem. After sending a configuration record to the subsystem over the network, the master waits for a positive acknowledgement, ACK, or a negative acknowledgement, NACK. The thread COMHANDLER services the events generated by the communication network. When entering the Waiting state of the state machine in Fig. 5.2, the master schedules a timeout, which prevents the master from getting stuck in the Waiting state in case the remote subsystem dies in the meantime. The thread TIMER is responsible for reacting to timeouts. There are two event handler threads, which perform transitions out of the state Waiting, the handler for the communication interface and the handler for the timeout timer. We have to synchronize execution of these transitions or else the implementation of the state machine will misbehave.

Notifying a processor of some event is supported by hardware. Event-triggered systems, therefore, can react very fast to an event. The multithreaded nature of these systems, however, renders it impossible to prove that the system satisfies its temporal requirements.

5.2.1 Implementing Event Triggered Systems

A processor, when fetching, decoding, and executing an instruction stream is oblivi-
ous of its surroundings and of the passage of time except for those instances when it
executes an instruction which probes either the environment or a timer. Such behavior
is hardly useful for general purpose systems and much less so for systems destined
for sensing and influencing the systems' surroundings.

5.2.1.1 Vectored Interrupt Systems

The key to making a processor responsive is a vectored interrupt system. Additional logic integrated into the processor, a so-called interrupt controller, monitors the
processor's environment. Interrupt request lines connect those clients of the processor
which need immediate service to the interrupt controller. Whenever such a client—an
interrupt source for short—needs service, it raises an interrupt request via its interrupt
line.

The interrupt controller represents each interrupt source using a small, unique
integer. For each interrupt source s, it maintains a unique interrupt priority ip_s, a
switch ie_s for enabling the source and a request bit ir_s. Depending on the capabilities
of the interrupt controller, the priorities may either be hardwired or may be software-
selectable. In each cycle of its processor's clock, the interrupt controller samples all
its interrupt sources and sets the request bit ir_s for all sources s having an active
interrupt request line. It determines whether there are sources s having both the
enable bit ie_s and the request bit ir_s set and chooses among those the source s' with
the highest priority. More precisely, it chooses the source s' satisfying $ie_{s'} = 1$, and
$ir_{s'} = 1$, and for all s, if $ie_s = 1$ and $ir_s = 1$ then $ip_s < ip_{s'}$. It passes s' to the
processor.

Provided it has interrupts enabled, the processor queries its interrupt controller
before each instruction fetch. If the interrupt controller indicates that there are interrupts to be serviced, the processor saves part of its state, including the program
counter, the program status word and any interrupt-related information. Next, it uses
s' as an index into the interrupt vector table in order to read the starting address
of the interrupt service routine for the source s'.[1] In order to prevent servicing the
same request twice, the processor disables interrupts. The interrupt service routine
clears the interrupt request bit $ir_{s'}$ and services the source s'. It finishes with a return
from interrupt instruction, which restores the processor to the state it was in before
servicing the interrupt. In effect, the interrupt controller and the processor together
preempt the executing program and schedule another thread of control, the interrupt service routine for s', for immediate execution. The interrupt service routine
typically saves those portions of the processor state which it overwrites and then
services the request, usually without switching runtime stacks. The delay between

---

[1] The actual content of the interrupt vector table depends on the processor's architecture.

when a request was raised and when the request is serviced is the request's interrupt latency.

Simple interrupt systems force the processor to execute interrupt service routines with interrupts disabled. Therefore, even an interrupt source with a priority higher than $ip_{s'}$ requesting service shortly after s' has to wait for the completion of the active service routine before receiving service by the processor. Nested interrupt systems allow lifting this restriction.

In a nested interrupt system, the interrupt controller not only provides the chosen interrupt source s' to the processor but also the source's interrupt priority $ip_{s'}$. The processor in turn maintains the interrupt priority level ipl it is executing at. When the processor has its interrupts enabled, it checks with its interrupt controller whether there are any interrupt requests pending and then fetches the next instruction. If there are, it compares ipl to the priority $ip_{s'}$. Provided the interrupt priority of the requesting source is greater than the priority it operates at, if $ip_{s'} > ipl$, it preempts whatever it is executing and switches to the interrupt service routine for the source s'. During the switch, it raises its own priority ipl to the priority $ip_{s'}$. As comparing, priorities prevents servicing the same request twice, there is no need for disabling interrupts while executing an interrupt service routine. When using nested interrupts, the programmer has to reserve more space for the saved processor states as there may be several unfinished invocations of interrupt service routines active at the same time.

Logically, an interrupt service routine constitutes a separate thread of control. Together with the processor's interrupt system it realizes an EVENTHANDLER, Fig. 5.1. The interrupt source the service routine handles acts as event generator g. The interrupt system adds the loop and the waiting operation, while the interrupt service routine handles the events e generated by the source g. In order to achieve a common goal, these threads cooperate by sharing resources—data and the available computing time in particular. They must synchronize for achieving mutual exclusion, or else data consistency will be compromised. When programming without being helped (or hindered!) by an operating system, we have to build this synchronization ourselves. As long as a single processor executes all threads which may access a shared data structure simultaneously, we can guarantee mutually exclusive access by disabling interrupts. Before enabling interrupts again, we have to ensure that all updates to the shared data structure really have taken place. In particular, we have to keep an optimizing compiler from optimizing away some write instructions. When several processors execute the threads which may simultaneously access the shared data structure we must, in addition to disabling interrupts, lock out the other processors from accessing the shared memory. Processors designed for shared memory parallel systems usually support hardware primitives for memory locks.

5.2.1.2 Software Organization

A digital feedback control system, for example, samples the output of the process it controls periodically. In order to preserve the phase margin designed into its control law it must produce a new input to the process in response to a measurement of the output with as little delay as possible. Usually, one sets up the hardware, so that a timer peripheral triggers the acquisition of a new sample periodically. The interrupt indicating that the new sample is available gets high priority. Its interrupt service routine computes the control law and updates the input of the process. Provided that the processor does not have to tend to even higher priority interrupts, this scheme delivers the shortest reaction time possible. In the extreme case, all computations are performed by interrupt service routines. The main procedure, after having set up the hardware, ends in the main loop. The body of this loop might put the processor into a low power sleep but does nothing else. Arrival of an enabled interrupt will eventually terminate this sleep phase and make the processor execute its interrupt service routine.

Operating a processor at a high priority for a long stretch of time, however, starves lower priority interrupts. Often we can split servicing a request into two parts, a short one, the so-called top half, that requires low latency, and a much longer one, the bottom half, that can be deferred to being processed by the main loop. An implementation of such a top half—bottom half arrangement requires a queue for storing the deferred bottom halves. Many communication protocols, for example, allow a liberal amount of time between reception of a message and response to the message. This makes servicing incoming messages a candidate for a top half—bottom half arrangement. Limited storage in the communication hardware used for implementing the protocol often forces us to react to a received message swiftly in order to prevent this message from being overwritten by newer data, but further processing of the message can wait till there is time in the main loop.

5.2.1.3 Direct Memory Access

While a peripheral device transmits or receives a block of data, the services it requests typically entail the transfer of a single item of data only. Invoking an interrupt service routine in order to perform just a single assignment is wasteful. Instead one delegates such services to direct memory access, DMA, units. Direct memory access units are processors dedicated to transferring blocks of data between memory and peripheral devices and between different ranges of memory. As such, they support a very limited programming model. They are able to execute procedures like the one in Fig. 5.3 consisting of a sequence of data transfers and of signaling completion by raising an interrupt. Direct memory access units for image processing applications support additionally transfers of two-dimensional subarrays to and from an enclosing array.

In order to turn DIRECTMEMORYTOMEMORYTRANSFER in Fig. 5.3 into a sequence of input transfers from a peripheral device, the source increment i_S has to be zero. Furthermore, the read access to the source S in line 4 has to be replaced with reading from the device's data register. This read access must block until the device has new data available. Likewise, for turning DIRECTMEMORYTOMEMORYTRANSFER into a sequence of output transfers to a peripheral device, the destination increment i_D has to be zero and the write to the destination D in line 4 has to be replaced with a write

```
1: procedure DIRECTMEMORYTOMEMORYTRANSFER(D, i_D, S, i_S, c)
2:     ι ← κ ← 0
3:     for i←0, c-1 do
4:         (D_ι, ι, κ) ← (S_κ, ι + i_D, κ + i_S)
5:     end for
6:     raise completion interrupt
7: end procedure
```

Fig. 5.3 A memory to memory transfer as supported by a typical direct memory access unit. Such a unit contains address generators both for the source address and the destination address in order to support transfers at full available bandwidth

to the device's data register. This write access must block until the device is able to accept new data.

5.2.2 Keeping Time in Event-Triggered Architectures

Event-triggered systems need a mechanism to react to the passing of time. Reliable communication, for example, requires that a communication partner confirms the reception of a message within a certain amount of time or else the sender has to assume that the message is lost. To be able to resend the message the sender must monitor this timeout.

Using individual hardware for each timing operation is too restrictive except for systems requiring very few such operations. Instead one resorts to software-based timers. The SCHEDULE operation takes a reference to a timer e, a delay Δt, a procedure a, and a parameter q for a. It schedules the timer referenced by e, so that a will be applied to q once Δt has elapsed. The CANCEL operation takes a reference to a timer e. Provided e has not elapsed yet, it removes e from the schedule. A timer has two fields, n and p, for referencing other timers, a field Δt representing a delay, a field referencing a procedure a, and a field for a parameter q. Unscheduled timers have both their n and their p field set to **nil**. A periodically reappearing interrupt, the so-called system tick, indicates the passage of time. In a naive solution, one keeps the scheduled timers in a circular, doubly linked list which is sorted according to the time to expiration. The sentinel *never* serves as the start into this list, where *never.*$\Delta t = \infty$. The reference *never.n* points to the timer which will expire the earliest. This timer's delay, *never.n.*Δt, is the number of system ticks that have to pass before the timer expires. In order to reflect the passing of time, the interrupt service routine PERTICK for the system ticks, Fig. 5.4, decrements this delay. For any other scheduled timer e, the delay $e.\Delta t$ is the timespan between its predecessor's expiration and its own expiration measured in system ticks. With this arrangement scheduling a timer, Fig. 5.5, amounts to an insert operation into the sorted list together with the required maintenance of the delay fields.

```
1: procedure PERTICK
2:     disable interrupts
3:     never.n.Δt ← never.n.Δt − 1
4:     while never.n.Δt = 0 do
5:         e ← never.n
6:         (never.n, e.n.p) ← (e.n, never)
7:         e.p ← e.n ← nil
8:         (a, q) ← (e.a, e.q)
9:         enable interrupts
10:        a(q)
11:        disable interrupts
12:     end while
13:     enable interrupts
14: end procedure
```

Fig. 5.4 Interrupt service routine for the system tick. Interrupts are disabled whenever the list is in an inconsistent state

```
1: procedure SCHEDULE(e, Δt, a, q)
2:     if Δt = 0 then
3:         Δt ← 1
4:     end if
5:     disable interrupts
6:     (e.a, e.q) ← (a, q)
7:     if e.n ≠ nil then
8:         (e.n.Δt, e.n.p, e.p.n) ← (e.n.Δt + e.Δt, e.p, e.n)
9:     end if
10:    i ← never.n
11:    while Δt > i.Δt do
12:        Δt ← Δt − i.Δt
13:        i ← i.n
14:    end while
15:    e.Δt ← Δt
16:    (e.p, e.n, i.p, i.p.n, i.Δt) ← (i.p, i, e, e, i.Δt − Δt)
17:    enable interrupts
18: end procedure
```

Fig. 5.5 Procedure for scheduling a timer. The list of scheduled timers is a critical section. Therefore, interrupts have to be disabled during the execution of SCHEDULE

Canceling a timer requires the removal of the timer from the doubly linked list and an update of the delay of its former successor. The list of scheduled timers is a critical section. Therefore, interrupts have to be disabled, while modifying this list. The CANCEL operation updating a few references only poses no problem. The interrupt service routine PERTICK expends a constant amount of work per expiring timer. As long as the priority of the system tick interrupt is low, its impact on the responsiveness of the system should be manageable. The SCHEDULE operation, however, requires time which on average grows linearly with the number of scheduled timers and, to make matters worse, it must disable interrupts during its operation. At best, this asks for painstaking attention to every single instruction of the implementation of SCHEDULE in order to minimize the time spent with interrupts disabled. For a sizable number of scheduled timers, however, we must expect every single call of SCHEDULE to potentially cause unacceptable loss of interrupt requests.

Timing wheels, see Varghese and Lauck (1987, 1997), help alleviate the problem. Instead of a single list of scheduled timers, one uses an array tw of n_w such lists. The interrupt service routine for the timer tick maintains the current slot σ. It 'turns' the wheel by one entry by setting $\sigma \leftarrow (\sigma + 1) \pmod{n_w}$ and then empties the list tw_σ using a loop similar to the one in PERTICK. A timer's Δt field now describes delays in terms of the number of revolutions of the timing wheel. When a timer is scheduled to elapse after d system ticks, it is entered into the list $tw_{\sigma'}$ where $\sigma' = (\sigma + d)$ $\pmod{n_w}$ and its Δt is set to $\lceil d/n_w \rceil$. Provided that n_w is large enough, most timers will expire within one revolution of the timing wheel. The expected effort for scheduling a timer then is constant too.

5.3 The Time Triggered Architecture

A time-triggered system maintains a common sense of time across all its processors. More precisely, each processor has access to a timer which ticks synchronously with the timers of the other processors in the system. A processor's work is partitioned into n tasks, which are invoked periodically. A task's invocation executes the program of the task. This restrictive model enables the computation of a schedule for the task invocations during development. For computing a schedule for a set of n tasks, we are given the period p_i of the task T_i, the temporal offset o_i of T_i, the deadline d_i of T_i, and an upper bound c_i for the execution time of the task's T_i invocation, where $1 \leq i \leq n$. Within a period of the task T_i, the task invocation for the period may start after the offset o_i has elapsed. The invocation must be finished before the deadline d_i. We have to provide a schedule table for a timespan of duration $\mathrm{lcm}(p_1, \ldots, p_n)$, which is called a hyperperiod. A schedule for a single hyperperiod suffices as the pattern formed by the periods, offsets and deadlines repeats after one hyperperiod. Each entry in this schedule consists of a time within the hyperperiod and a scheduling action to be performed at that time. The scheduling action specifies which task invocation to start or which task invocation to resume. As long as the tasks are independent, the earliest deadline first policy, for example, will compute a schedule provided there exists one. For cooperating tasks, finding a schedule is much more complicated, typically involving an exhaustive search. Bounds for the execution time of an invocation of a task can be acquired by measurement. Alternatively, these bounds can be derived analytically using tools for worst-case execution time estimation.

Provided a task that is responsible for reaction to some stream of external events samples this stream only once during its period, the task's period bounds the task's worst case response time tightly from above. A required short reaction time therefore results in a short period, putting high demands on the system. If the events occur rarely, the system will be very busy with doing nothing useful most of the time. Provided one is able to give good upper bounds for the execution times, however, time-triggered systems allow proving formal statements about their temporal behavior.

5.3.1 Maintaining a Common Sense of Time

When we build the system as a single clock domain all timers operate necessarily from the same clock, thereby establishing a common sense of time automatically. This approach seems alluringly simple, but it introduces severe problems. The clock generator becomes a single point of failure. Loss of the clock generator stops operation of such a system immediately. Therefore, the system cannot enter a save state. Quartz crystal oscillators, for example, might fail when subjected to severe shock. In addition, distributing a high-frequency clock signal over any appreciable distance creates its own set of problems. When the connections from the clock generator to the processors have different lengths one and the same clock edge will arrive at the processors at different times, the distribution network introduces clock skew. While we might be able to compensate for the skew doing so introduces additional complexity into the system.

Large systems always consist of several clock domains. In order to maintain a common sense of time, each processing node in such a system is equipped with a timer. The nodes collaborate in synchronizing these timers to a certain precision using a clock synchronization protocol.

When no relation to some universal timescale is required, internal clock synchronization will do. In this case, the precision of synchronization bounds the disagreement between the values exhibited by any two timers of the system at any single point in time. An event generator periodically generates events, which all nodes observe. At each such event, each node records the value of its timer and exchanges this value with all the other nodes. The precision achievable is limited by the uncertainties with which the nodes recognize the events and read their timers. Each node computes the nominal value all timers should have exhibited and the deviation of its timer from this nominal value. Finally, each node makes adjustments in order to minimize the deviation of its timer. Simply setting the timer will lead to a nonincreasing timescale in some nodes, which will make the timer unsuitable as a source of time stamps. Therefore, it is better to adjust the rates of the timers to make the slow ones tick faster and the fast ones tick slower. Controlling the length of the timer's tick can be achieved in two ways. A relatively long tick can be produced from the clock domain's clock using a counter, which divides the clock by a programmable divisor. By adjusting this divisor, the rate of the timer can be controlled. In order to enable small adjustments, the frequency of the domain's clock must be high. Alternatively, for ticks with short duration, the timer can be equipped with its own adjustable oscillator.

External synchronization protocols strive to synchronize the system's timers with some universal timescale like the International Atomic Time TAI. For external

synchronization protocols, the precision of synchronization is a bound for the deviation of any timer in the system from the universal timescale. In a system based on the
Precision Time Protocol, (IEEE 2008), all participating nodes are called clocks. The
clocks periodically execute a best master clock algorithm in order to determine the
clock representing the universal timescale best. In the course of the best master clock
algorithm, the clocks arrange themselves in a spanning tree with the best clock, the
grandmaster clock, at the root. The tree's edges represent direct network links. By
sending timestamped messages back and forth the clocks estimate the delay introduced by each network link. For best accuracy, the messages must be time stamped
at the very moment a message's time stamp point, the temporal reference point of
the message, is put onto the network medium. The clocks cascade synchronization
messages down the spanning tree. A clock uses the information in the incoming synchronization messages for adjusting the rate of its timer. The Precision Time Protocol
can use several protocols, most prominently Ethernet, for the underlying network.

5.3.2 Implementing Time Triggered Systems

When we operate an embedded processor or a digital signal processor in a time
triggered fashion, the timer representing the common sense of time becomes the
only interrupt source the processor has to service. The timer interrupt service routine
processes the schedule table, creating, suspending, and resuming task invocations in
its course.

Tasks which have a very short period are candidates for being implemented in
synchronous logic. Effectively, we create a peripheral processor for such a task. As
long as the clock for the peripheral processor is derived from the timer which represents the global sense of time, the peripheral processor will operate synchronously
with the rest of the system. Synchronous logic allows us to set the time for a task
invocation with perfect accuracy.

5.4 The Polling Architecture

In a polling-based system, the processors check for events whenever they feel like
doing so. Implementing polling is easy. Each event generator is attached to a flip-flop.
An event sets the flip-flop and the processor servicing this event generator can reset
it. In this processor's program, one introduces checks for events in the main loop at
places appropriately fitting. When the processor encounters a set flip-flop indicating
that one or several events have occurred, it resets the flip-flop and processes the event.
The timing behavior of a polling-based system is highly irregular. Moreover, it will
change fundamentally with every change of the software. Polling is attractive for
acquiring information that changes very slowly, such as temperatures of objects with
large heat capacity. It is of no use for anything else.

5.5 Hardware Organization

We focus again on systems processing sensor inputs and controlling actuators. The hardware of these systems consists of one or more processor cores for executing the programs of specialized peripheral devices and of memories. Communication infrastructure, mostly buses, connects these parts. A core can either be an embedded microprocessor or a digital signal processor. In realizing such a system's function, the peripherals are at least as important as the microprocessors; ignoring them is a major mistake. A peripheral device performs its task with a certain amount of autonomy. Therefore, we better think of it as a special purpose processor. Each system contains some peripherals for enabling the exchange of data with the clients the system serves. Together, the parts form a heterogeneous parallel computing system.

Highly autonomous subsystems tend to exchange a limited amount of data with the rest of the system. If such a subsystem in addition allows for some latency when communicating, it is loosely coupled to the rest. A loosely coupled subsystem may form its own clock domain and may connect to the rest via some serial bus. A subsystem exchanging large amounts of data or demanding short latencies is tightly coupled with the rest. Such a subsystem better resides in the same clock domain connecting with the rest via a parallel bus.

A microcontroller integrates one or a few cores, volatile and nonvolatile memories, direct memory access units, an assortment of tightly coupled peripheral devices, and a number of connecting buses onto a single chip. The on-chip memories usually are multiported allowing several concurrent accesses. Some machines augment each word in memory with an error-detecting or error-correcting code improving reliability. In microcontrollers with several cores, the memories usually form a hierarchy. Each core has its own local memory providing the highest bandwidth and the lowest latency. If both code and data reside in a core's local memory, the core should be able to operate at full speed. The memories shared between the cores, the direct memory access units, and the peripheral processors provide lower bandwidth. The on-chip volatile memories usually are static random access memories; their combined size ranges from a few kilobytes to about one megabyte. The peripheral devices and the cores communicate via device registers. A device acts on the contents of its registers. The device registers of tightly coupled peripherals also appear in the cores' address spaces. When programming, we have to make sure that accesses to device registers are not optimized away and that these accesses are executed in the order defined by the program. We have to prevent compilers from optimizing too much and we have to be particularly careful when programming processors which have caches or access memory out of order (Patterson and Hennessy 2013; Hennessy and Patterson 2011).

Control systems controlling a few actuators or measurement systems processing the inputs of a few simple sensors usually can do with a microcontroller's on-chip memory resources. Chips for such applications dispense with an external memory interface in favor of reduced cost and a smaller package. Putting caches into such chips is pointless. The applications are better served by providing the fast memory that would make up the cache as addressable memory instead.

Some systems have to collect a considerable amount of data before they can start to process these data. An embedded vision system, for example, has to collect a complete frame before it can analyze an image. The size of this data precludes it from being stored on-chip. A suitable microcontroller must have an interface for external memory. The relatively small on-chip memories offer high bandwidth and low latency, while the external memory contributes sheer volume to the system at a lower bandwidth and longer latency; the external memory adds another layer to the memory hierarchy. The internal memories serve as scratchpad for frequently used data, while the external memory provides bulk storage. In order to ease managing the different bandwidths and the different latencies in such an elaborate memory hierarchy, these microcontrollers allow configuring some internal memories as caches.

5.6 Peripherals

Peripheral devices provide the processors with connections to the world. Many peripherals operate autonomously and in parallel with a system's microprocessors freeing the processors from tasks that require short cycle times or clock-cycle accurate timing. For requesting service, peripherals generate events. A processor may process such an event in an interrupt service routine. Alternatively another peripheral, triggered by an event, might provide the requested service. Such cascaded peripherals interact with their surroundings with precise timing. Once all the action has taken place, the processors come into play again. They evaluate the results and adjust the cascade for the next round.

5.6.1 General Purpose Input and Output

A general purpose input and output peripheral puts a microcontroller's pins under direct program control. Programs can configure each pin either as input or as output. They can set and reset output pins and read input pins.

5.6.2 Counters and Timers

A counter-timer module in counter mode counts events generated by some external event generator. Software can at any time read the number of events and process it. Alternatively, it can set up the counter to produce an event once a certain programmable limit is reached. When operated in timer mode, a clock generator provides the clock for the module. Operated in single shot mode, a timer generates an event upon having reached a certain preset value and stops counting. Operated in continuous mode, a timer generates an event upon having reached the preset value and continues counting from zero. This way the timer generates a stream of periodic events. Some counter-timer modules also contain so-called capture registers. Upon recognizing a capture event, the timer module stores the value of the counter in the capture register, which is assigned to the event. Once the capture register is written, the counter-timer module produces another event for requesting service. Such capture functions allow for the precise measurement of the time of occurrence of the capture event.

A watchdog timer resets the system whenever it expires. As long as it is alive, software must periodically reset the watchdog timer in order to keep it from expiring. When the software crashes, the watchdog timer will expire soon thereafter and, by generating a reset, will bring the system back to live again.

5.6.3 Analog Data Acquisition

An analog-to-digital converter, ADC for short, converts a voltage at its input into a number as its output. The converters integrated into microcontrollers often are paired with an analog multiplexer allowing selection from several inputs. In such a combination, a single converter can acquire data from several analog sources. Periodic measurements can be achieved by cascading a continuously operating timer with an analog-to-digital converter.

5.6.4 Waveform Generation

When embedded systems act on their environments, they have to employ actuators. Many actuators must be supplied with precise voltage or current waveforms. For an actuator which requires appreciable power for its operation, pulse width modulated switching allows generating waveforms with low power losses. A pulse-width modulation unit, PWM unit for short, Fig. 5.6, consists of a counter operating from a high-frequency clock, a compare register and a shadow register. The counter counts continuously. It wraps around periodically, the period is the period of the pulse-width modulated signal the unit generates. The value of the noninverting output is one as long as the content of the counter is less than the content of the compare register; it is zero otherwise. Every time the counter wraps around the PWM unit updates the compare register with the content of the shadow register and generates an event. Updating the compare register only when the counter wraps around prevents the generation of two pulses per single period. The frequency f_{clk} of the clock determines the resolution r and the frequency f_{pwm} of the pulse-width modulated signal, $f_{clk} = r f_{pwm}$.

A digital-to-analog converter, DAC for short, generates output using a combination of switches and resistors which form a configurable voltage divider. A digital-to-analog converter can synthesize signals with much higher frequency than is possible

Fig. 5.6 A single channel
PWM unit

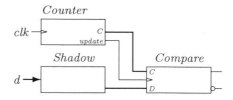

with a pulse-width modulation unit. The electrical power a DAC can deliver, however, is limited.

5.6.5 Communication

A communication peripheral is a protocol processor, which implements the lower layers of a protocol. Software, a so-called protocol stack, is responsible for the upper layers.

The serial peripheral interface bus, SPI bus for short, is a serial bus protocol for connecting devices with moderate bandwidth requirements, up to about $10\,\mathrm{Mbit\,s^{-1}}$, to a processor. An SPI bus operates in a master-slave fashion. During a transfer, the master and the selected slave exchange data; the master sends data to and receives data from the selected slave simultaneously. The bus consists of three shared lines. The serial-clock line, SCLK, driven by the master defines the timing of the transfer. The 'master-out-slave-in' line, MOSI, transports data from the master to the slave. The 'master-in-slave-out' line, MISO, carries data in the reverse direction. In addition, the master has a separate chip select output CS for each slave. In order to enable sharing, the MISO line between slaves, a slave's MISO output has a tri-state buffer which is controlled by the slave's chip select input. For each transfer, the master has to program the number of bits to be exchanged, the frequency of the SCLK signal, and the mode of the transfer, see Fig. 5.7. Slaves such as analog-to-digital and digital-to-analog converters exchange a few bytes at most in a single transfer; SPI-based mass storage devices exchange blocks of data up to a few kilobytes in length. Often general purpose outputs provide the chip select signals for the slaves. Software must then toggle the respective output. Typically, an SPI bus connects components on a few circuit boards. It rarely provides connectivity outside an instrument's enclosure.

An I^2C bus consists of the clock line SCL and the data line SCA. The I^2C standard, (NXP 2012), specifies modes with transfer rates of $100\,\mathrm{kbit\,s^{-1}}$, $400\,\mathrm{kbit\,s^{-1}}$, $1\,\mathrm{Mbit\,s^{-1}}$ and $3.4\,\mathrm{Mbit\,s^{-1}}$ for bidirectional transfers. For unidirectional transfers it also specifies a mode with a rate of $5\,\mathrm{Mbit\,s^{-1}}$. The I^2C bus uses a master-slave protocol which allows for multiple masters. The slaves have addresses which are either seven or ten bits long. An I^2C message consists of an address field, a direction bit and the data bytes together with an acknowledgement for each byte. Message

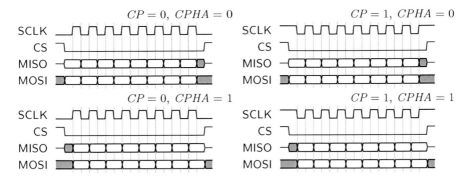

Fig. 5.7 The four modes of operation of SPI. The mode bit *CP* sets the polarity of the SCLK signal. If the mode bit *CPHA* is reset, the participants in a transfer must sample the bit on the first clock edge and shift to the next bit on the second edge. If the mode bit *CPHA* is set, they have to shift first and sample second

transmissions are initiated by masters only. In case two masters start transmitting simultaneously, a lossless arbitration procedure determines a winner.

The oldest communication standard still in use is the asynchronous serial port. A universal asynchronous receiver transmitter, UART for short, converts single bytes into a stream of bits. While the serial port has vanished from today's general-purpose computers, UARTs still serve as building blocks for many communication systems. In factory automation, several bus protocols use asynchronous transmission for transferring individual bytes, as does the LIN protocol, used for networking automotive body electronics (LIN 2013; Paret and Riesco 2007). Serial port communication requires little software which makes it attractive.

The universal serial bus, USB for short, is a protocol for connecting a host and several peripheral devices (USB-IF 2013). It specifies transfer rates between $1.5\,\mathrm{Mbit\,s^{-1}}$ and $10\,\mathrm{Gbit\,s^{-1}}$. Several classes of USB devices are defined, for example, audio devices, human interface devices, and communication devices. A USB host provides power via the USB cables to its peripheral devices, allowing us to build instruments that are host-powered only. We can achieve USB connectivity either by attaching a USB to serial converter chip to a UART or by using a USB interface integrated into a microcontroller. Using a dedicated USB to serial converter frees us from dealing with the intricate details of the USB protocol. All we have to provide is the program operating the UART. We have no control, however, over the device class and similar parameters, as these are hardwired into the converter chip. Using a USB interface integrated into a microcontroller gives us full control. It requires, however, a substantial protocol stack.

Ethernet-based networks not only find use in general-purpose computers but also in automation equipment (IEEE 2012). The hardware support for Ethernet consists of two parts, the media access control, MAC for short, and the physical layer interface, PHY for short. Some microcontrollers integrate media access control for 10 and $100\,\mathrm{Mbit\,s^{-1}}$ Ethernet. The physical layer interface is a separate chip. Some media access control peripherals support time stamping of Ethernet frames for the Precision Time Protocol. For even higher precision time stamping, some PHY chips support time stamping too. Again, any protocol based on Ethernet requires a substantial protocol stack.

Control area network CAN for short, (CAN 2003–2013), is a bus intended originally for automotive applications. Due to its low cost, it has found widespread use, in factory automation and in aerospace to name a few. It offers a bandwidth of up to $1\,\mathrm{Mbit\,s^{-1}}$. The nodes on a CAN bus have no addresses, instead each message has a message identifier, which describes the content of a message and defines the priority of the message. The lossless arbitration mechanism used in CAN limits the physical length of the bus for a given data rate. The maximum length for a CAN bus operating at $1\,\mathrm{Mbit\,s^{-1}}$ is 40 m. The CAN interface services a send request as soon as the message wins arbitration. In contrast, operating a CAN bus requires a very moderate protocol stack only.

FlexRay is a bus intended for automotive applications (FlexRay 2010). FlexRay uses one or two buses (or active stars) operating at $10\,\mathrm{Mbit\,s^{-1}}$ each. The dual bus configuration provides either redundancy or doubled bandwidth. Communication over FlexRay is cyclical. A communication cycle consists of a static segment for real-time data and an optional dynamic segment. The static segment consists of slots, the dynamic segment of minislots. Each node has one or several slots assigned, in which it can send its messages. In order to keep message transmission coordinated, FlexRay provides internal synchronization for the node's clocks. The software executed by a FlexRay node must run in sync with the communication cycle. Data to be sent must be available before the slot for this data appears on the bus, forcing the software to operate in a time-triggered fashion. Likewise, data to be processed becomes available only after the data's slot has passed.

A simple communication link can be established by converting a digitally represented signal with a digital-to-analog converter at the sender. The receiver digitizes the signal again using an analog-to-digital converter. The 4–20 mA current loop used in many industrial automation systems encodes zero as a current of 4 mA flowing through the communication line and the largest possible value as 20 mA. Assuming that the receiver has an input impedance of $500\,\Omega$, it will dissipate between 8 and 200 mW. In order to become noticeable, electromagnetic interference must inject significant power into the communication line. As most sources of interference are too weak, the current loop is well suited for electrically noisy environments.

5.7 Memory

The systems we consider need memory for storing programs and data. After being powered, a processor needs a program. This program must be stored in nonvolatile memory, which retains its content even when not powered. As programs are seldom if ever changed write performance of the nonvolatile memory is of secondary importance. The program's variables on the other hand need not retain their values, while the system is not powered. The memory for storing data must allow fast read and fast write accesses. Processors can only execute instructions after they have fetched them from memory and they can only process data they have loaded first. Therefore, the temporal properties of memory determine the timing of our programs to the largest extent.

5.7.1 Volatile Memory

Random access memory, RAM for short, can be read and written with speeds comparable to those of current processors. Random access memory comes in two varieties, static RAM and dynamic RAM. The time required for accessing a word in static RAM is constant. This allows us to calculate a precise timing for a sequence of memory accesses. The time required for accessing a word in dynamic RAM, however, depends on the preceding memory accesses. This makes predicting the time required for a sequence of memory accesses a formidable task. The memory size available in a single package is much larger for dynamic RAM than for static RAM, therefore dynamic RAM is the technology of choice for memory intensive applications such as embedded vision systems, despite the timing uncertainties it introduces.

5.7.1.1 Static RAM

Static RAMs, SRAMs for short, use a flip-flop for each bit they store.

Access to an asynchronous SRAM is timed by the edges of the control and address signals. Access to a synchronous SRAM is timed by a dedicated clock input.[2] Asynchronous static RAM chips, Fig. 5.8, come in sizes up to two megawords and a word size of 2 bytes. More often, we find this type of memory in microcontroller chips. The chip enable line CE enables the memory. The output enable line OE activates the output buffers for the data lines. The write enable line WE indicates a write access. During a write access, the byte low enable BLE enables writing to the low byte of the addressed word, while the byte high enable BHE enables writing to the high byte. All control signals are low-active. The data lines D_0–D_{15} are bidirectional.

When Asynchronous SRAM is supplied with an address during a read cycle, Fig. 5.9, it responds with the data stored in the addressed word after the memories access time. In a write cycle, the processor has to drive the data onto the data lines some time before it drives the WE line low for the memory to accept the data. Each access to an asynchronous SRAM completes within a single memory cycle, whose duration is determined by the memory's access time. This uniformity simplifies the temporal analysis of programs considerably.

---

[2] Synchronous static RAM is beyond the scope of this book.

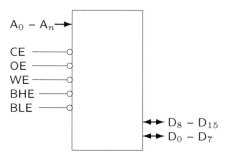

Fig. 5.8 Sixteen bit wide asynchronous static RAM. The address lines A_0–A_n select 16 bit wide words. For word wide accesses, the data lines D_0–D_{15} are used. For byte-wide accesses BLE indicates a transfer via the lines D_0–D_7 while BHE indicates a transfer via the lines D_8–D_{15}

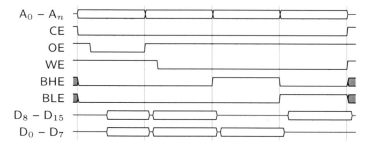

Fig. 5.9 A read access followed by three write accesses. The first write is word wide, the second one effects the low byte and the third one effects the high byte

5.7.1.2 Dynamic RAM

Dynamic RAMs, DRAMs for short, use a cell consisting of a capacitor and a transistor for storing one bit. Positive charge in the capacitor represents a logic one, while no charge represents a zero. These capacitors loose their charges over time, therefore their content has to be refreshed periodically, typically every 64 ms. Nowadays, the interface to DRAM is synchronous, the full designation for these memories is synchronous dynamic RAM, SDRAM for short. Synchronous DRAM samples the command signals, the address lines A, the bank select lines BA, the data lines DQ, and the data mask signals DQML and DQMH all at the rising edge of the memory clock CLK. The command signals chip select CS, write enable WE, column address strobe CAS and row address strobe RAS encode commands to the SDRAM. The clock enable signal CKE, when one, enables the memory to react to commands on the command signals. When zero it places the memory into a low power mode, in which the memory retains data. The memory is organized in a number of banks. Each bank comprises a matrix of memory cells, a read data latch and column select logic, Fig. 5.10. Four banks are typical for SDRAMs, while double data rate SDRAMs, DDR-SDRAMs, typically have eight or more banks. The memory matrix in a single

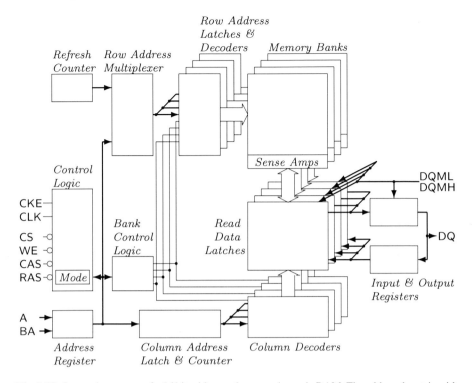

Fig. 5.10 Internal structure of a 16 bit wide synchronous dynamic RAM. The address bus A is wide enough for addressing either a row or a column in a bank. Two bank select lines BA address four banks. The sixteen data lines DQ are used together for word-wide accesses. For byte-wide accesses either the lower eight lines or the upper eight lines are used. The data mask signal DQML selects the lower eight lines for a transfer, while DQMH selects the upper eight lines

bank can contain 8,192 or more rows, and a single row can contain 8,192 or more bits. For storing data into and reading data out of a bank, the protocol sketched by the state machine in Fig. 5.11 has to be adhered to. Usually, this necessitates separate memory control logic for translating the memory accesses issued by the processors. This logic is responsible also for keeping the memory and an accessing processor synchronized, usually by stalling the processor until the requested data is delivered during a read access.

After initialization of the memory, all banks are in the Idle state. When the memory control logic issues an ACTIVE command together with a row address via the address lines A, the memory transfers the content of the whole row from the bank's memory array into the bank's read data latch using the bank's sense amplifiers. This transfer destroys the information stored in the addressed row in the memory matrix. After the ACTIVE to READ or WRITE delay t_{RCD} has passed the bank is in the Active state, and the memory control logic can read data from and write data to the row stored in the bank's read data latch. For reading, the memory control logic issues a READ command together with the column address it wants to read from. The memory delivers the

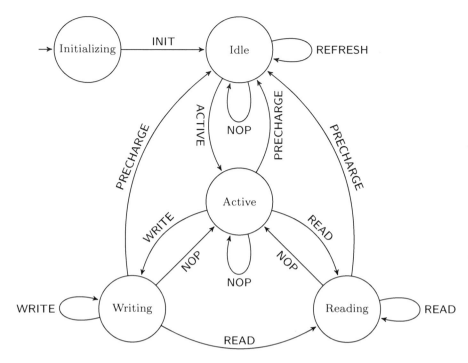

Fig. 5.11 Simplified state machine for one bank of an SDRAM

data a CAS latency *CL* later via the data lines DQ, Fig. 5.12. The memory control logic selects whether to read the lower byte, the upper byte, or both bytes of a word using the data mask signals DQML and DQMH. The memory controller may issue subsequent READ commands without intervening NOP commands. When the memory control logic issues a WRITE command, it has to supply a column address, the data mask signals and the data to be written, all in the same clock cycle. In order to allow the data transfers initiated by preceding READ commands to clear the data lines, it has to introduce additional NOP commands before the WRITE. All banks can be in the Active state simultaneously, allowing access to several address ranges without the overhead of repeated ACTIVE commands. A bank interprets commands addressed to another bank as NOP commands. For changing the row in a bank, the memory control logic has to first issue a PRECHARGE command. After the PRECHARGE command period t_{RP} has passed, the bank has transferred the data stored in the bank's read data latch back into the bank's memory array. Then, the memory controller may issue a new ACTIVE command, in order to address a new bank. The ACTIVE to PRECHARGE parameter, t_{RAS}, specifies a minimum and a maximum time between these two commands when both commands go to the same bank. For initializing the SDRAM, the memory control logic has to program the memory's mode register with the CAS latency. Without proper initialization, the memory system will not work!

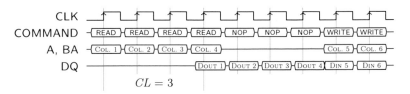

Fig. 5.12 Four read accesses to an active row in a bank of an SDRAM, followed by two write accesses to the same row. The memory control logic has to issue NOP commands for the data lines to become free

The capacitors storing the bits are not ideal. Leakage currents make these capacitors forget. In order to keep this memory loss at bay, the memory control logic has to issue REFRESH commands repeatedly. A REFRESH command effects all banks simultaneously. It causes the bank to read one row from the bank's memory array into the bank's read data latch and write the row back to the memory array. The SDRAM's refresh counter provides the rows' address to all banks. As a refresh operation overwrites the read data latch in each bank, all banks must be brought into the Idle state by a PRECHARGE command addressing all banks simultaneously before the memory control logic issues a REFRESH command. The AUTO REFRESH period, t_{RFC}, specifies the duration of one refresh operation, while the refresh period t_{REF} specifies the maximum time that may pass between two refreshes to the same row.

Double data rate synchronous dynamic RAMs transfer blocks of data in bursts. They use only half a clock cycle for a single data transfer, therefore doubling the transfer rate. Commands, however, still require one complete clock cycle.

5.7.2 Memory Hierarchy

When designing systems which require a sizable amount of memory, such as embedded vision systems, we most probably have to resort to SDRAM in order to keep price and size of the system down. This forces us to design our programs with the idiosyncrasies of SDRAM in mind. In general-purpose computers, caches help by speeding up memory accesses on average, see for example, Patterson and Hennessy (2013), Hennessy and Patterson (2011). While well-designed caches are very effective in hiding the latency of the memory from the processor and from the programs they make it impossible to give tight bounds for the worst case execution times of the programs. A few microcontrollers and a few digital signal processors allow the programmer to operate some of the on-chip memory as cache. Managing the memory hierarchy in the programs explicitly, however, can result in programs which are both faster and easier to analyze. When dealing with issues of memory hierarchy, we have several possibilities.

Naively, we can ignore the difference between on-chip memory and external DRAM in terms of latency and bandwidth. Dynamic RAM will be accessed mostly

Fig. 5.13 Basic structure of a
block by block computation.
The inputs, referenced by I,
and the outputs, referenced
by O, reside in external
memory. The procedure uses
two buffers, referenced by a
and d, in on-chip memory

```
1: procedure BLOCKBYBLOCK(I, O)
2:    for i ← 0, n − 1 do
3:        LOADBLOCK(I_i, a)
4:        DOBLOCK(a, d)
5:        STOREBLOCK(d, O_i)
6:    end for
7: end procedure
```

in the order dictated by the program. Operations which simultaneously access several data items located far apart in memory, such as matrix multiplication, will be penalized. While this is an acceptable strategy in the early phases of development, when we strive to arrive at a functionally correct prototype, this will hardly give us the execution times required. If we are forced to stop development at this stage, either by lack of time or lack of ideas, the best we can do is turn on all available caches and hope for the best.

Alternatively, we can try to partition the data to be processed into n blocks, sized so that the processing of one block fits comfortably into on-chip memory, and perform the computation on a block by block basis, Fig. 5.13. The procedures LOADBLOCK and STOREBLOCK transfer a block between external memory and on-chip static RAM. Not being bound by the sequence of memory accesses dictated by the program, they can address external memory in an order matched to the memory. The procedure DOBLOCK addresses on-chip memory only. It benefits from the fast access time and the high bandwidth the on-chip RAM offers. As the access time of the on-chip memory does not depend on history, we need not arrange the memory accesses in DOBLOCK to fit the memory. Besides reducing the run-time, this transformation tends to ease computation of the worst-case execution time. When analyzing DOBLOCK, the idiosyncrasies of the external memory can be disregarded. Likewise, when analyzing LOADBLOCK and STOREBLOCK, the algorithm performing the computation has no influence.

We can move the execution of the procedures LOADBLOCK and STOREBLOCK to a direct memory access unit, Fig. 5.14. Doing so allows us to build a pipeline, provided we have space for two more buffers in the on-chip RAM. The pipeline's first stage loads blocks from external into on-chip memory, the second stage processes blocks, and the third stage writes back results from on-chip into external memory.

5.7.3 Nonvolatile Memory

Nonvolatile memory stores the code which is executed immediately after a reset. Usually, this is a bootstrap loader, which after initializing and testing the hardware loads the operating system or the application program from some storage medium. In systems which do not have additional storage media, the operating system and the application must reside in nonvolatile memory.

```
 1: procedure PIPELINEDBLOCKBYBLOCK(I, O)
 2:     LOADBLOCK(I₀, a)
 3:     fork LOADBLOCK(I₁, b)
 4:     DOBLOCK(a, c)                                          LOADBLOCK
 5:     for i ← 2, n − 1 do
```

```
 6:         (a, b, c, d) ← (b, a, d, c)
 7:         fork STOREBLOCK(d, Oᵢ₋₂)
 8:         fork LOADBLOCK(Iᵢ, b)
                                                               STOREBLOCK
 9:         DOBLOCK(a, c)
                                                               LOADBLOCK
10:     end for
```

```
11:     fork STOREBLOCK(c, Oₙ₋₂)
12:     DOBLOCK(b, d)                                          STOREBLOCK
```

```
13:     STOREBLOCK(d, Oₙ₋₁)
14: end procedure
```

Fig. 5.14 Basic structure of a pipelined block by block computation. In parallel with the processor, *left side*, a direct memory access unit, *right side*, executes the transfers between on-chip and external memory. Two more buffers referenced by b and c are required. Horizontal lines indicate join operations, which wait till the direct memory unit has finished all previously forked tasks

5.7.3.1 ROM, PROM, and EPROM

These three types of memory cannot be erased and reprogrammed within the circuit. Read only memory, ROM for short, is programmed during the manufacture of the chip. Its content cannot be modified after production. Programmable read only memory, abbreviated PROM, represents each bit with a fuse. By selectively blowing fuses with some programming fixture, one determines the content of a PROM once and for all. Erasable programmable read only memory, EPROM for short, uses electrical charges trapped in special structures on the semiconductor chip. Irradiating the chip with ultraviolet light releases the charges and erases the memory. Typically, these memories have interfaces resembling the interface of asynchronous static RAM.

5.7.3.2 EEPROM and Flash

Electrically erasable programmable read only memories, EEPROMs for short, allow byte by byte read and write accesses. They need no erase operation before being written. EEPROMs typically interface to a serial bus, most often SPI. With capacities ranging from 256 bytes to several hundred kilobytes, these memories serve as storage for parameters and as bootstrap device.

Flash memory is organized in blocks. Before writing a word to Flash, one must erase the complete block the word resides in. Flash memory comes in two variants. NOR Flash allows word by word read and write accesses. Its main use is to store program code. Most NOR Flash devices have an interface similar to asynchronous static RAM. Chips with a capacity of up to 64 megabytes are available. The blocks of

NAND Flash are further subdivided into pages. Read and write accesses must transfer whole pages. Besides the space dedicated to data each page contains some room for organizational information and an error correcting code. Chips with a capacity of several gigabytes are available. NAND Flash serves in mass storage devices.

Repeated write and erase operations wear out the memory cells of EEPROM and Flash devices. A memory cell is able to endure up to a million updates before becoming dysfunctional. This endurance is exhausted easily by, for example, writing every second to the same memory cell. When used as mass storage, Flash devices need a special software layer providing wear leveling. Wear leveling prevents hot spots, which fail prematurely, from forming.

5.8 Embedded Microprocessors

Embedded microprocessors are designed to support a wide range of tasks including communication stacks, control system procedures and lightweight signal processing. In order to achieve good performance, modern embedded microprocessors execute their instructions using pipelines usually consisting of three or four stages. The processors' register files typically consist of sixteen general purpose registers. Most processors are load-store architectures; before they can operate on a data item it has to reside in a register. Only a few of their instructions access memory, the load instructions for moving data from memory into registers and the store instructions for moving data back from registers into memory. These processors usually have separate buses for fetching instructions and for transferring data, they constitute Harvard architectures. When the processor is integrated on a chip together with several memory modules or with multiported memory, this allows simultaneous instruction fetches and data transfers. In order to maximize the use of the on-chip nonvolatile memory, the instruction sets of embedded microprocessors favor code density, while maintaining acceptable execution speed.

Communication stacks typically have to handle byte-wide data. Therefore, embedded microprocessors are byte-addressing machines. Modern embedded processors have word sizes of 16 or 32 bits. Thirty-two bit wide arithmetic and a 32 bit wide addresses allow for a large linear address space, while 16 bit architectures require some segmentation scheme in order to extend their address spaces beyond 64 kB. When having to deal with large data structures, a linear address space results in simple address arithmetic requiring few instructions only. In comparison, a segmented address space results in a much more complicated address arithmetic requiring substantially more code. Programs which frequently handle large data structures in segmented address spaces often exhibit performance problems attributable to this added complexity.

Once built into a device, an embedded microprocessor executes only the few programs necessary for the operation of the device. Therefore, a memory management unit translating logical into physical addresses is not required. Instead some embedded processors include a memory protection unit. Such a unit allows real-time operating systems to confine programs or parts of programs to well-defined memory regions.

Embedded processors are designed for short interrupt latency. The ARM® Cortex® M3 and Cortex® M4 processors, for example Yiu (2014), allow mutually exclusive read modify write operations for locks without having to disable interrupts by providing a mechanism resembling atomic transactions. First, the processor executes an exclusive load instruction for reading the lock. If it finds the lock taken, it terminates the exclusive sequence. Otherwise, it executes an exclusive write instruction for obtaining the lock. This instruction updates the lock variable provided the processor has not executed an interrupt service routine between the exclusive read and the exclusive write and no other processor has accessed memory in the same timespan. In case of success the lock has been acquired; else the program may retry the locking operation creating a so-called spin lock.

Most physical quantities are real-valued. The majority of embedded microprocessors supports integer arithmetic only. Emulating floating point arithmetic in software is too costly; we need a way to do with integer arithmetic. The key to processing real-valued quantities on an integer processor is fixed-point arithmetic. Any actuator can realize physical stimuli only within a certain range, and any sensor is restricted to a limited measurement range. Let us assume that the realizable or observable values x of some physical quantity are restricted to $-f < x < f$ by either an actuator or a sensor for some full-scale value $f \in \mathbb{R}$, $f > 0$. By scaling the value x we get $y = x/f$ with $-1 < y < 1$. Therefore, we only have to approximate values between -1 and 1. Normally, we interpret the w bit word, Fig. 5.15, using two's complement representation as the signed integer n

$$n = \left(\sum_{i=0}^{w-2} d_i 2^i \right) - s \left(2^{w-1} \right).$$

Instead, we may interpret it as the fraction y

$$y = \left(\sum_{i=0}^{w-2} d_i 2^{i-w+1} \right) - s,$$

by putting the binary point in front of d_{w-2}. This is a fixed-point number in $0.(w-1)$ format. The product of two such numbers will be $2w$ bits long. There will be no overflow. When we convert the result back to w bits, we will loose information. We can either truncate the result by shifting the result to the right by $w-1$ positions and then dropping the w most significant bits, or we can round the result by adding 2^{2-w} to the result before shifting. Addition or subtraction of two fixed-point numbers may result in an overflow or an underflow. The standard behavior of truncating the overflowing bits results in errors much larger than necessary. The best possible fixed-point approximation to an overflowing result is almost 1, the one for an underflowing result is almost -1. This saturation behavior even results in a crude antiwindup when

s	d_{w-2}		$\cdots$		d_1	d_0

Fig. 5.15 A signed machine word consisting of w bits

we implement a proportional and integral controller using fixed-point arithmetic. Division is better avoided as very few processors support it in hardware. A number of microprocessors have instructions for 0.15 or 0.31 fixed-point arithmetic. The Cortex® M4 processor supports saturation arithmetic with a special instruction taking the number of bits w as parameter.

Some embedded microprocessors include single-precision floating-point operations in their instruction sets. Often a separate floating-point unit together with a separate register file is added. Applications requiring substantial numerical computations benefit from this unit. Interrupt service routines performing floating-point operations have to save and restore floating-point registers in addition to the general-purpose registers, affecting interrupt latency negatively.

5.9 Digital Signal Processors

Digital signal processors, DSPs, when compared to embedded microprocessors are optimized for fast execution of numerical algorithms necessary for analyzing signals. These algorithms include scalar products arising in the computation of digital filters and the discrete Fourier transform. Typically, digital signal processors exploit parallelism within an instruction. They consist of several functional units working in parallel. When computing the scalar product of two vectors, for example, a computational block will perform multiply-accumulate operations. For the computational block to operate during every single clock cycle, the block has to be supplied with one entry from each of the two vectors in each clock cycle. This necessitates two buses between the processor and the memory, and two address generators, one for each of the two buses. Furthermore, the loop necessary for stepping through the vectors' entries must not introduce any additional clock cycles. For being able to command all functional units at the same time, the instruction word must contain fields for controlling each functional unit, resulting in long or even very long instruction words. For programs which do not allow for instruction level parallelism, however, the long instruction words require a lot of memory containing mostly no-operations. Some signal processors support floating-point operations mostly avoiding numerical overflow and underflow problems.

Besides one or several cores and some multiported memory, a typical digital signal processor chip contains interfaces for analog-to-digital and digital-to-analog converters, and a direct memory access engine for supporting the processor with transferring blocks of data between memory and peripheral units. The direct memory access units found in signal processor chips targeted for audio and instrumentation applications typically handle one-dimensional blocks. Processor chips targeted for image and video processing typically contain direct memory access engines capable of copying a two-dimensional subimage out of a larger image. These chips may contain dedicated hardware for low-level operations such as edge-detection in image frames.

5.9.1 The SHARC® Family of Digital Signal Processors

The SHARC® signal processors by Analog Devices are 32 bit floating-point machines designed for general instrumentation and audio applications. A SHARC® chip, Fig. 5.16, consists of a core processor, an input-output processor, on-chip multiported static RAM, and peripherals. The core processor executes the signal processing programs and controls operation of the other units. The input-output processor transfers blocks of data back and forth between on-chip memory and external memory or between on-chip memory and peripherals. The peripherals include interfaces for analog-to-digital and digital-to-analog converters.

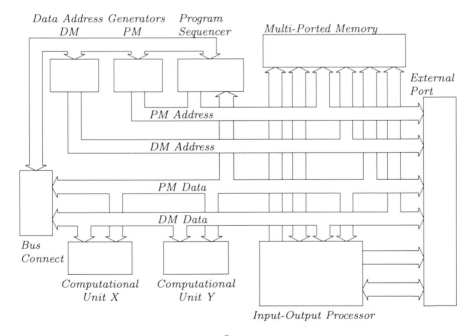

Fig. 5.16 Simplified structure of a SHARC® signal processor without peripherals. The abbreviations PM and DM stand for program memory and data memory respectively. The core's two computational units, its two data address generators, its program sequencer, and the input-output processor work in parallel

5.9.1.1 The Core Processor

The core processor is a word-addressing long instruction word 32 bit architecture. An instruction for the core specifies operations for all the core's parts, the computational blocks, the data address generators and the program sequencer. This allows all these units to work simultaneously.

Each of the core processor's two computational blocks consists of a register file with 16 primary registers and 16 alternate registers, an arithmetic-logic unit, a barrel shifter, and a multiplier. All registers in the register files are 40 bits long. Usually, the primary registers are active and the alternate registers are inactive. Software can switch to the alternate registers, for example, in an interrupt service routine for avoiding the overhead of saving register content to memory. A register can either hold a 32 bit integer, a 32 bit floating point number, or a 40 bit float. The 40 bit float data type has a 31 bit mantissa. Operands for and results of computational operations must reside in one of the registers. Besides addition, subtraction, and comparison for integer and floating-point data types, the arithmetic-logic unit can compute the sum and difference of the same two inputs simultaneously. Saturating integer arithmetic is available for supporting fixed-point data. The bit-wise logical operations are available for integers only. When operating on integer data the multiplier performs multiply-accumulate operations. In order to be able to store the sum of several 64 bit products, the accumulator is 80 bits long. When operating on floating-point data, the multiplier computes products only. The barrel shifter shifts data arithmetically or logically by an arbitrary number of bits. It also deposits bit-fields into and extracts bit-fields out of integers and performs conversions between integers and floating-point data. The arithmetic unit and the multiplier can work simultaneously. When both computational blocks are activated, both of them perform the same operations, each on the data stored in its register file, they operate in single instruction multiple data mode. Each of the computational blocks produces its own set of condition flags describing the outcome of the last operation of its arithmetic-logic unit, barrel shifter, and multiplier.

The program sequencer consists of the instruction pipeline, the program counter and a program counter stack, a selective instruction cache, a stack for the condition flags, a loop stack and a loop counter stack, a timer, and a vectored interrupt controller. It fetches instructions from memory via the PM bus. As the PM bus is also used for data transfers, instruction fetches may conflict with data transfers on the PM bus commanded by instructions down the instruction pipeline. In order to resolve these conflicts, the sequencer caches those instructions whose fetch conflicts with a data transfer over the PM bus. Doing so allows programs to perform two memory accesses per instruction in inner loops without being hindered by instruction fetches. In order to avoid costly jumps over short sections of code, most instructions specify a condition. If the condition is met when the sequencer executes the instruction, the core processor updates its state, otherwise executing the instruction has no effect. Procedure calls push the return address automatically onto the program counter stack, a return instruction pops the return address from the program counter stack. Branching instructions, calls and jumps and returns, can be delayed, where the sequencer fetches and executes the two instructions following the branch, or can be immediate where the sequencer sends two no-operations through its pipeline instead. For an indirect branch, the PM data address generator can provide the branch's target address. When the sequencer executes a loop instruction, it pushes the address of the loop top onto the program counter stack; it also pushes the address of the loop bottom and the termination condition onto the loop stack. For counting loops, it additionally pushes the number of iterations onto the loop counter stack. When the sequencer executes the instruction two addresses above the loop bottom, it decrements the loop counter at the top of the loop counter stack for counting loops and tests the loop's termination condition. If the loop terminates, the sequencer pops the respective stacks and fetches the instruction following the loop bottom, otherwise it fetches the instruction at the top of the loop. The vectored interrupt controller monitors the interrupt signals. Each clock cycle, it determines the interrupt sources which are both enabled and signal an outstanding request. Provided such sources exist, it selects the one with the highest priority. The sequencer then pushes the program counter onto the program counter stack and fetches the next instruction from the address associated with the selected source. The timer counts clock cycles. It serves as one of the interrupt sources.

Each of the two data address generators contains a primary register set comprising eight base registers, eight index registers, eight modify registers, and eight length registers. Each one also contains a software-selectable alternate set. The PM data address generator provides addresses for accessing memory via the PM bus, while the DM data address generator provides addresses to the DM bus. The PM data bus and the DM data bus each are 64 bit wide. The data address generators support two kinds of memory accesses, one which updates an index register and one which does not. Both the DM data address generator and the PM data address generator support the same types of accesses. In case only the first computational unit is active, a read access without index register update via the DM bus amounts to the assignment

$$f_i \leftarrow \mathrm{DM}_{\iota+\mu},$$

where f_i, $0 \leq i \leq 15$ is one of the registers of the first computational unit, ι is an index register of the DM data address generator, and μ is a modify register of the DM data address generator. When both computational blocks are active such a load amounts to the parallel assignment of two consecutive words

$$(f_i, g_i) \leftarrow (\mathrm{DM}_{\iota+\mu}, \mathrm{DM}_{\iota+\mu+1}),$$

where g_i is the register associated with f_i in the other computational unit. Write accesses work in an analogous manner. In the following code pieces, we omit the second computational block when it is understood that it is active too. A load with index register update via the DM bus amounts to the parallel assignment

$$(f_i, \iota) \leftarrow (\mathrm{DM}_\iota, \iota + \mu).$$

Fig. 5.17 Circular
addressing, equivalent
program

$$(\mathrm{DM}_\iota, \iota) \leftarrow (f_i, \iota + \mu)$$
$$\textbf{if } \iota < \beta \textbf{ then}$$
$$\iota \leftarrow \iota + \lambda$$
$$\textbf{else if } \iota \geq \beta + \lambda \textbf{ then}$$
$$\iota \leftarrow \iota - \lambda$$
$$\textbf{end if}$$

In both types of accesses, the address sent to the memory can be bit-reversed.
A load with index register update and bit-reversed addressing amounts to the parallel
assignment

$$(f_i, \iota) \leftarrow (\mathrm{DM}_{\mathrm{rev}\iota}, \iota + \mu).$$

A broadcast load loads a single value into the register f_i and the associated register g_i
in the other computational unit,

$$(f_i, \iota) \leftarrow (\mathrm{brDM}_\iota, \iota + \mu).$$

For supporting circular buffers, accesses with index register updates can be modified
with circular addressing. The store operation

$$(\mathrm{DM}_\iota, \iota) \leftarrow (f_i, \mathrm{circ}(\iota + \mu))$$

is equivalent to the program in Fig. 5.17, where β is the base register associated with
the index register ι and λ is the length register associated with ι. The update of the
index register completes in a single clock cycle. When both computational blocks
are active transfers between four registers, two in the first computational block and
two in the second, are possible in each instruction.

5.9.1.2   The External Port and the Input-Output Processor

The external port interfaces external memory to the signal processor. It allows the core
processor to access this external memory either via the PM bus or the DM bus. It also
allows accesses by the input-output processor. The input-output processor comprises
a number of direct memory access channels, some capable of transfers between on-
chip and external memory, others for transfers between the on-chip memory and a
peripheral. A data structure in internal memory, a so-called transfer control block,
controls a transfer. It specifies the start address of the block of data in internal memory
to be transferred, the address increment and the number of data items. For memory
to memory transfers, it also specifies the start address and the address increment for
the external memory. Further, it specifies whether the input-output processor will
deliver an interrupt to the core upon completion of the transfer. The input-output
processor handles linked lists of transfer control blocks without intervention by the

core. Circular linked lists of transfer control blocks describe transfers which have to keep going indefinitely, for example, between memory and analog-to-digital and digital-to-analog converters.

5.9.2 Case Study—Fast Fourier Transform

The discrete Fourier transform forms the basis of many digital signal processing applications. In 1965 Cooley and Tukey (Cooley and Tukey 1965), discovered an efficient algorithm, the fast Fourier transform, for computing the discrete Fourier transform for a discrete-time signal s. The fast Fourier transform for signals with period $p = 2^l$ for some $l \in \mathbb{N}$ and $l > 0$ is particularly simple. Given the sequence of samples of one period

$$s = \langle s_0, \ldots, s_{p-1} \rangle$$

we have to compute the transform

$$S = \mathcal{F}_p(s) = \langle S_0, \ldots, S_{p-1} \rangle,$$

where

$$S_k = \sum_{n=0}^{p-1} s_n e^{-i\frac{2\pi}{p}nk} = \sum_{n=0}^{p-1} s_n W_p^{nk}, \tag{5.1}$$

for $W_p^{nk} = e^{-i\frac{2\pi}{p}nk}$ and $0 \le k < p = 2^l$. This suggests an algorithm for computing the sequence S requiring a number of complex multiplications and a number of additions, where both numbers depend quadratically on p. By splitting the sum in this equation we can do better,

$$
\begin{aligned}
S_k &= \sum_{r=0}^{p/2-1} s_{2r} W_p^{2rk} + \sum_{r=0}^{p/2-1} s_{2r+1} W_p^{(2r+1)k} \\
&= \sum_{r=0}^{p/2-1} s_{2r} W_p^{2rk} + W_p^k \sum_{r=0}^{p/2-1} s_{2r+1} W_p^{2rk} \\
&= \sum_{r=0}^{p/2-1} s_{2r} W_{p/2}^{rk} + W_p^k \sum_{r=0}^{p/2-1} s_{2r+1} W_{p/2}^{rk}.
\end{aligned}
\tag{5.2}
$$

Note that both the first and the second sum in (5.2) are periodic in k with period $p/2$ because

$$W_{p/2}^{n(k+p/2)} = e^{-i\frac{4\pi}{p}n(k+p/2)} = e^{-i\frac{4\pi}{p}nk} = W_{p/2}^{nk}.$$

Therefore, for computing the Fourier transform S of s, we have to compute two Fourier transforms, one of the subsequence of s comprising the entries with even index and one of the subsequence of s comprising the entries with odd index, and combine the two transforms. Employing this observation repeatedly suggests a divide and conquer approach resulting in a run-time proportional to $p \log p$. Besides splitting up the input, the divide phase does nothing. Therefore, we can skip the recursive divide phase and build the solution nonrecursively.

Before we can start, we have to analyze how the repeated decompositions reorder the input sequence s. One decomposition splits its input sequence into two subsequences according to whether the index of an entry is even or odd. An odd index has a least significant bit of 1, an even index one of 0. The first divide brings the entries of s with indexes having 0 as least significant bit into the first half, and the other entries into the second half. The second divide brings the entries with indexes having least significant bits 00 into the first quarter, the entries with 10 into the second quarter, the entries with 01 into the third quarter and the entries with 11 into the forth quarter. Continuing this pattern, we see that the divide phase reorders the entries of the input sequence s into bit-reversed order.

Expanding the divide and conquer scheme in (5.2) results in the dataflow graph in Fig. 5.18. Dots indicate summation operations. A number under an edge indicates multiplication with that number. Factors of 1 are omitted from the diagrams. Close inspection shows that the computation shown on the left side of Fig. 5.19 is the primitive operation in Fig. 5.18. Transforming it into the computation shown on the right side of Fig. 5.19 saves one multiplication. The resulting operation is called a butterfly and the complex number W_p^r is called a twiddle. The two outputs of the butterfly, C and D, are related to the two inputs, A and B, by

$$C = A + W_p^r B$$
$$D = A - W_p^r B.$$

Using butterflies and reordering we redraw the dataflow diagram in Fig. 5.18. The columns in the new arrangement, due to Singleton (1967), all have the same structure, Fig. 5.20. A program computing the transform for $p = 2^l$ points based on this arrangement uses two arrays of complex numbers. It processes the butterflies column by column, retrieving the inputs of a column from one array and storing the outputs in the other. For the next column it switches arrays. The program stores output C of the ith butterfly in the jth column, $0 \leq i < p/2$ and $0 \leq j < l$, at position i of the output array, output D at $i + p/2$. This butterfly uses the twiddle $W_p^{\lfloor i/2^{l-j-1}\rfloor 2^{l-j-1}}$.

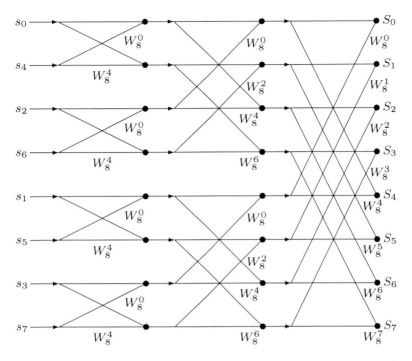

Fig. 5.18 Dataflow diagram resulting from expanding the divide and conquer scheme for eight inputs. Dots indicate summation operations. A number under an edge indicates multiplication with that number

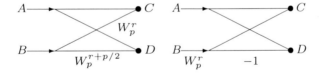

Fig. 5.19 The primitive operation in the fast Fourier transform computation. Scrutiny saves one multiplication

5.9.2.1 Fast Fourier Transform of Long Input Sequences

We wrote the following program for a SHARC® floating-point signal processor equipped with word-wide external SDRAM. For dealing with long inputs, too long to fit into the on-chip memory of a signal processor, we split the input vector into blocks of size 2^m, m chosen according to the size of the available on-chip memory. The core processor operates only on data located in the on-chip memory. For computing the discrete Fourier transform, Fig. 5.21 with $m = 3$, the input-output processor loads the first two blocks, one block for each of the core's computational units, from external into on-chip memory. The core computes those butterflies of the first m columns depending solely on the data in these two blocks. The simple structure of

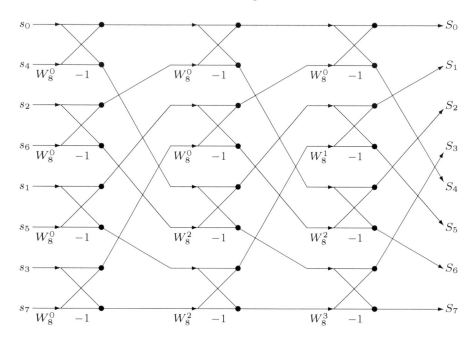

Fig. 5.20 Dataflow diagram of a fast Fourier transform for eight points. All columns have the same structure

the dataflow diagram in Fig. 5.20 allows us to analyze the flow of data through several columns. The input-output processor writes back the results into an auxiliary vector placed in the external memory and loads the next two blocks. Once all input data has been processed, the auxiliary vector contains the intermediate result after m columns. Further similar passes over the data complete the transform.

The need to address the input vector with bit-reversed indexes complicates the first pass. We perform bit-reversed indexing in two steps. First, we gather the data of the ith block by finding all the indexes which have the bit-reverse of i in their lower bits. These indexes are spaced 2^{l-m} apart, and the first index is just the bit-reverse of i. Once the entries of the ith block have been transferred into a vector in on-chip memory, we address this vector with bit-reversed indexes completing the second step.

By having the input-output processor write back results from on-chip memory to external memory and read inputs from external memory into on-chip memory in parallel with computing two blocks of butterflies we end up with a pipeline like the one in Fig. 5.14.

The procedure PIPELINEDFFT, Fig. 5.22, uses this scheme for two passes, limiting the size of the input to $2^l \leq 2^{2m}$. It accepts the input data s in two arrays, R for the real parts and I for the imaginary parts, such that $s_n = R_n + iI_n$ for $0 \leq n < 2^l$. Both arrays must reside in external memory. It places the results in these two arrays

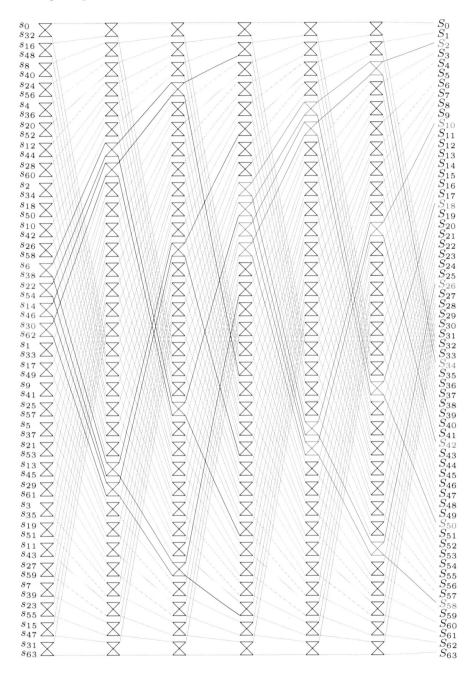

Fig. 5.21 Dataflow of a block by block decimation-in-time fast Fourier transform for 64 points. The transform is computed in two passes over the data. One block for the first pass and one for the second are highlighted in different colors. The twiddle factors have been omitted

1: **procedure** $\text{PipelinedFft}(R, I, l, T, T')$
2:    $(r, r's, b_0, b_1) \leftarrow (\text{rev } 2^{l-m-2}, 2^{2m-l}, 2^{l-m}, 0, \text{rev } 2^{l-m-1})$
3:    **create** RdBlocks1
4:    **fork** $\text{RdBlocks1}(R, I, s, \text{rev } b_0, \text{rev } b_1, N, N')$
5:    **create** WrBlocks1 RdBlocks1

6:    $(b_0, b_1) \leftarrow (b_0 + r, b_1 + r)$
7:    **fork** $\text{RdBlocks1}(R, I, s, \text{rev } b_0, \text{rev } b_1, D, D')$
8:    $(N, N', F, F') \leftarrow \text{Fft1}(N, N', F, F')$ RdBlocks1
9:    **for** $b \leftarrow 2, 4, 6, \ldots, 2^{l-m} - 4$ **do**

10:      $(N, N', D, D', b_0, b_1) \leftarrow (D, D', N, N', b_0 + r, b_1 + r)$
11:      **fork** $\text{WrBlocks1}(D, D', s, b - 2, A, A')$
12:      **fork** $\text{RdBlocks1}(R, I, s, \text{rev } b_0, \text{rev } b_1, D, D')$
13:      $(N, N', F, F') \leftarrow \text{Fft1}(N, N', F, F')$ WrBlocks1
                                                        RdBlocks1
14:    **end for**

15:    $(N, N', D, D', t') \leftarrow (D, D', N, N', 2^{m+1} - 2^{2m-l+1})$
16:    **fork** $\text{WrBlocks1}(D, D', s, s - 4, A, A')$
17:    **fork** $\text{RdTwiddles}(T, T', 0, t', U, U')$
18:    $(N, N', F, F') \leftarrow \text{Fft1}(N, N', F, F')$ WrBlocks1
19:    **create** RdBlocks2 RdTwiddles

20:    **fork** $\text{WrBlocks1}(N, N', s, s - 2, A, A')$
21:    **fork** $\text{RdBlocks2}(A, A', 0, N, N')$
22:    **create** WrBlocks2 WrBlocks1
23:    $t \leftarrow t'$ RdBlocks2

24:    **fork** $\text{RdBlocks2}(A, A', 2^{m+1}, D, D')$
25:    **fork** $\text{RdTwiddles}(T, T', t, t', V, V')$
26:    $(N, N', F, F') \leftarrow \text{Fft2}(N, N', F, F', U, U')$ RdBlocks2
                                                        RdTwiddles

---

27:    **for** $b \leftarrow 2, 4, 6, \ldots, 2^{l-m} - 4$ **do**
28:      $(N, N', D, D') \leftarrow (D, D', N, N')$
29:      $(U, U', V, V') \leftarrow (V, V', U, U')$
30:      $t \leftarrow t + t'$
31:      **fork** $\text{WrBlocks2}(D, D', r', s, (b - 2)r', R, I)$
32:      **fork** $\text{RdBlocks2}(A, A', (b + 2)2^m, D, D')$
33:      **fork** $\text{RdTwiddles}(T, T', t, t', V, V')$
                                                        WrBlocks2
34:      $(N, N', F, F') \leftarrow \text{Fft2}(N, N', F, F', U, U')$ RdBlocks2
                                                        RdTwiddles

---

35:    **end for**
36:    **fork** $\text{WrBlocks2}(N, N', r', s, (s - 4)r', R, I)$
37:    $(D, D', F, F') \leftarrow \text{Fft2}(D, D', F, F', V, V')$ WrBlocks2

38:    **fork** $\text{WrBlocks2}(D, D', r', s, (s - 2)r', R, I)$
                                                        WrBlocks2

39: **end procedure**

Fig. 5.22 Pipelined block by block fast Fourier transform computation, main procedure. The code for the core processor is shown on the *left side*, the direct memory transfers from external to internal memory and back on the *right side*. Horizontal lines indicate join operations too. The core processor executes the left column in Fig. 5.22, while the input-output processor executes the right column. Horizontal lines indicate join operations which are realized as spin locks. The input-output processor executes operations which have been forked in the order of the forks. A join operation waits till the operation that has been forked last completes. The procedure uses three pairs of references to arrays located in the on-chip memory of the signal processor. Each array is large enough to hold either the real parts or the imaginary parts of two blocks. In the two loops, which constitute the main parts of the two passes, after line 10 and after line 28 the pair (N, N') references new data that has just been loaded from external memory, the pair (D, D') references dirty data that has just been computed and must be written back to external memory, and the pair (F, F') references two arrays which are empty and serve as auxiliary storage for processing the next two blocks. The procedures FFT1 for processing two blocks during the first pass and FFT2 for the second pass return references to the two arrays containing the results and the two arrays whose content is irrelevant.

During the first pass, the same 2^{m-1} twiddles are required for each block. The array W located in the on-chip memory, contains the real parts of these twiddles, the array W' the imaginary parts. The start-up code initializes these two arrays. For the second pass, however, each block requires its own set of

$$t'/2 = 2^m - 2^{2m-l} \tag{5.3}$$

different twiddles. These twiddles have to be brought into on-chip memory in the order in which they are required. The pairs (U, U') and (V, V') reference four arrays located in on-chip memory. After line 29, the pair (U, U') references the real parts and the imaginary parts of the twiddles required for processing the next two blocks. The input-output processor loads these arrays from two arrays in external memory which are referenced by the two parameters T for the real parts and T' for the imaginary parts.

The input-output processor cannot execute arbitrary sequences of instructions. Instead it executes groups of transfers described by possibly circular linked lists of so-called transfer control blocks. In the procedure PIPELINEDFFT, the core processor creates these lists on the fly.

Each invocation of the procedure RDBLOCKS1, Fig. 5.23, executed by the input-output processor, reads two blocks of input data from the arrays referenced by R and I and located in external memory into the two arrays referenced by $\hat{R}$ and $\hat{I}$ in on-chip memory. In order to allow for processing two blocks of data simultaneously using the signal processor's two computational units, it stores the data of the first block in entries with even indexes and data of the second block in entries with odd indexes. It addresses the external arrays with stride $s = 2^{l-m}$. The parameter b_0 is the index in R and I of the first entry of the first block, b_1 is the index of the first entry of the second block.

Each invocation of the procedure WRBLOCKS1($\hat{R}, \hat{I}, s, b, R, I$) writes back two blocks located in on-chip memory to two consecutive blocks in external memory. The real parts of the two blocks are stored interleaved in the array referenced by $\hat{R}$, while

Fig. 5.23 Transfer of two blocks from external to internal memory for the first pass. The procedure RDBLOCKS1 is executed by the input-output processor

```
 1: procedure RDBLOCKS1(R, I, s, b₀, b₁, R̂, Î)
 2:     (i, i′) ← (b₀, 0)
 3:     for j ← 0, 2ᵐ − 1 do
 4:         (R̂ᵢ′, i, i′) ← (Rᵢ, i + s, i′ + 2)
 5:     end for
 6:     (i, i′) ← (b₀, 0)
 7:     for j ← 0, 2ᵐ − 1 do
 8:         (Îᵢ′, i, i′) ← (Iᵢ, i + s, i′ + 2)
 9:     end for
10:     (i, i′) ← (b₁, 1)
11:     for j ← 0, 2ᵐ − 1 do
12:         (R̂ᵢ′, i, i′) ← (Rᵢ, i + s, i′ + 2)
13:     end for
14:     (i, i′) ← (b₁, 1)
15:     for j ← 0, 2ᵐ − 1 do
16:         (Îᵢ′, i, i′) ← (Iᵢ, i + s, i′ + 2)
17:     end for
18: end procedure
```

the imaginary parts are stored in the array referenced by $\hat{I}$. The parameter b is the index of the first entry of the first block in the arrays R and I located in external memory. The first entry of the second block has index $b + 1$. The procedure addresses R and I with stride $s = 2^{l-m}$. Like RDBLOCKS1 the procedure WRBLOCKS1 consists of four loops.

Each invocation of the procedure RDBLOCKS2($R, I, b, \hat{R}, \hat{I}$) reads two blocks of data from the two auxiliary arrays located in external memory referenced by R and I into two arrays referenced by $\hat{R}$ and $\hat{I}$ in on-chip memory. it reads the arrays R and I sequentially. The parameter b is the index of the first entry of the first block in the arrays referenced by R and I. For allowing the simultaneous processing of both blocks using the signal processor's two computational units RDBLOCKS2 stores the data it reads in interleaved fashion.

Each invocation of the procedure RDTWIDDLES(T, T', t, c, V, V') transfers twiddles from the arrays referenced by T and T' in external memory into the arrays referenced by V and V' located in on-chip memory. The parameter t is the index of the first entries to be transferred and the parameter $c = 2^{m+1} - 2^{2m-l+1}$ is the number of twiddles.

Each invocation of the procedure WRBLOCKS2 writes back two blocks from the arrays referenced by $\hat{R}$ and $\hat{I}$ located in on-chip memory to the arrays referenced by R and I located in external memory. The second pass computes $l - m$ columns of butterflies for each block of 2^m data points considerably complicating writing back the results. The contents belonging to one of the two blocks of the arrays referenced by $\hat{R}$ and $\hat{I}$ have to be placed in the arrays R and I in $c = 2^{l-m}$ groups of contiguous data, each of length $r = 2^{2m-l}$ and spaced 2^m apart. The parameter b is the index of first entry of the first group in the arrays referenced by R and I. When creating the procedure WRBLOCKS2, we expand the outer loop. The program that the input-output processor executes consists actually of $4r$ loops. The alternative

Fig. 5.24 Transfer of two blocks from internal to external memory for the second pass

```
 1: procedure WRBLOCKS2(R̂, Î, r, c, b, R, I)
 2:     for j ← 0, r − 1 do
 3:         (i, i′) ← (b + j, 0)
 4:         for j′ ←, 1, c do
 5:             (Rᵢ, i, i′) ← (R̂ᵢ′, i + 2ᵐ, i′ + 2r)
 6:         end for
 7:         (i, i′) ← (b + j, 0)
 8:         for j′ ←, 1, c do
 9:             (Iᵢ, i, i′) ← (Îᵢ′, i + 2ᵐ, i′ + 2r)
10:         end for
11:         (i, i′) ← (b + j + r, 1)
12:         for j′ ←, 1, c do
13:             (Rᵢ, i, i′) ← (R̂ᵢ′, i + 2ᵐ, i′ + 2r)
14:         end for
15:         (i, i′) ← (b + j + r, 1)
16:         for j′ ←, 1, c do
17:             (Iᵢ, i, i′) ← (Îᵢ′, i + 2ᵐ, i′ + 2r)
18:         end for
19:     end for
20: end procedure
```

implementation, iterating over the groups in the outer loop and iterating over the entries in a group in the inner loops, is more efficient for $l < 3m/2$, for larger l the alternative is less efficient and requires a larger number of transfer control blocks than the code in Fig. 5.24.

5.9.2.2 Simultaneous Computation of the Transforms of Two Blocks

Provided the twiddle is already available in the register pair w and w', computing a single butterfly, Fig. 5.25, requires ten instructions. Only four of these instructions use the core processor's hardware multipliers and only four transfer data. Data dependencies inherent in the computation prevents one from increasing resource utilization within a single butterfly.

In order to better use the resources available in the core processor one has to compute several butterflies at the same time in a pipelined fashion. In the trace in Fig. 5.26, six butterflies are computed simultaneously, three on each computational unit. While the pipeline is fully loaded every singe instruction uses the multiply units and transfers data via both buses. The procedure BUTTERFLYCOLUMN, Fig. 5.27, implements this approach. For switching twiddles it has to leave the inner loop. During the last iteration of the outer loop it overshoots its input arrays and performs some superfluous computations. Avoiding this behavior, however, is too costly in terms of code size as well as run time, considering that it does no harm.

The procedure SECONDLASTCOLUMN($R, I, R̂, Î$) computes the second to last column. It changes the twiddle every other butterfly. For writing SECONDLASTCOL-UMN one modifies BUTTERFLYCOLUMN. One must set $t = 2$ and replace the inner

$$
\begin{aligned}
&1:\ (b, b', \iota) \leftarrow (R_\iota, I_\iota, \iota - 2)\\
&2:\ f_1 \leftarrow bw\\
&3:\ f_2 \leftarrow b'w'\\
&4:\ (a, a', \iota, f_1') \leftarrow (R_\iota, I_\iota, \iota + 6, bw')\\
&5:\ (f_2', f_3) \leftarrow (b'w, f_1 + f_2)\\
&6:\ f_3' \leftarrow f_2' - f_1'\\
&7:\ (c', d') \leftarrow (a' + f_3', a' - f_3')\\
&8:\ (c, d) \leftarrow (a + f_3, a - f_3)\\
&9:\ (\hat{R}_\kappa, \hat{I}_\kappa, \kappa) \leftarrow (c, c', \kappa + 2)\\
&10:\ (\hat{R}_\nu, \hat{I}_\nu, \nu) \leftarrow (d, d', \nu + 2)
\end{aligned}
$$

Fig. 5.25 Code for computing a single butterfly. The pair (a, a') of floating point registers holds the complex input $A = a + ia'$, the pair (b, b') holds the complex input B, the pair (c, c') the complex output C, the pair (d, d') the complex output D. The pair (w, w') holds the twiddle. The pairs (f_1, f_1'), (f_2, f_2'), and (f_3, f_3') hold complex intermediate values. The registers ι, κ and ν are index registers. Data dependencies inherent in the butterfly computation prohibit better utilization of the available hardware

loop with two copies of the inner loop's body. Furthermore one adds an instruction for changing the twiddle in the middle of the first copy. Last, the part of the outer loop's body after the inner loop has to go. The procedure LASTCOLUMN(R, I, $\hat{R}$, $\hat{I}$), which loads a new twiddle for each butterfly, computes the last column.

The first two columns deserve special attention. The butterflies in these two columns use two peculiar twiddles only, $W_p^0 = 1$ and $W_p^{p/4} = i$. These twiddles are so peculiar in fact that no multiplications are required for computing these butterflies. This allows one to forgo the single butterflies and compute four butterflies at once, two from the first column and two from the second. Such a four-point transform, Fig. 5.28, computes the outputs

$$
\begin{aligned}
E &= A + B + C + D\\
F &= A - B + i(C - D)\\
G &= A + B - C - D\\
H &= A - B - i(C - D).
\end{aligned}
$$

The procedure FIRSTCOLUMNS, Fig. 5.29, computes the first two columns of butterflies of two blocks in Fig. 5.21. It is supplied with the input vector R comprising the real parts of the inputs, the input vector I comprising the imaginary parts, and the output vectors, $\hat{R}$ for the real parts and $\hat{I}$ for the imaginary parts. It addresses the input vectors using the index register ι' with bit-reversed addressing. Incrementing ι' is somewhat tricky, the counting has to take place in the appropriate number of *most* significant bits. The following bit-reverse operation, revι', then moves the bits into the least significant positions. The loop is pipelined; the computation of one four-point transform takes two iterations through the loop. A four-point transform is started each iteration.

$(b, b', \iota) \leftarrow (R_\iota, I_\iota, \iota - 2)$
$f_1 \leftarrow bw$
$f_2 \leftarrow b'w'$

$(a, a', \iota) \leftarrow (R_\iota, I_\iota, \iota + 6)$	$f_1' \leftarrow bw'$	
$(b, b', \iota) \leftarrow (R_\iota, I_\iota, \iota - 2)$	$(f_2', f_3) \leftarrow (b'w, f_1 + f_2)$	
$f_1 \leftarrow bw$	$f_3' \leftarrow f_2' - f_1'$	
$f_2 \leftarrow b'w'$	$(c', d') \leftarrow a' \pm f_3'$	

$(a, a', \iota) \leftarrow (R_\iota, I_\iota, \iota + 6)$	$f_1' \leftarrow bw'$	$(c, d) \leftarrow a \pm f_3$
$(b, b', \iota) \leftarrow (R_\iota, I_\iota, \iota - 2)$	$(f_2', f_3) \leftarrow (b'w, f_1 + f_2)$	
$f_1 \leftarrow bw$	$f_3' \leftarrow f_2' - f_1'$	$(\hat{R}_\kappa, \hat{I}_\kappa, \kappa) \leftarrow (c, c', \kappa + 2)$
$f_2 \leftarrow b'w'$	$(c', d') \leftarrow a' \pm f_3'$	$(\hat{R}_\nu, \hat{I}_\nu, \nu) \leftarrow (d, d', \nu + 2)$

$(a, a', \iota) \leftarrow (R_\iota, I_\iota, \iota + 6)$	$f_1' \leftarrow bw'$	$(c, d) \leftarrow a \pm f_3$
$(b, b', \iota) \leftarrow (R_\iota, I_\iota, \iota - 2)$	$(f_2', f_3) \leftarrow (b'w, f_1 + f_2)$	
$f_1 \leftarrow bw$	$f_3' \leftarrow f_2' - f_1'$	$(\hat{R}_\kappa, \hat{I}_\kappa, \kappa) \leftarrow (c, c', \kappa + 2)$
$f_2 \leftarrow b'w'$	$(c', d') \leftarrow a' \pm f_3'$	$(\hat{R}_\nu, \hat{I}_\nu, \nu) \leftarrow (d, d', \nu + 2)$

$(a, a', \iota) \leftarrow (R_\iota, I_\iota, \iota + 6)$	$f_1' \leftarrow bw'$	$(c, d) \leftarrow a \pm f_3$
	$(f_2', f_3) \leftarrow (b'w, f_1 + f_2)$	
	$f_3' \leftarrow f_2' - f_1'$	$(\hat{R}_\kappa, \hat{I}_\kappa, \kappa) \leftarrow (c, c', \kappa + 2)$
	$(c', d') \leftarrow a' \pm f_3'$	$(\hat{R}_\nu, \hat{I}_\nu, \nu) \leftarrow (d, d', \nu + 2)$

	$(c, d) \leftarrow a \pm f_3$
	$(\hat{R}_\kappa, \hat{I}_\kappa, \kappa) \leftarrow (c, c', \kappa + 2)$
	$(\hat{R}_\nu, \hat{I}_\nu, \nu) \leftarrow (d, d', \nu + 2)$

Fig. 5.26 Trace of the pipelined computation of four butterflies. The pipelined inner loop has three stages. The pair (a, a') of floating-point registers holds the complex input $A = a + ia'$, the pair (b, b') holds the complex input B, the pair (c, c') the complex output C, the pair (d, d') the complex output D. The pair (w, w') holds the twiddle. The pairs (f_1, f_1'), (f_2, f_2'), and (f_3, f_3') hold complex intermediate values. The registers ι, κ and ν are index registers. Each line corresponds to the execution of a single instruction. Horizontal lines indicate separate iterations of the pipelined butterfly loop

The procedure FFT1 computes the fast Fourier transform of two blocks with 2^m points each simultaneously, Fig. 5.30. It is supplied with the two input arrays R and I, and two auxiliary arrays $\hat{R}$ and $\hat{I}$. It passes back the two arrays R and I containing the outputs and the two auxiliary arrays $\hat{R}$ and $\hat{I}$. The procedure FFT2, which processes the blocks during the second pass, is similar to FFT1 except for the bit-reversed addressing of the inputs and the optimized computation of the first two columns.

5.9.2.3 Overall Timings

Table 5.1 gives measurements for the execution times in clock cycles of parts of the procedure PIPELINEDFFT in terms of the size of the input. The hardware on which these measurements were taken consists of an ADSP-21121 signal processor equipped with word-wide external SDRAM. Both the processor and the memory run off the same clock. The block size is 256, that is $m = 8$, the row size of the

```
 1: procedure BUTTERFLYCOLUMN(R, I, t, R̂, Î)
 2:     ι ← 2
 3:     κ ← ω ← 0
 4:     ν ← 2^m
 5:     (w, w', ω) ← (br W_ω, br W'_ω, ω + t)
 6:     (b, b', ι) ← (R_ι, I_ι, ι − 2)
 7:     f_1 ← bw
 8:     f_2 ← b'w'
 9:     (a, a', ι, f'_1) ← (R_ι, I_ι, ι + 6, bw')
10:     (b, b', ι, f'_2, f_3) ← (R_ι, I_ι, ι − 2, b'w, f_1 + f_2)
11:     (f_1, f'_3) ← (bw, f'_2 − f'_1)
12:     (f_2, c', d') ← (b'w', a' + f'_3, a' − f'_3)
13:     for i ← 1, 2^{m−1}/t do
14:         for j ← 1, t − 2 do
15:             (a, a', ι, f'_1, c, d) ← (R_ι, I_ι, ι + 6, bw', a + f_3, a − f_3)
16:             (b, b', ι, f'_2, f_3) ← (R_ι, I_ι, ι − 2, b'w, f_1 + f_2)
17:             (f_1, f'_3, R̂_κ, Î_κ, κ) ← (bw, f'_2 − f'_1, c, c', κ + 2)
18:             (f_2, c', d', R̂_ν, Î_ν, ν) ← (b'w', a' + f'_3, a' − f'_3, d, d', ν + 2)
19:         end for
20:         (a, a', ι, f'_1, c, d) ← (R_ι, I_ι, ι + 6, bw', a + f_3, a − f_3)
21:         (b, b', ι, f'_2, f_3) ← (R_ι, I_ι, ι − 2, b'w, f_1 + f_2)
22:         (w, w', ω) ← (br W_ω, br W'_ω, ω + t)
23:         (f_1, f'_3, R̂_κ, Î_κ, κ) ← (bw, f'_2 − f'_1, c, c', κ + 2)
24:         (f_2, c', d', R̂_ν, Î_ν, ν) ← (b'w', a' + f'_3, a' − f'_3, d, d', ν + 2)
25:         (a, a', ι, f'_1, c, d) ← (R_ι, I_ι, ι + 6, bw', a + f_3, a − f_3)
26:         (b, b', ι, f'_2, f_3) ← (R_ι, I_ι, ι − 2, b'w, f_1 + f_2)
27:         (f_1, f'_3, R̂_κ, Î_κ, κ) ← (bw, f'_2 − f'_1, c, c', κ + 2)
28:         (f_2, c', d', R̂_ν, Î_ν, ν) ← (b'w', a' + f'_3, a' − f'_3, d, d', ν + 2)
29:     end for
30: end procedure
```

Fig. 5.27 Procedure for computing one column of butterflies. The real parts of the input are passed in the array R, the imaginary parts in I. The real parts of the outputs are returned in $\hat{R}$, the imaginary parts in $\hat{I}$. The parameter t indicates how may butterflies share the same twiddle. The pair (a, a') of floating-point registers holds the complex input $A = a + ia'$ to a butterfly, the pair (b, b') holds the complex input B, the pair (c, c') the complex output C, the pair (d, d') the complex output D. The pair (w, w') holds the twiddle. The pairs (f_1, f'_1), (f_2, f'_2), and (f_3, f'_3) hold complex intermediate values. The registers ι, κ and ν are index registers. Each line represents a single machine instruction

SDRAM is 512 words, the CAS latency is $CL = 3$ clock cycles, the PRECHARGE command period is $t_{RP} = 2$ cycles, the ACTIVE to READ or WRITE delay is $t_{RCD} = 2$ cycles, and the ACTIVE to PRECHARGE delay is $t_{RAS} = 5$. The time for executing the procedure FFT1 measured at 4,165 clock cycles. We have no good way for measuring the times for RDBLOCK2 and for RDTWIDDLES individually, so we measured the time for executing the transfers performed in the second pass, WRBLOCK2 and RDBLOCK2 and RDTWIDDLES, together. In all but one case, the transfers to and from external memory take longer than the computations executed by the core processor. The nonsequential accesses to the external SDRAM in the procedures RDBLOCK1, WRBLOCK1 and WRBLOCK2 are particularly costly. The signal processor's memory

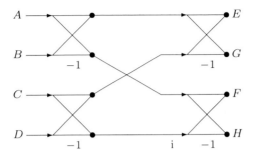

Fig. 5.28 Four-point fast Fourier transform

```
 1: procedure FIRSTCOLUMNS(R, I, R̂, Î)
 2:     ι' ← 0
 3:     κ ← 0
 4:     κ' ← 2^{m-1}
 5:     ν ← 2^m
 6:     ν' ← 3(2^{m-1})
 7:     (a, a', ι') ← (R_{rev ι'}, I_{rev ι'}, ι' + rev 2^m)
 8:     (b, b', ι') ← (R_{rev ι'}, I_{rev ι'}, ι' + rev 2^m)
 9:     (e, a, c, c', ι') ← (a + b, a - b, R_{rev ι'}, I_{rev ι'}, ι' + rev 2^m)
10:     (e', f', d, d', ι') ← (a' + b', a' - b', R_{rev ι'}, I_{rev ι'}, ι' + rev 2^m)
11:     (g, d) ← (c + d, c - d)
12:     (g', h') ← (c' + d', c' - d')
13:     for i ← 1, 2^{m-2} do
14:         (f, h, a, a', ι') ← (a + h', a - h', R_{rev ι'}, I_{rev ι'}, ι' + rev 2^m)
15:         (e', g', b, b', ι') ← (e' + g', e' - g', R_{rev ι'}, I_{rev ι'}, ι' + rev 2^m)
16:         (e, g, c, c', ι') ← (e + g, e - g, R_{rev ι'}, I_{rev ι'}, ι' + rev 2^m)
17:         (h', f', d, d', ι') ← (f' + d, f' - d, R_{rev ι'}, I_{rev ι'}, ι' + rev 2^m)
18:         (e, a, R̂_κ, Î_κ, κ) ← (a + b, a - b, e, e', κ + 2)
19:         (e', f', R̂_{κ'}, Î_{κ'}, κ') ← (a' + b', a' - b', f, f', κ' + 2)
20:         (g, d, R̂_ν, Î_ν, ν) ← (c + d, c - d, g, g', ν + 2)
21:         (g', h', R̂_{ν'}, Î_{ν'}, ν') ← (c' + d', c' - d', h, h', ν' + 2)
22:     end for
23: end procedure
```

Fig. 5.29 The procedure FIRSTCOLUMNS computes the first two columns of butterflies of two blocks during the first pass of a block by block fast Fourier transform. The pair (a, a') of floating point registers holds the complex input $A = a + ia'$ to a four-point transform, the pair (b, b') holds the complex input B, the pair (c, c') the complex input C, the pair (d, d') the complex input D, the pair (e, e') holds the complex output E, the pair (f, f') the complex output F, the pair (g, g') the complex output G, and the pair (h, h') the complex output H. These registers are used for intermediate values too. The register $ι'$ is the index for reading the inputs, while $κ$, $κ'$, $ν$ and $ν'$ are indexing the outputs

Fig. 5.30 The procedure
FFT1 computes the fast
Fourier transform of two
blocks simultaneously

```
 1: procedure FFT1(R, I, R̂, Î)
 2:     FIRSTCOLUMNS(R, I, R̂, Î)
 3:     t ← 2^(m-3)
 4:     for i ← 1, m − 4 do
 5:         BUTTERFLYCOLUMN(R̂, Î, t, R, I)
 6:         (R, I, R̂, Î, t) ← (R̂, Î, R, I, t/2)
 7:     end for
 8:     SECONDLASTCOLUMN(R̂, Î, R, I)
 9:     LASTCOLUMN(R, I, R̂, Î)
10:     return (R̂, Î, R, I)
11: end procedure
```

Table 5.1 Measured execution times in clock cycles of PIPELINEDFFT and parts of it with respect to the size of the input $p = 2^l$

l	RDBLOCKS1	WRBLOCKS1	FFT2	Transfers 2nd pass	WRBLOCK2	PIPELINEDFFT
12	6431	1304	2481	6876	4662	121253
13	6625	1438	3035	6368	4146	236597
14	6943	1756	3581	6154	3916	482655
15	7565	2418	4135	6010	3764	1036549
16	8869	3702	4675	5960	3708	2394901

Note that the time spent for accessing external memory dominates the total execution time

controller penalizes nonsequential reads further, it waits for the completion of each single read access instead of issuing several in a pipelined fashion.

5.10 Programmable Logic

Using programmable logic we can design hardware fitting our requirements and our imagination perfectly. We want to achieve high performance by putting a computational task directly into digital hardware. Often clock cycle accurate timing is a secondary goal. A task well suited for implementation directly into logic should exhibit simple control structure without many special cases. Repeated operations, with a fixed number of repetitions, will result in heavily used hardware blocks, favoring an implementation in logic. The handling of exceptional cases, however, will result in hardware blocks sitting idle most of the time. Tasks consisting mostly of dataflow with many repeated operations and a regular network structure lend themselves very well to an implementation in logic. Repeated operations getting their inputs from only a few sources often can be executed in parallel. We may be able to exploit this parallelism either by building pipelines or by replicating parts of the hardware. Many operations not available in general-purpose instruction sets can be realized in logic at little cost. Take computing a fixed permutation of a word's bits as an example; it requires painfully many instructions when implemented in software. When realized

Fig. 5.31 Two bit Gray code output by an incremental rotary encoder. For the first one and a half rotations the encoder's shaft rotates clockwise, for the second one and a half rotations it rotates counterclockwise

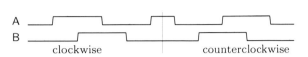

in hardware it requires interconnecting wires crossing in the correct pattern only. Memory requirements, however, are a deal breaker. Large memories usually must be realized using off-the-shelf dynamic RAM. The resulting limited memory bandwidth tends to become the performance bottleneck annihilating all previous gains. In addition, memory accesses render the analysis of the precise timing behavior difficult. The design of such a special-purpose processor quickly turns into a serious undertaking by itself. Hardware description languages such as VHDL and Verilog[3] together with logic simulators and hardware synthesis tools aid in such a design effort (IEEE 2009, 2006).

When implementing an input-output task directly in logic we want to achieve clock cycle accurate timing. We can realize autonomous peripherals for example for data acquisition or waveform generation using a moderate number of flip-flops (as available in complex programmable logic devices).

5.10.1 Case Study—Rotary Encoder Interface

In many mechanical devices rotary encoders sense the position of assemblies moved by some actuators in order to provide feedback concerning the positions of these assemblies. Rotating knob user interfaces are often realized with rotary encoders. A rotary encoder measures the angular position of its shaft. An absolute encoder reports the shaft's angle with respect to a well defined home position. An incremental encoder does not have a home position. It reports increments of the shaft's angle. In order to represent the sense of direction it uses 2 bit Gray code for representing the increments. An increment in a Gray code, Fig. 5.31, changes a single bit only. This prevents reading outrageously wrong values during the transition from one count to the next. Together with a sensor for an assembly's home position an incremental encoder can provide absolute position information. During the device's startup the actuator moves the assembly without the help of the encoder until the sensor for the home position detects the assembly. Then the incremental encoder takes over and provides position information.

---

[3] Hardware description languages are beyond the scope of this book.

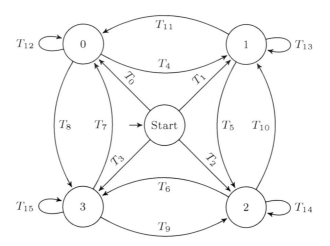

	Condition		Condition		Condition		Condition
T_0	$(A, B) = (0, 0)$	T_4	$(A, B) = (1, 0)$	T_8	$(A, B) = (0, 1)$	T_{12}	**default**
T_1	$(A, B) = (1, 0)$	T_5	$(A, B) = (1, 1)$	T_9	$(A, B) = (1, 1)$	T_{13}	**default**
T_2	$(A, B) = (1, 1)$	T_6	$(A, B) = (0, 1)$	T_{10}	$(A, B) = (1, 0)$	T_{14}	**default**
T_3	$(A, B) = (0, 1)$	T_7	$(A, B) = (0, 0)$	T_{11}	$(A, B) = (0, 0)$	T_{15}	**default**

State	A	B	cnt	up	State	A	B	cnt	up	State	A	B	cnt	up	State	A	B	cnt	up
0	0	0	0	0	1	0	0	1	0	2	0	0	0	0	3	0	0	1	1
0	1	0	1	1	1	1	0	0	0	2	1	0	1	0	3	1	0	0	0
0	1	1	0	0	1	1	1	1	1	2	1	1	0	0	3	1	1	1	0
0	0	1	1	0	1	0	1	0	0	2	0	1	1	1	3	0	1	0	0
Start	–	–	0	0															

Fig. 5.32 State transition diagram, transitions, and output function of a state machine for decoding the 2 bit Gray code. See also Sect. 4.4.3

The state machine in Fig. 5.32 converts a rotary encoder's outputs A and B into the count strobe cnt and the direction signal up. We may implement this state machine in software and call it periodically.

For handling high angular velocities, we build an interface for an incremental encoder directly in digital hardware. The device provides a 12 bit count representing the position of the encoder's shaft relative to the shaft's position when the device was reset. It autonomously updates this count whenever the encoder's shaft moves. In addition, it provides 12 digital inputs and 12 digital outputs. To the outside, the device appears as SPI slave. It communicates with its master using 16 bit words sending and receiving data in most significant bit first order. The SPI mode

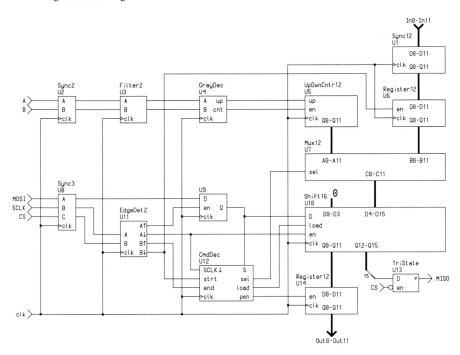

Fig. 5.33 Schematic of the counter and parallel input-output peripheral. *Wide lines* indicate parallel wires. Reset circuitry has been omitted for clarity

is $(CP, CPHA) = (0, 0)$. The data sent by the master consists of a 4 bit long command followed by 12 bits of data. When reset the least significant bit of the command selects the digital inputs and outputs. When set, it selects the encoder position. The data sent by the device consists of four undefined bits followed by 12 data bits.

The circuit in Fig. 5.33 is fully synchronous. All inputs are synchronized to the local high-frequency clock clk. The decoder U4 decodes the inputs from the rotary encoder and controls the up-down counter U5. The optional filter U3 eliminates the bouncing introduced by mechanical contacts in the encoder. It changes its output only after having seen for a specific time an input opposite to its output. The input register U1 freezes the inputs In0-In11 at the falling edge of CS. The flip-flop U9 samples MOSI whenever the edge detector U11 has detected a rising edge on SCLK during a transfer. After having received 4 bits from the master, the command decoder U12, a state machine, switches the multiplexer U7 to the selected source. During the next cycle of the clock clk, it commands the shift register U8 to perform a parallel load of the selected data. At the end of a transfer involving the digital inputs and outputs, it commands the transfer to the output register U14.

5.11 Tools

When developing software for embedded microprocessors or digital signal processors the target's instruction set is different from that of the development system. Cross compilers execute on the development system but produce code for the target. Often the system we are developing has to do without an operating system. Start-up code must establish a system state from which a program written in a higher programming language, mostly C, can take off.

5.11.1 Cross Compiler and Linker

The output of a compiler consists of the descriptions of several memory sections. The code section contains executable instructions and the interrupt vectors. The constant section contains preinitialized data that is not changed by the program such as string constants. Data that has to be initialized at start-up but will be changed later occupies another section. Yet another section contains the variables that have to be initialized to zero at start-up. The runtime stacks occupy a separate section. A final section contains the heap for dynamic storage allocation.

The linker places these sections in the processors' address spaces. When we cannot rely on an operating system, we must instruct the linker to place each section into the best-fitting memory. The code section must be mapped to nonvolatile memory. The interrupt vectors must end up at the addresses the processors expect them to be. The constant section has to reside in nonvolatile memory too. The data that will be initialized at start-up must reside in volatile memory. In order to enable the start-up code to initialize this data, the initial values have to reside in nonvolatile memory. The remaining sections all reside in volatile memory. The linker reports the memory map, containing the addresses of each of the program's variables and procedures. Before trying to run a program you should check the memory map for correct placement of the sections.

5.11.2 Debugger and In-Circuit Emulator

Embedded programs sense and influence their environment. For observing this interaction with any precision in a simulation, we need not only faithful models of the embedded system's peripheral components but also models of the behavior of the system's environment. Often these models are difficult to obtain; forcing us to relegate observations onto prototypes of the system.

In-Circuit emulators attached to a microcontroller or a digital signal processor allow downloading programs into the processors' memories. They allow setting breakpoints, and accessing processor registers and memories. Together with a debugger executing on the development system they provide the developer with some limited control over the execution of the embedded program. Statistical profilers, analyzing data gathered by an emulator without slowing down the processor, help identify bottlenecks in embedded programs. The JTAG interface is widely used for connecting emulators to programmable devices (IEEE 2001).

5.11.3 Logic Design Tools

Logic design tools support the simulation of schematics and of programs written in hardware design languages. They translate designs for the target chips and fit the results to the resources available on the chip. The tools download device configurations into the programmable logic chips using JTAG.

5.12 Bibliographic Notes

Patterson and Hennessy (2013), Hennessy and Patterson (2011) cover computer architecture from a modern point of view. The material presented there helps in understanding the choices the designers of embedded microprocessors and digital signal processors have to make. The book by Liu (2000), is a comprehensive source for scheduling in real-time systems. Kopetz (2011) discusses time-triggered systems extensively. The book by Marwedel (2010) gives a broad overview over skills and knowledge relevant when designing embedded systems. Lee and Seshia (2011) cover techniques for modeling, designing and analyzing embedded systems. Wolf (2014) covers processor architectures and system software for high-performance embedded systems, such as smart phones and multimedia devices. The C programming language (Kernighan and Ritchie 1988), is the lingua franca for writing embedded programs. For time-triggered systems, synchronous programming languages are a better alternative, see Halbwachs (1993) for an early reference. A comprehensive treatment of techniques for estimating the worst-case execution time of programs can be found in Wilhelm et al. (2008). The book by Oppenheim and Schafer (1975), is the classic text on digital signal processing. Our introduction of the fast Fourier transform in Sect. 5.9.2 follows their presentation. The processing of a single block in Sect. 5.9.2.2 is derived from the procedure `cfftf`, which is part of the signal processing library delivered with Analog's development tools for the SHARC® family, (Analog Devices 2012). For in-depth material on digital circuit design, see for example Kaeslin (2008). For a rarely used type of memory see Signetics (1972).

5.13 Exercises

Exercise 5.1 Design an event driven, master side software framework organizing transfers via an SPI bus. The framework shall queue transfer requests when the SPI framework is busy with some other transfer. The framework shall be able to support several slaves. Design the framework for an SPI interface which starts a transfer as soon as it is supplied with the data to be transmitted, provided it is not busy with another one. The interface signals completion of a transfer by raising an interrupt. Try minimizing the knowledge the framework must have about the slaves attached to the bus.

Exercise 5.2 Argue that (5.3) specifies the number of twiddles required for processing one block during the second pass of procedure PIPELINEDFFT in Fig. 5.22.

Exercise 5.3 Design a procedure for initializing the arrays T and T' of twiddles in Fig. 5.22.

Exercise 5.4 Design a time-triggered program for providing position information from the 2 bit Gray code produced by an incremental rotary encoder.

Exercise 5.5 The circuit in Fig. 5.33 uses a state machine for decoding the commands received from the master. Design this state machine.

Exercise 5.6 Design the filter for cleaning the outputs of the rotary encoder used in Fig. 5.33.

5.14 Lab Exercises

Exercise 5.7 Implement the procedures PERTICK and SCHEDULE in Sect. 5.2.2. Implement a timing wheel. Make performance comparisons.

Exercise 5.8 Implement the framework you designed in Exercise 5.1.

Exercise 5.9 Implement the time triggered program you designed in Exercise 5.4.

Exercise 5.10 Implement the input-output subsystem described in Sect. 5.10.1. Test it with the framework you implemented in the lab Exercise 5.8.

References

Analog Devices (2012) CrossCore®
Braess H, Seiffert U (2005) Handbook of automotive engineering. SAE International, Warrendale
CAN (2003–2013) ISO 11898 Road vehicles—controller area network (CAN). ISO

Cooley JW, Tukey JW (1965) An algorithm for the machine calculation of complex fourier series. Math Comput 19(90):297–301. doi:10.2307/2003354

FlexRay (2010) ISO 10681 Road vehicles—Communication on FlexRay. ISO

Halbwachs N (1993) Synchronous programming of reactive systems. Kluwer Academic Publishers, Norwell

Hennessy JL, Patterson DA (2011) Computer architecture, fifth edition: a quantitative approach, 5th edn. Morgan Kaufmann Publishers Inc., San Francisco

IEEE (2001) IEEE Standard test access port and boundary scan architecture. IEEE Std 1149.1-2001. doi:10.1109/IEEESTD.2001.92950

IEEE (2006) IEEE standard for verilog hardware description language. IEEE Std 1364–2005 (Revision of IEEE Std 1364–2001). doi:10.1109/IEEESTD.2006.99495

IEEE (2008) IEEE standard for a precision clock synchronization protocol for networked measurement and control systems. IEEE Std 1588–2008 (Revision of IEEE Std 1588–2002) pp c1–269. doi:10.1109/IEEESTD.2008.4579760

IEEE (2009) IEEE standard VHDL language reference manual. IEEE Std 1076–2008 (Revision of IEEE Std 1076–2002). doi:10.1109/IEEESTD.2009.4772740

IEEE (2012) IEEE standard for ethernet. IEEE Std 802.3-2012 (Revision of IEEE Std 802.3-2008). doi:10.1109/IEEESTD.2008.4579760

Kaeslin H (2008) Digital integrated circuit design: from VLSI architectures to CMOS fabrication, 1st edn. Cambridge University Press, New York

Kernighan B, Ritchie D (1988) The C programming language. Prentice-Hall software series, Prentice Hall, Upper saddle River

Kopetz H (2011) Real-time systems: design principles for distributed embedded applications. Real-time systems series, Springer, Berlin

Lee EA, Seshia SA (2011) Introduction to embedded systems, A cyber-physical systems approach. http://LeeSeshia.org

Leveson NG (1995) Safeware: system safety and computers. ACM, New York

Leveson NG (2011) Engineering a safer world: systems thinking applied to safety. Engineering systems, MIT Press, Cambridge

LIN (2013) ISO/DIS 17987 Road vehicles—Local interconnect network (LIN). ISO

Liu JWSW (2000) Real-time systems, 1st edn. Prentice Hall PTR, Upper Saddle River

Marwedel P (2010) Embedded system design: embedded systems foundations of cyber-physical systems. Embedded systems, Springer

NXP (2012) I2C-bus specification and users manual

Oppenheim AV, Schafer R (1975) Digital signal processing. Prentice-Hall, Upper Saddle River

Paret D, Riesco R (2007) Multiplexed networks for embedded systems: CAN, LIN, flexray, safe-by-wire. Wiley, London

Patterson DA, Hennessy JL (2013) Computer organization and design: the hardware/software interface, 5th edn., The Morgan Kaufmann series in computer architecture and design, Elsevier Science

Reddy T (2010) Linden's handbook of batteries, 4th edn. Mcgraw-hill, New York

Signetics (1972) 25120 final specification

Singleton RC (1967) A method for computing the fast Fourier transform with auxiliary memory and limited high-speed storage. IEEE Trans Audio Electroacoustics 15(2):91–98. doi:10.1109/TAU.1967.1161906

USB-IF (2013) Universal Serial Bus Revision 3.1 specification. Hewlett-Packard Company and Intel Corporation and Microsoft Corporation and Renesas Corporation and ST-Ericsson and Texas Instruments

Varghese G, Lauck A (1997) Hashed and hierarchical timing wheels: Efficient data structures for implementing a timer facility. IEEE/ACM Trans Netw 5(6):824–834. doi:10.1109/90.650142

Varghese G, Lauck T (1987) Hashed and hierarchical timing wheels: Data structures for the efficient implementation of a timer facility. In: Proceedings of the eleventh ACM symposium on operating systems principles, SOSP '87, ACM, New York, NY, USA, pp 25–38. doi:10.1145/41457.37504

Wilhelm R, Engblom J, Ermedahl A, Holsti N, Thesing S, Whalley D, Bernat G, Ferdinand C, Heckmann R, Mitra T, Mueller F, Puaut I, Puschner P, Staschulat J, Stenström P (2008) The worst-case execution-time problem—overview of methods and survey of tools. ACM Trans Embed Comput Syst 7(3):36:1–36:53. doi:10.1145/1347375.1347389, http://doi.acm.org/10.1145/1347375.1347389

Wolf M (2014) High-performance embedded computing: architectures, applications, and methodologies, 2nd edn. Morgan Kaufmann Publishers Inc., San Francisco

Yiu J (2014) The definitive guide to the Arm® Cortex®-M3 and Cortex®-M4 Processors. Elsevier Science and Technology Books

Chapter 6
Analog Circuits—Signal Conditioning and Conversion

The world is analog. The detectors in embedded systems often pick up analog information and convert this information into feeble analog continuous-time signals. These signals have to be amplified and filtered before being converted into digital discrete-time signals, streams of numbers, which we can process using digital computers. This signal conditioning and conversion determines the performance of an embedded system: even with the most clever algorithm we cannot recover any information destroyed during this process! When an embedded system acts on the physical world it is embedded in, it has to either produce analog voltages and currents or generate other analog physical quantities like sounds, pressures, or positions using actuators. For being able to design the architecture of such a system we have to understand the properties of analog signal conditioning and conversion methods.

6.1 Sampled Data Systems

Continuous-time signals do not lend themselves easily to processing by a digital computer. Instead we approximate such a signal by measuring the value of the signal at uniformly spaced points in time. This process, the creation of a discrete-time signal from a continuous-time signal, is called sampling. Sampling introduces several kinds of artifacts. The type of problem we have to tackle determines which types of artifacts limit the performance of the solution most. We look at these issues in some detail by considering three prototypical types of systems, digital controllers, measurement systems, and communication systems.

In a control system a controller governs the input of a process in such a way that the output of the process approaches a given set point. Most control systems are feedback systems. Conceptually the controller measures the process' output continuously and computes the error, the difference between the set point and the measurement, and sets the process' input according to a control law in order to minimize the error. In a digital control system, Fig. 6.1, an analog-to-digital converter samples the output of the process, which has been picked up by a detector and filtered, with a constant rate. It provides these samples to the digital controller. For each sample the controller

© Springer International Publishing Switzerland 2015
P. Hintenaus, *Engineering Embedded Systems*, DOI 10.1007/978-3-319-10680-9_6

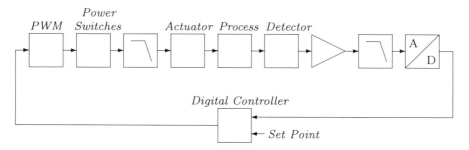

Fig. 6.1 Block diagram of a digital feedback control system

computes the error and, according to a control algorithm, a new input value for the process. Whenever one has to exert appreciable power to achieve the value at the process' input it is advisable to pursue a switching approach in order to achieve good energy efficiency. Thereby, the controller passes the input value to a pulse width modulation unit. The pulse width modulated outputs activate power switches which drive a transducer. If one has to apply only little power to set the process' input we may use a digital-to-analog converter for driving the actuator. In both cases the lowpass filter evens out the switching pulses. For the digital controller to function we have to minimize the time between an excursion of the process' output and the reaction of the controller to this excursion. Therefore, we have to understand the delays introduced by the filters and the data converters. Furthermore, monotonicity of the analog-to-digital converter is crucial for control systems applications. A monotonic converter produces increasing conversion values for increasing inputs and decreasing conversion values for decreasing inputs. Non-monotonous converters can introduce erratic behavior of the control system.

In many measurement systems the target to be measured is subject to a predefined excitation via some source. The source can produce an acoustical stimulus, as in ultrasonic range finders, or an optical one or any other physical stimulus the target reacts to. The measurement system picks up the target's response to the excitation using a detector. By relating the response to the stimulus it computes the measurement. In a digital realization of such a system, Fig. 6.2, the digital signal processor produces the excitation waveform and feeds a stream of samples to the digital-to-analog converter.

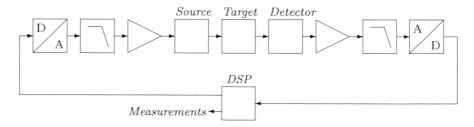

Fig. 6.2 Block diagram of a digital measurement system

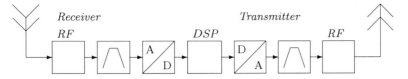

Fig. 6.3 Block diagram of a wireless digital communication system

The lowpass filter at the output of the digital-to-analog converter removes unwanted frequency content, which is introduced by the conversion. The source impinges the amplified excitation onto the target. The detector picks up the target's response. After amplification the lowpass filter at the input of the analog-to-digital converter removes frequency content which would mess up the result of the subsequent conversion. The converter supplies the digital signal processor (DSP) with a stream of data, which the processor analyzes in order to produce the measurement values. In measurement applications the fidelity of the excitation presented to the target and the degree to which the digital stream produced by the analog-to-digital converter represents the response of the target determine the accuracy of the measurements.

In a wireless communication system, Fig. 6.3, the analog radio frequency block in the receiver bandpass filters and amplifies the radio signal picked up by the receiving antenna in order to select the band of frequencies of interest. It down converts the signal into some intermediate frequency band. A bandpass filter removes the frequency content which would destroy the information contained in the signal during the subsequent analog-to-digital conversion. By sampling with a rate that depends on the bandwidth of the signal and not on the highest frequency in the signal the analog-to-digital converter produces a stream of samples that lends itself to being demodulated by the digital signal processing engine. Due to the nature of the signals digital-to-analog converters can produce, the digital signal processing engine has to synthesize the modulated sample stream for transmission at a very high rate. A digital-to-analog converter capable of such rates converts the sample stream and produces a signal in some intermediate frequency band. The bandpass filter rids the signal of unwanted frequency content. The analog radio frequency block in the transmitter up converts the signal to the transmission frequencies, amplifies it, and passes it on to the antenna for transmission. In communication applications the converters must have ample analog bandwidth. Furthermore nonlinearities, which introduce additional frequency content, have to be low.

6.2 Analog Filters

Filters attenuate signal content at certain frequencies while passing content at other frequencies. Besides crystal filters, surface acoustic wave filters, and ceramic filters, which employ mechanical waves and are beyond the scope of this treatise, analog electronic filters use the frequency dependent impedance of capacitors and inductors. Passive filters consist of capacitors, resistors, and of inductors only. Active filters contain amplifiers too. The number of components with frequency dependent impedance determines the effectivity of a filter. This number is called the filter's order.

A lowpass filter passes low frequencies up to its corner frequency $\hat{f}$ and attenuates high frequencies. The filter's corner frequency $\hat{f}$ is set so that when the lowpass filter is subjected to a sinusoidal signal with frequency $\hat{f}$ and amplitude A it responds with a sinusoidal signal with frequency $\hat{f}$ and an amplitude, which is 3 dB down from A. The filter's passband spans the frequencies between 0 Hz, and $\hat{f}$, the filter's stopband is the range of frequencies in which the filter's attenuation is above a required, application dependent, stopband attenuation. The area around $\hat{f}$ is called the filter's corner or the filter's knee. The transition from the passband into the stopband is gradual, the filter's order determines the sharpness of this transition. The attenuation increases with about 6 dB per order per octave in the frequency region above $\hat{f}$.

A highpass filter passes high frequencies down to its corner frequency $\hat{f}$ and attenuates low frequencies. It attenuates a sinusoidal signal with frequency $\hat{f}$ by 3 dB. The filter's passband spans the frequencies from $\hat{f}$ upwards, the filter's stopband is the range of frequencies in which the filter's attenuation is above a required stopband attenuation. The stopband includes 0 Hz. The filter's attenuation below the filter's corner frequency $\hat{f}$ increases with 6 dB per order per octave.

A bandpass filter passes frequencies between its lower corner frequency $\hat{f}_1$ and its upper corner frequency $\hat{f}_2$. It can be understood as the series combination of a lowpass filter with corner frequency $\hat{f}_2$ and a highpass filter with corner frequency $\hat{f}_1$.

A band reject filter passes all frequencies but those between its lower corner frequency $\hat{f}_1$ and its upper corner frequency $\hat{f}_2$. It can be understood as a lowpass filter and a highpass filter, which both filter the same input, and whose outputs are added together.

6.2.1 Filter Characteristics

A filter's transfer function $\hat{H}(s)$, the Laplace transform of the filter's impulse response, describes the behavior of the filter. The filter's frequency response $H(f) = \hat{H}(2\pi i f)$ is commonly shown as a Bode plot, consisting of a frequency versus gain plot and a frequency versus phase plot, with logarithmic frequency scale, logarithmic gain scale and linear phase scale. An equivalent representation of the filter's response in the time domain is the filter's step response, the output of the filter when subjected to a unit step. In the step response the filter's settling time, the time it takes the filter to arrive and stay in a specified band around its final output, the filter's overshoot, the amount by which the filter exceeds its final output, and whether the filter exhibits ringing, a dampened oscillation at its output in response to the unit step, can be read off readily.

An ideal filter has an instantaneous transition from its passband into its stopband and lets all frequencies within its passband pass completely unchanged and annihilates all frequencies within its stopband completely. A practical filter exhibits a transition from its passband into its stopband that is not perfectly sharp, a gain in the passband that is not constant, and it introduces a frequency dependent nonlinear phase shift between its input and its output. A phase shift linear with frequency amounts to a frequency independent delay introduced by the filter, which can be compensated for easily. When designing a filter we can only optimize one of these parameters but we have to compromise the others. The characteristics of a filter defines which parameter has been optimized and which have been compromised.

Butterworth filters exhibit a maximally flat passband. The transition from the passband into the stopband is reasonable but these filters introduce a phase shift which is nonlinear in the passband, causing overshoot and ringing in the step response. Butterworth filters are called for when the information carried by a signal is mostly contained in the amplitudes of the signal's spectral components, Figs. 6.4 and 6.5.

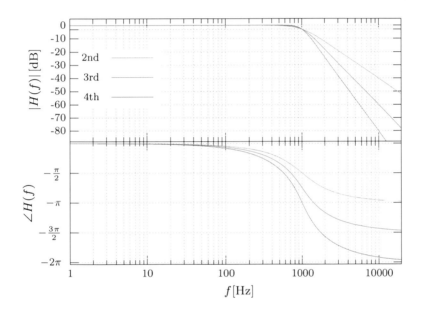

Fig. 6.4 Bode plots of a second, a third and a fourth order Butterworth filter

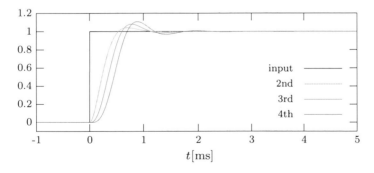

Fig. 6.5 Step responses of a second, a third, and a fourth order Butterworth filter, all having corner frequency $\hat{f} = 1\,\mathrm{kHz}$

Chebyshev filters (of the first kind) exhibit the sharpest transition for a given order and a given ripple of the gain in their passband. Such a filter introduces a phase shift between its input and its output which is nonlinear with frequency in the filter's passband, causing considerable overshoot and ringing in the filter's step response. Chebyshev filters are called for when a design requires a sharp transition from the passband into the stopband, Figs. 6.6 and 6.7.

A Bessel filter, also called a Thomson filter, exhibits the best approximation to a linear dependency in the filter's passband of the phase shift $\phi(f)$ the filter introduces between its input and its output and frequency. The group delay of this

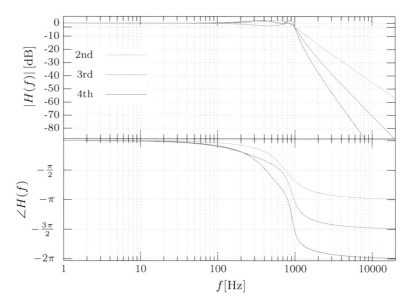

Fig. 6.6 Bode plots of a second, a third, and a fourth order Chebyshev filter, all having a maximum passband ripple of 2 dB

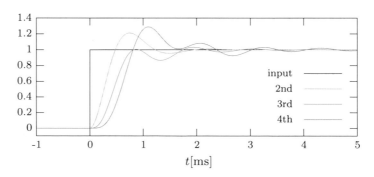

Fig. 6.7 Step responses of a second, a third, and a fourth order Chebyshev filter, all having corner frequency $\hat{f} = 1$ kHz and a maximum passband ripple of 2 dB

filter, the negative of the derivative of the phase with respect to angular frequency $\omega = 2\pi f$, $-\mathrm{d}\phi(f)/\mathrm{d}\omega$, is almost constant in the filter's passband, therefore a Bessel filter preserves the temporal relationship between a signal's spectral components. The step response of a lowpass Bessel filter shows no overshoot and almost no ringing. The transition from the filter's passband into the filter's stopband, however, is very gradual. Bessel filters are called for when the temporal relationship between the spectral content of a signal or between signals has to be preserved as much as possible, Figs. 6.8 and 6.9.

The transfer functions of these filters are tabulated, for example, in Tietze et al. (2008). For computing the transfer function $H'(s)$ of a highpass filter with certain order, certain characteristics, and certain corner frequency, one takes the transfer

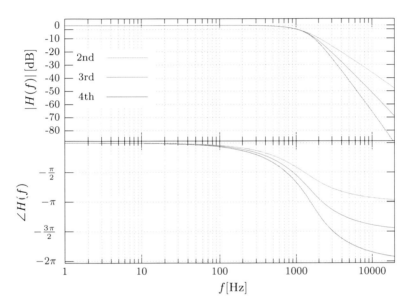

Fig. 6.8 Bode plots of a second, a third, and a fourth order Bessel filter

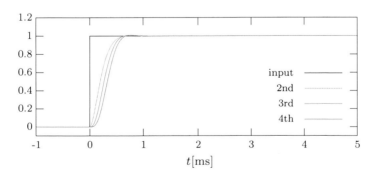

Fig. 6.9 Step responses of a second, a third, and a fourth order Bessel filter, all having corner frequency $\hat{f} = 1\,\mathrm{kHz}$

function $H(s)$ of the lowpass filter with same order, same characteristics, and same corner frequency and substitutes $H'(s) = H(1/s)$.

Chebyshev filters of the second kind optimize the sharpness of their transition for a given maximum ripple in their stopbands. Elliptic filters, also called Cauer filters, optimize the sharpness of their transition for a given maximum simultaneous ripple in both their passbands and their stopbands. These and further filters are beyond our scope.

6.3 Sampling and Reconstruction

Sampling has been studied extensively by Shannon (1948) based on earlier work by Nyquist (1924a, 1928), and Hartley (1928). All three have been with Bell Labs.

Let $s : \mathbb{R} \to \mathbb{C}$ be a complex-valued continuous-time signal. Let the signal s be band limited such that s has frequency content only between $f_c - \frac{b}{2}$ and $f_c + \frac{b}{2}$ and between $-f_c - \frac{b}{2}$ and $-f_c + \frac{b}{2}$ for some center frequency $f_c > 0$ and some bandwidth $b > 0$. Let the Fourier transform of s be $S : \mathbb{R} \to \mathbb{C}$ such that

$$s(t) = \frac{1}{2\pi} \int_{-\infty}^{\infty} S(\omega) e^{i\omega t} \, d\omega = \frac{1}{2\pi} \int_{2\pi\left(f_c - \frac{b}{2}\right)}^{2\pi\left(f_c + \frac{b}{2}\right)} \left(S(-\omega) e^{-i\omega t} + S(\omega) e^{i\omega t} \right) d\omega.$$

When we sample the signal s with sample rate f_s and sample period $t_s = 1/f_s$ the sampled version is the discrete-time signal $s' : \mathbb{Z} \to \mathbb{C}$ with

$$s'_n = s(t_s n).$$

The discrete-time Fourier transform $S' : \mathbb{R} \to \mathbb{C}$ of s' is according to (2.20)

$$S'(\omega) = \frac{1}{t_s} \sum_{k=-\infty}^{\infty} S\left(\frac{\omega - 2\pi k}{t_s}\right). \tag{6.1}$$

When the signal $e^{i2\pi g t}$ with frequency $g \geq f_s$ is sampled with rate f_s the sampled signal will have content with frequency g' where $0 \leq |g'| < f_s$. This shift of frequencies is called aliasing. Sampling s puts bands, each f_s wide, of the frequency content of s on top of each other in the sampled signal s'. This process can destroy the information contained in s. In order to preserve this information during sampling we must impose conditions on the center frequency f_c of s, on the bandwidth b, and the sampling frequency f_s such that for all ω only a single summand in (6.1) is nonzero. If there is a positive integer $\hat{k}$ such that

$$(\hat{k} - 1)\frac{f_s}{2} \leq f_c - \frac{b}{2} \quad \text{and} \quad f_c + \frac{b}{2} < \hat{k}\frac{f_s}{2} \tag{6.2}$$

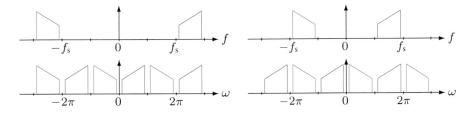

Fig. 6.10 Fourier transforms of two different continuous-time signals (*top*), Fourier transforms of the respective sampled signals (*bottom*)

it is guaranteed. This is the generalized Nyquist condition. It allows aliasing, as long as no information is lost due to sampling. The transform S' is then

$$S'(\omega + (k-1)\pi) = \begin{cases} \frac{1}{t_s} S\left(\frac{\omega + \pi(\hat{k}-1)}{t_s}\right) & \text{both } \hat{k}, k \text{ odd or both even} \\ \frac{1}{t_s} S\left(-\frac{\omega + \pi(\hat{k}-1)}{t_s}\right) & \text{otherwise} \end{cases}$$

for $0 \le \omega < \pi$ and $k \in \mathbb{Z}$, see Fig. 6.10.

For reconstructing the continuous-time signal s from the discrete-time signal s' we consider the periodic signal p, an impulse train with repetition rate f_s,

$$p(t) = \sum_{k=-\infty}^{\infty} \delta(t - kt_s).$$

The Fourier series coefficients P_n of p are

$$P_n = f_s \int_{-t_s/2}^{t_s/2} \left(\sum_{k=-\infty}^{\infty} \delta(t - kt_s)\right) e^{-i2\pi f_s n t} \, dt$$

$$= f_s \int_{-t_s/2}^{t_s/2} \delta(t) e^{-i2\pi f_s n t} \, dt$$

$$= f_s.$$

We can drop the sum, as $k = 0$ delivers the only summand that is nonzero within the integration bounds. The rest follows from the sifting property (2.4) of the Dirac δ distribution. Using (2.19), the continuous-time transform $P(\omega)$ of p is

$$P(\omega) = 2\pi f_s \sum_{k=-\infty}^{\infty} \delta(\omega - 2\pi k f_s).$$

The continuous-time Fourier transform W of the signal $w : \mathbb{R} \to \mathbb{C}$,

$$w(t) = s(t)p(t) = \sum_{k=-\infty}^{\infty} s'_k \delta(t - kt_s), \qquad (6.3)$$

is $W(\omega) = \frac{1}{2\pi}(S * P)(\omega)$ the convolution of the Fourier transform S and the Fourier transform P. Expanding the convolution we get

$$W(\omega) = f_s \int_{-\infty}^{\infty} S(\Omega) \left(\sum_{k=-\infty}^{\infty} \delta(\omega - \Omega - 2\pi k f_s) \right) d\Omega$$

$$= f_s \sum_{k=-\infty}^{\infty} \int_{-\infty}^{\infty} S(\Omega) \delta(\omega - \Omega - 2\pi k f_s) \, d\Omega$$

$$= f_s \sum_{k=-\infty}^{\infty} S(\omega - 2\pi k f_s). \qquad (6.4)$$

Assuming that s satisfies (6.2) we have to apply a suitable bandpass filter to the signal w in order to recover the original signal s. The frequency response $R : \mathbb{R} \to \mathbb{C}$ of the ideal reconstruction filter, which reconstructs the signal s without error, is

$$R(\omega) = \begin{cases} 1 & \text{for} - f_c - \frac{b}{2} < \frac{\omega}{2\pi} \le -f_c + \frac{b}{2} \text{ or } f_c - \frac{b}{2} \le \frac{\omega}{2\pi} < f_c + \frac{b}{2} \\ 0 & \text{otherwise.} \end{cases}$$

The impulse response $r : \mathbb{R} \to \mathbb{C}$ of this filter is

$$r(t) = \frac{\sin(\pi t(2f_c + b)) - \sin(\pi t(2f_c - b))}{\pi t}.$$

This filter is not causal. In practical applications the ideal bandpass filter is replaced with a causal one. Moreover, instead of producing the impulse train w a digital-to-analog converter holds the analog value of a sample at its output until being supplied with and having converted a new sample.

6.3.1 Sampling Real-Valued Signals

If the signal $s : \mathbb{R} \to \mathbb{R}$ is real-valued instead of complex-valued its Fourier transform obeys $S(-\omega) = S^*(\omega)$, therefore the transform for $\omega < 0$ contains no additional information which is not contained in the transform for $\omega \ge 0$ already. Once we have agreed upon the sampling frequency f_s we divide the positive frequency scale into intervals

$$N_i = \left\{ f \in \mathbb{R} : (i-1)\frac{f_s}{2} \leq f < i\frac{f_s}{2} \right\},$$

for $i \in \mathbb{N}, i > 0$, the Nyquist zones. Let us sample a signal s having all its frequency content in a single Nyquist zone. The information contained in s is also contained in the frequencies $f \in N_1$ of the sampled signal. Note that sampling a signal in an even-numbered Nyquist zone reverses the frequency scale in the sampled signal for frequencies in the first Nyquist zone N_1, see Fig. 6.10.

6.4 Antialiasing and Reconstruction Filters

For a signal in the first Nyquist zone any content outside the first zone, be it signal or be it noise, will be folded back into the first zone during sampling. The folded content will be indistinguishable from the signal of interest. Therefore, we have to introduce an analog antialiasing filter which attenuates the unwanted content. When designing such an antialiasing filter we start out with a specification of the frequency band of interest and a specification of the fidelity, that is, the resolution, or the signal-to-noise ratio, with which the digital representation shall approximate the analog signal. The signal-to-noise ratio required for achieving n bits resolution using a perfect analog-to-digital converter is $(6.02n + 1.76)$ dB, see Bennet (1948). We have to select a suitable sampling rate, the corner frequency of the filter, the characteristics of the filter, and the sharpness, that is the order of the filter. Let us assume that the highest frequency of interest is f_a, see Fig. 6.11. Furthermore, there shall be signal content up to much higher frequencies with amplitudes of appreciable magnitude. The frequencies between $f_s/2$ and $f_s - f_a$ is imaged into the range between f_a and $f_s/2$ by sampling, where it can be removed by digital filtering. The content above $f_s - f_a$, however, is imaged into the frequency range of interest, where it cannot

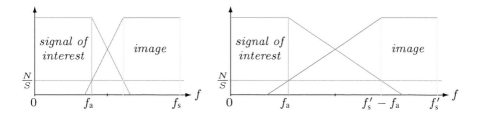

Fig. 6.11 Frequency domain description of the sampling process of a signal with content between 0 and f_a with a maximum noise content N/S. On the *left* a small sampling rate f_s is chosen, necessitating an antialiasing filter with a rather sharp transition from the passband to the stopband. On the *right* a larger sampling rate f_s' is chosen allowing for a more gradual transition

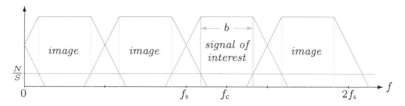

Fig. 6.12 Frequency domain description of a signal in the third Nyquist zone with center frequency f_c and bandwidth b that is undersampled. A bandpass filter is required for attenuating content folded back into the third zone

be removed by any means, analog or digital. We will set the corner frequency of the filter to f_a or slightly above. In order to achieve the maximum allowable noise content N/S we have to design the antialiasing filter such that it attenuates content above $f_s - f_a$ down to below N/S. When we choose the sampling frequency f_s close to its theoretical minimum of $2 f_a$ the antialiasing filter has to have high order for achieving the required sharpness. Such filters, requiring precision resistors and capacitors, are hard to design and costly to manufacture. When choosing a higher sampling frequency f_s' than the theoretically required one, by oversampling the signal, a filter of lower order will do. The increased sampling rate will, however, put a larger computational burden on the digital part of the system. We determine the characteristics of the filter depending on which features of the signal have to be preserved the most.

Converting a signal in a Nyquist zone N_2 and up is called undersampling. Then a bandpass filter will be required. In order to achieve the required signal-to-noise ratio we not only have to attenuate frequency content above the Nyquist zone but also the one below, Fig. 6.12. Making the Nyquist zones wider than theoretically mandated, oversampling while undersampling, eases the sharpness requirements for the bandpass filter. Undersampling when converting from analog-to-digital amounts to shifting down the frequency of the input signal. It is used mainly in the receiving section of digital communication devices.

6.5 Analog-to-Digital Converters

An ideal analog-to-digital converter with n bits resolution compares the voltage at its signal input V_i to the voltage of a reference V_r, and produces the digital measurement m,

$$m = \left\lfloor 2^n \frac{V_i}{V_r} + \frac{1}{2} \right\rfloor.$$

Depending on the converter the converter's signal input might either be presented with voltages between $0\,\mathrm{V}$, and V_r or between $-V_r/2$ and $V_r/2$. Input voltages outside the allowable range cause the converter to clip. The converter then outputs either the largest possible value or the smallest possible one. The discrete nature of the digital measurement introduces an error, the quantization error. This error limits the signal-to-noise ratio achievable with a converter with n bits resolution to $(6.02n + 1.76)$ dB. The reference voltage can either be generated by the converter itself or provided externally.

In general, the accuracy of the reference determines the absolute accuracy of the measurements. With some detectors however, we can forgo the reference completely and even gain precision. Such a detector, called ratiometric, requires an excitation voltage V_x in order to operate and produces an output voltage V_o, which is proportional both to the physical quantity p to be measured and to its excitation,

$$V_o = pV_x.$$

With a ratiometric detector a converter with external reference shines. When we connect the excitation voltage to the reference input of the converter too, the digital output becomes

$$m = \lfloor 2^n p + 1/2 \rfloor,$$

which is independent of the excitation voltage. We can use any voltage source producing a voltage of the right magnitude and free of high-frequency content as simultaneous source for the detector's excitation and the converter's reference.

6.5.1 Single-Shot Conversion

Many microcontrollers feature an integrated analog-to-digital converter with a resolution of about 12 bits. Usually such a converter is equipped with an analog multiplexer allowing it to convert the signals produced by several sources. The software has to provide a start conversion command to the analog-to-digital converter for every single conversion. Some microcontrollers, however, allow a timer to provide the start conversion command, enabling either precise delays, or periodic measurements with a precise period. When the converter receives a start conversion command, it freezes the analog value at its input in a track and hold circuit. Using a built in digital-to-analog converter it performs a binary search for the digital value. Every iteration of the binary search loop produces an additional bit of resolution in principle, but the finite accuracy of the track and hold limits the resolution achievable. This type of converter, the successive approximation type, requires a couple of microseconds for producing a result, making it well suited for control systems.

At the time of writing the shortest conversion times for successive approximation converters housed in separate packages is about 100 ns, and the highest resolution is 18 bits. These converters operate in the single-shot mode too. For monitoring slowly changing analog signals tiny successive approximation converters with moderate resolution are available. These converters communicate via a serial bus such as SPI or I^2C allowing them to be placed at the signal origin.

6.5.2 Periodic Conversion

Once started, a periodically converting analog-to-digital converter converts the signal at its input periodically with a fixed sampling rate. Without further intervention it presents a continuous stream of data at its outputs to the digital signal processing functions.

A Σ-Δ analog-to-digital converter oversamples its input by a large factor compared to its output rate. It digitally decimates the input stream in order to achieve the output rate, while increasing resolution of the samples. Resolutions of 16–24 bits are common. At the time of writing output rates of five mega samples per second are possible. Their high resolution makes Σ-Δ analog-to-digital converters prime candidates for measurement instruments and digital audio systems. The digital signal processing, inherent in their functional principle, introduces a considerable group delay, which precludes them from being used in most control systems. When using these converters in conjunction with analog multiplexers one has to skip a number of conversions after changing the input channel as the converter needs time for settling.

An n bit flash converter compares the input signal with $2^n - 1$ voltage levels in parallel, using $2^n - 1$ comparators. A high precision voltage divider with the corresponding number of taps produces the required voltage levels. The converter's digital logic converts the resulting code into binary. Flash converters can be made very fast but the huge number of precision resistors and comparators and the associated power consumption renders them impractical for all but low resolutions.

A pipelined analog-to-digital converter consists of several stages. The first stage produces the most significant bits, the subsequent stages the bits of lesser, and lesser significance until the last stage, which produces the least significant bits. Each stage uses an analog-to-digital converter with low resolution for producing its bits, and each stage but the first uses a digital-to-analog converter with assorted analog circuitry for producing and amplifying the error leftover by the preceding stages. Track and hold circuits decouple the individual stages from each other. While the first stage in such a converter converts a sample, the second stage converts the preceding sample, and so on. The low-resolution analog-to-digital converters in each stage can be flash converters. Pipelined converters can be very fast. They are suited for digital radio applications and for measurement instrumentation, which has to cope with signals having high bandwidth.

6.5.3 Effects of Sampling Clock Uncertainty

Periodically converting analog-to-digital converters need a clock signal for determining the sample points. Variations of the lengths of the clock cycles, clock jitter, make the converter sample the input at uneven intervals, which introduces errors in the samples and limits the signal-to-noise ratio achievable, see Fig. 6.13. The signal- to-noise

Fig. 6.13 Error introduced
by uncertain sampling times
when sampling the signal s

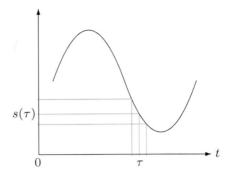

ratio in decibel achievable when sampling a sinusoid s with frequency f using a
sampling clock with root-mean-square jitter t_j is

$$20 \log \frac{1}{2\pi f t_j},$$

see, for example Zumbahlen and Analog Devices (2008) or Smith (2004). When
sampling a sinusoid with a frequency of 1 MHz ,and a sampling clock jitter of 1 ns
the achievable signal-to-noise ratio 44 dB, so the resolution is around 7 bits only. The
precision required from the sampling clock precludes the clock's generation under
software control. Instead, the clock should be considered an important analog signal,
which must be kept away from fast-switching logic as much as possible even at the
cost of introducing an additional clock domain for a system's data converters.

6.6 Digital-to-Analog Converters

An ideal digital potentiometer with n bits resolution converting the input i produces
the ratio of resistances

$$\frac{R_w}{R} = \frac{i}{2^n - 1}.$$

between its three outputs. The input i may range between $0 \le i < 2^n$. Usually,
the total resistance R is not controlled very well, but the ratio is accurate. Such a
potentiometer consists of a string of $2^n - 1$ equal-valued resistors forming a voltage
divider with 2^n taps, a Kelvin divider, and 2^n switches where a single switch conducts
at any time. Such a string converter is inherently monotonic. The large number of
matched resistors and switches required limits string converters to low resolutions.
Typically digital potentiometers communicate via a serial bus such as SPI or I^2C.
Housed in small packages this allows them to be placed where a digitally controlled
voltage divider is required.

A digital-to-analog converter with n bits resolution converting the input i produces the voltage

$$V_\text{o} = V_\text{r} \frac{i}{2^n}$$

at its output. Depending on the converter the input may either range between $0 \le i < 2^n$ or between $-2^{n-1} \le i < 2^{n-1}$. In some converters an internal reference supplies the voltage V_r, in others V_r must be supplied externally. A multiplying digital-to-analog converter allows an arbitrary analog signal as reference, it multiplies its digital input with its analog reference in a very elegant way. The basic digital-to-analog converter consists of a number of resistors and a number of switches arranged in ingenious ways to minimize the number of elements required while providing high precision and easy manufacture, see, for example, Kester (2004). These converters can be very fast; update rates of several hundred megahertz and more are possible.

One group of digital-to-analog converters do not impose any restrictions on the timing of the data passed to them for conversion. Such a converter produces a new analog output whenever it is supplied with a new digital input. Provided the digital host provides the digital samples with sufficiently low jitter such a converter can convert low-frequency signals with medium resolution satisfactorily. Digital-to-analog converters intended for converting high-frequency signals at high-resolution demand a continuous stream of samples at a fixed sample rate. Such converters are often segmented, a low-resolution converter converts the most significant bit, another one the less significant bits and yet another one the least significant ones. Fast analog circuitry combines the conversion for the analog output. Making a segmented converter monotonous requires analog art well beyond our scope.

The Σ-Δ principle can be used for digital-to-analog converters too. A digital Σ-Δ modulator translates the digital input stream into a stream of samples with a much higher sampling rate but with lower resolution. A low-resolution converter produces the analog output. The high sampling rate eases the requirements for the reconstruction filter considerably. Such a converter demands a continuous stream of samples from its digital host. The digital signal processing involved in the data conversion process introduces a large group delay making Σ-Δ converters unsuitable for most closed-loop controls systems. The main application of these converters is digital audio, but the very high resolution they offer, about 24 bits, makes them attractive for measurement instruments too.

6.7 Analog Building Blocks

The advances in semiconductor technology provide us with more and more huge analog subsystems integrated on single chips. This trend notwithstanding, we still have to build an occasional signal conditioning circuit or an occasional filter, using analog building blocks such as amplifiers or voltage references. The circuits in this section and the next will show up again in the following chapters as parts of larger systems.

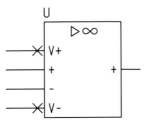

Fig. 6.14 Circuit symbol for an operational amplifier. The connections V+ and V− are the positive and negative supply, the input labeled + is the non-inverting input, the one labeled − is the inverting input. The single non-inverting output labeled + is on the *right*

6.7.1 Operational Amplifiers

A voltage feedback operational amplifier,[1] Fig. 6.14, is a device with five terminals. The terminals V+ and V− are the positive and the negative power supply, usually +15 and −15 V for precision amplifiers. Let the voltage at the amplifier's non-inverting input, labeled +, be V_+ and the voltage at the amplifier's inverting input, also called the summing junction, labeled −, be V_-. An ideal operational amplifier amplifies the difference $V_+ - V_-$ with gain infinity and presents the result at its output, labeled + too. Moreover, the ideal amplifier does not introduce any delay between its inputs and its output, its inputs do not draw any current, and its output is able to drive any current.

The key to putting such as concept to work is negative feedback. Should the difference $V_+ - V_-$ be nonzero in the circuit in Fig. 6.15 the amplifier will instantly change the output such that the difference becomes zero again. The circuit is a unity gain follower. While it passes its input signal without amplification it provides a zero impedance drive without loading the signal source at its input. Positive feedback on the other hand is useless, should there be any nonzero difference $V_+ - V_-$ positive feedback increases this difference, driving the output to plus or minus infinity.

Fig. 6.15 Voltage follower. The input's impedance is high while the output's impedance is low

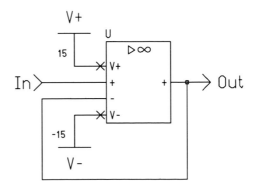

---

[1] We do not cover current feedback operational amplifiers, but see for example Jung (2005).

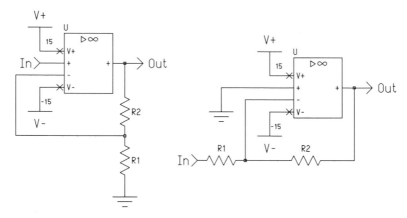

Fig. 6.16 Non-inverting amplifier to the *left*, inverting amplifier to the *right*

These observations give us the three rules for analyzing circuits including operational amplifiers: First, loading the output does not matter, second no currents flow into or out of the inputs, and third the amplifier will do anything with its output to make the voltages at its two inputs equal, provided there is negative feedback.

Armed with these three rules we are able to study some circuits, starting with the left one in Fig. 6.16. In this circuit diagram and in the remaining ones in this chapter we let the resistance of R1 be R_1, of R2 be R_2, and so on. We let the capacitance of C1 be C_1 and so on. Further, the voltage at In is V_i and at Out is V_o. The resistors R1 and R2 form an unloaded voltage divider. Using (3.5) we get

$$V_o = V_i \left(1 + \frac{R_2}{R_1} \right).$$

The circuit is a non-inverting amplifier, its input impedance is infinity. On to the circuit to the right, the voltage at the summing junction must be zero, therefore $V_i R_1 + V_o R_2 = 0$, or

$$V_o = -V_i \frac{R_2}{R_1}.$$

The circuit is an inverting amplifier, its input impedance is R_1.

Next the circuit in Fig. 6.17, left side. Let V_1 be the voltage at In1, V_2 the one at In2. The voltage at the non-inverting input is

$$V_+ = (R_1 \parallel R_2) \left(\frac{V_1}{R_1} + \frac{V_2}{R_2} \right).$$

The rest of the circuit is an non-inverting amplifier, therefore

$$V_o = V_+ \left(1 + \frac{R_4}{R_3} \right) = (R_1 \parallel R_2) \left(1 + \frac{R_4}{R_3} \right) \left(\frac{V_1}{R_1} + \frac{V_2}{R_2} \right).$$

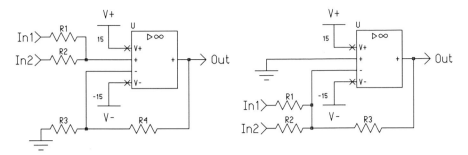

Fig. 6.17 Non-inverting summing amplifier to the *left*, inverting one to the *right*

Choosing $R_1 = R_2$ this simplifies to

$$V_o = \frac{1}{2}\left(1 + \frac{R_4}{R_3}\right)(V_1 + V_2),$$

therefore the circuit is a summing amplifier. What remains for us to do is choosing the resistances. You might have already guessed, there is no such thing as an ideal operational amplifier. Each input of a real operational amplifier does draw a little current, lower than 1 pA for some amplifiers, the input bias current. These bias currents cause voltage drops in the resistors connected to the amplifier's inputs. We first chose the gain-setting resistances R_3 and R_4. Choosing these resistances too low waste energy, while choosing them too high increases the thermal noise introduced by the resistors themselves and drives the summing junction with high impedance, making it susceptible to picking up interference. Resistances between a couple 100 Ω to about 100 kΩ form an acceptable compromise. The impedance the summing junction sees is $R_3 \parallel R_4$ and the one the non-inverting input sees is $R_1 \parallel R_2$.

The circuit in Fig. 6.17 to the right is an inverting summing amplifier. Its output voltage is

$$V_o = -R_3\left(\frac{V_1}{R_1} + \frac{V_2}{R_2}\right). \tag{6.5}$$

Speaking of deviations from the ideal, the output of a real operational amplifier can only drive a finite current and the output voltage will always stay within the limits set by the supply voltages. Furthermore, a real amplifier does not drive its output to zero when both inputs are at the same potential, only when there is a small voltage—the amplifier's input offset voltage—between them. The assumption that the ideal operational amplifier has infinite gain and introduces no delay is overly optimistic too: the gain of a real operational amplifier without feedback, its open-loop gain, is finite and, after being flat for low frequencies, drops off with 6 dB per octave. An amplifier's gain-bandwidth product is the product the open loop gain for a given frequency, it is constant. The phase shift introduced by an operational amplifier varies with frequency too. To aid in analyzing a circuit's stability an amplifiers phase

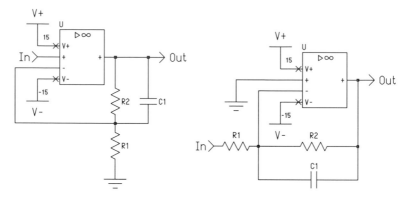

Fig. 6.18 Non-inverting lowpass filter to the *left*, inverting to the *right*

margin is given. Amplifiers having a positive phase margin are unity gain stable, they operate reliably with gain 1. A decompensated amplifier on the other hand needs a gain greater than 1 for stable operation. It has however a higher gain-bandwidth product than a unity stable amplifier of similar design.

The circuit in Fig. 6.18, left side, is basically a non-inverting amplifier with lowpass characteristics. For frequencies between $\frac{1}{2\pi R_2 C_1}$ and $\frac{1}{2\pi (R_1 \| R_2) C_1}$ its gain, V_o/V_i, drops off with 6 dB per octave but for frequencies above $\frac{1}{2\pi (R_1 \| R_2) C_1}$ the capacitor C1 resembles a short, the amplifier becomes a follower and its gain becomes one. The gain of the inverting lowpass, Fig. 6.18 to the right, at frequency zero is $-R_2/R_1$. It drops off with 6 dB for frequencies above the circuits corner frequency

$$\hat{f} = \frac{1}{2\pi R_2 C_1}. \tag{6.6}$$

When designing a circuit containing a large portion of digital functions having to provide two additional supply voltages for the analog part is a major nuisance. Instead one tries to power the analog functions with the filtered digital supply. In these single supply circuits the amplifiers' V- terminals are connected to the circuit's ground and a voltage somewhere between ground and the positive supply serves as the virtual ground for the analog section. In the following circuits virtual ground is the 1.65 V rail. In order to keep the signals between the amplifier's supply rails signals are typically AC coupled between amplifiers, see Sect. 3.6. The coupling capacitor forms a highpass filter with the input impedance of a following inverting stage, therefore it must be large enough. When the next amplifier operates in a non-inverting configuration on the other hand we have to provide a direct current path for the bias current of the amplifier's non-inverting input. Connecting the coupling capacitor's terminal, which provides the input to the next stage, to virtual ground via a large resistance will do.

The circuit in Fig. 6.19 is a second-order lowpass filter with Butterworth characteristics, while the one in Fig. 6.20 is a second-order highpass. Filter design software

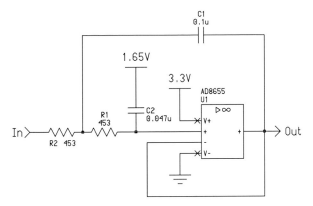

Fig. 6.19 Second-order lowpass filter, Butterworth characteristics, corner frequency $\hat{f} = 5\,\text{kHz}$, Sallen-Key topology

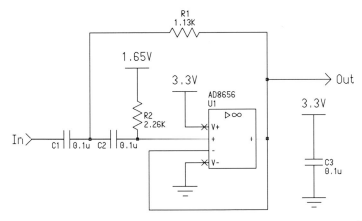

Fig. 6.20 Second-order highpass filter, Butterworth characteristics, corner frequency $\hat{f} = 1\,\text{kHz}$, Sallen-Key topology

available form several semiconductor manufactures helps with choosing a circuit and finding the component values for a given specification.

Putting a digital potentiometer into the feedback path of an inverting amplifier creates a variable gain amplifier whose gain can be set via SPI, Fig. 6.21. The digital potentiometer introduces some parasitic capacitances between the terminals of the variable resistor and ground necessitating capacitor C1 for stabilizing the amplifier.

Last we consider a low-side current sense amplifier, Fig. 6.22, for measuring currents that change fast. The circuit operates off a single 5 V supply. It measures the current flowing through a load's ground connection. The minimum current is $-10\,\text{A}$ resulting in a output of 0 V, while the maximum current is 10 A resulting in an output of 5 V. The circuit produces a voltage drop proportional to the current with the shunt R2. A resistance of $20\,\text{m}\Omega$ for keeping the losses low results in a

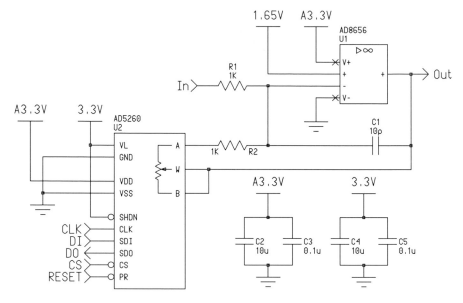

Fig. 6.21 Inverting amplifier with digitally programmable gain ranging from 1.5 to 200

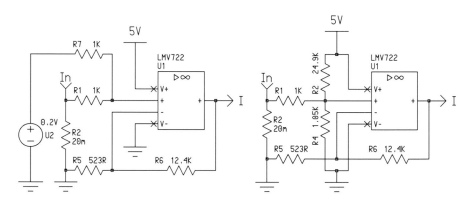

Fig. 6.22 Low-side current sense amplifier, current range −10 A to 10 A. Principle circuit to the *left*, practical realization to the *right*

voltage swing of ±200 mV across the shunt. To handle the negative voltage we add 200 mV from a voltage source and amplify the sum in the non-inverting summing amplifier in Fig. 6.22 to the left. We choose the resistances R_1 and R_7 relatively low for driving the amplifiers non-inverting input with a relatively low impedance, in order to decrease the circuit's sensitivity to interference. We choose the gain-setting resistances R_5 and R_6 for a gain of 25. Further, the amplifier's summing junction sees the same impedance as the non-inverting input. In the practical realization of this circuit, Fig. 6.22 right side, we replaced the voltage source U2 and the resistor U7 with the equivalent voltage divider R2 and R4, see (3.5). This makes the circuit's output dependent on the power supply voltage, but we expect the errors to be dominated by interference from the highly dynamic current. The gain-bandwidth product of the chosen amplifier is 10 MHz so the bandwidth of this circuit is 400 kHz. The amplifier can operate with both it's inputs at -0.3 V so operating them between 0 and 0.4 V is fine.

6.7.2 Voltage References

Ratiometric circuits offer the advantage of high precision without relying on some precise voltage. The same principle holds true when we have to produce a precise ratio of some voltage. By wiring this voltage to the reference input of the digital-to-analog converter, the output of the converter will, within the limits of the converter's accuracy, assume the digitally programmed ratio. When we have to make absolute measurements or produce analog outputs on some absolute scale we have to rely on voltage references. Voltage references exhibiting a long term stability in the range of 50.0×10^{-6} per 1,000 h and a typical temperature drift of $1.0 \times 10^{-6}\,°C^{-1}$ are available. Despite this precision references limit the absolute accuracy achievable.

6.7.3 Comparators

A comparator, Fig. 6.23, has two inputs X and Y, both drawing negligible current, and a logic level output. It compares the voltage V_X at the input X with the voltage V_Y at the input Y and outputs 1 if $V_X < V_Y$ and 0 otherwise. Essentially a comparator is a single bit analog-to-digital converter. When V_X and V_Y are very close noise picked up by the inputs causes the output to switch back and forth repeatedly and rapidly. Such behavior may cause havoc in any logic downstream of the comparator's output. To alleviate this problem we can provide some hysteresis by introducing some positive

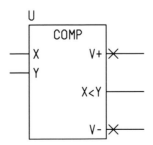

Fig. 6.23 Circuit symbol for a comparator. The connections V+ and V− are the positive and negative supply, X and Y are the inputs. If the voltage at X is less than the voltage at Y the comparator drives its output to a logic one, otherwise it drives the output to a logic zero

Fig. 6.24 Positive feedback
introduces hysteresis

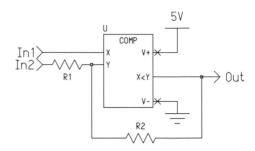

feedback, Fig. 6.24. Assuming, that the output swings between the supply rails and
that V_S is the supply voltage the hysteresis is

$$V_H = V_S \frac{R_1 + R_2}{R_1}. \tag{6.7}$$

At first sight the behavior of a comparator and that of an operational amplifier
when operated open loop are similar. Comparators are designed for fast speeds,
while operational amplifiers are designed for precise amplification, therefore using
an operational amplifier in the place of a comparator is a bad idea.

To facilitate interfacing a comparator to different logic families many comparators
have an additional ground connection and an open-drain or open-collector output.
A pull-up to the correct voltage for a logic 1 then constitutes the interface to the
downstream logic.

6.8 Discrete Active Elements

Diodes come in a bewildering variety. It seems any effect a semiconductor material
exhibits can be put to use in some specialized diode. Therefore, diodes play major
roles in many circuits not only for rectifying current but also as detectors or as protective elements. Integrated circuits on the other hand have mostly replaced discrete
transistors. Transistors still show up as switches or in conjunction with an integrated
circuit for boosting the circuit's drive capabilities.

6.8.1 Diodes

Diodes, Fig. 6.25, are devices with two terminals, the anode A and the cathode C.
The diodes we cover here are made out of semiconductor materials, most often
silicon. Besides being used for rectifying alternating current, diodes can be tuned for
purposes, such as detectors for light, resulting in a vast variety of types of diodes.

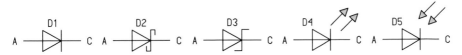

Fig. 6.25 Symbols for different types of diodes, D1 ordinary diode, D2 Schottky diode, D3 Zener diode, D4 light emitting diode (LED), and D5 photodiode

When a positive voltage larger than the diodes forward voltage is applied between an ordinary diode's anode and cathode, when the diode is forward-biased, the diode conducts. The current through the diode grows exponentially with the voltage across it. For silicon diodes the forward voltage is about 0.7 V. When the voltage between a diode's anode and cathode is negative, the diode is reverse-biased, only a very small current, the diode's reverse current, flows through the diode. If the voltage across a reverse-biased diode exceeds the diode's peak reverse voltage the diode breaks through and allows current to flow. Unless this current is limited it will destroy the diode. If the voltage across a diode changes suddenly from forward-biasing the diode to reverse biasing, the diode requires some time, the diode's reverse recovery time, until it blocks current. Diodes for signal applications have a reverse current of 20 nA at room temperature and 50 μA at 150 °C, a peak reverse voltage of about 100 V and a reverse recovery time of a few nanoseconds. Diodes for rectifying alternating current at the mains frequency have a reverse current between 5 and 50 μA, a peak reverse voltage up to 1 kV and a reverse recovery time so bad, no one wants to specify it. The diode circuits in Fig. 6.26 rectify single phase and three phase alternating currents. These circuits are available packaged together for use in off-line power supply applications.

Schottky diodes have a forward voltage of about 0.3–0.4 V, a reverse current of a few μA and a peak reverse voltage of up to 50 V. They switch almost instantly. As long as their low peak reverse voltage does not pose a problem Schottky diodes are a good choice for high-frequency applications, both for rectifying signals and power.

A Zener diode is designed to break through at a well-defined voltage, the diode's Zener voltage. In the circuit Fig. 6.27 the Zener diode D is operated in breakdown. The resistor R limits the current through the diode. Provided the voltage at the input rail is higher than the diode's Zener voltage and the output rail Vo is loaded lightly

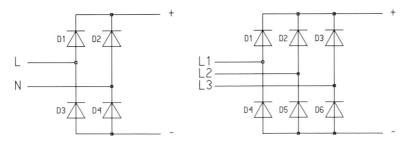

Fig. 6.26 Diode rectifiers, single phase on the *left*, three-phase on the *right*

Fig. 6.27 Zener diode used in a crude reference circuit

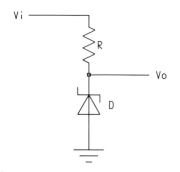

the voltage at Vo depends mostly on the Zener voltage of D and only very little on the input voltage. Therefore, this circuit forms a crude voltage reference. Power Zener diodes are used as transient voltage suppressors. They absorb the energy from electrostatic discharges in order to protect delicate electronics from being zapped.

Light emitting diodes, LEDs, turn electrical energy into light. They are not suitable as rectifiers, their maximum reverse voltage is about 5 V. For a given voltage across a LED the current through the LED rises with the LED's temperature. This behavior may lead to thermal destruction of the LED when operated with constant voltage. Increasing temperature leads to increasing losses and, ultimately, to destruction. Constant current operation on the other hand leads to a decrease in total power consumption with temperature, therefore the LED will reach an equilibrium. LEDs suitable for indicators have a forward voltage between 2.2 V for red ones and 3.1 V for blue ones at a forward current of a couple mA. LEDs designed for lighting applications have a forward voltage of about 3 V at a total power consumption of up to 10 W. LEDs emitting in the near-infrared part of the electromagnetic spectrum serve in various applications such as communications. A wide variety of diode lasers is also available.

A photodiode, used as a detector for light, acts as a current source. It delivers a current that is proportional to the intensity, the power per unit area, of the light that hits it. Silicon photodiodes are sensitive to visible light, indium gallium arsenide photodiodes are sensitive to infrared radiation with a wavelength between $1.0\,\mu m$ and $1.9\,\mu m$ and mercury zinc telluride photodiodes are sensitive between $2.5\,\mu m$ and $12\,\mu m$.

The circuit in Fig. 6.28 is an amplifier for a photodiode. It turns the current sourced by the photodiode D1 into a voltage at the output L. Such an amplifier, which turns a current into its input into a voltage at its output is called a transimpedance amplifier. The summing junction is at $V_0 = 1.65\,V$. When we consider the currents into the summing junction we get $I_P + (V_o - V_0)/R_1 = 0$, where I_P is the current out of the photodiode, and V_o is the voltage at the output L. The output voltage, therefore, is

$$V_o = R_1 I_P + V_0.$$

There is a twist however, any practical photodiode exhibits a parasitic capacitance in parallel to the diode. The feedback resistor R1 together with the parasitic capacitance of D1 turns the feedback circuit into a lowpass filter and therefore the whole circuit

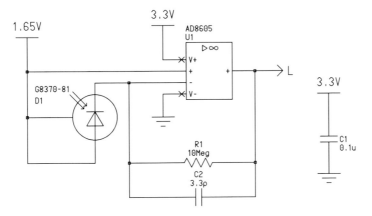

Fig. 6.28 Photodiode amplifier

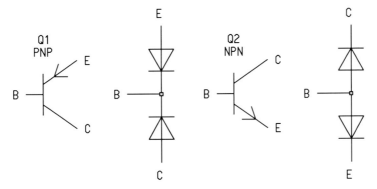

Fig. 6.29 Bipolar transistors and their diode equivalents, Q1 is a PNP, Q2 is an NPN

has highpass characteristics and is unstable.[2] To make it stable we have to reduce the circuits gain at high frequencies. The capacitor C2 does exactly that.

6.8.2 Transistors

A bipolar transistor, Fig. 6.29, has three terminals, the collector C, the base B, and the emitter E. Bipolar transistors come in two varieties, PNP and NPN. As indicated in Fig. 6.29 transistors look like two diodes, but there is more to the story. In the circuit in Fig. 6.30 the current I_C flowing from the collector to the emitter of the NPN transistor Q1 grows monotonically with the current I_B flowing from the base to the emitter. As long as the voltage V_{BE} across the base emitter diode is below 0 V

---

[2] Intuitively the operational amplifier has to work harder at high frequencies to compensate for the attenuation of the feedback network at these frequencies.

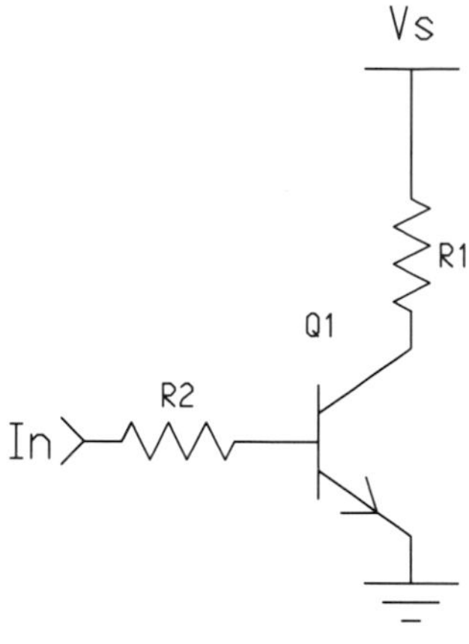

Fig. 6.30 Behavior of an
NPN transistor. The voltage at
In sets the current through R2.
The current through R1
increases monotonically with
the current through R2

the base and the collector current are $I_B = 0$ and $I_C = 0$, the transistor is turned
off. An increase of V_{BE} past 0 V leads to an exponential increase of the current I_B,
which in turn leads to a monotonous increase of the current I_C. The small current I_B
controls the larger current I_C. The dependence of I_C on I_B is usually approximated
by the linear relationship $I_C \approx B I_B$ where the DC current gain B ranges between
100 and 800 for the small signal transistor (BC848C) we use in this book. Eventually
the resistor R1 limits the current I_C and increases of I_B do not lead to increases of
I_C any more, the transistor has saturated. The voltage drop from the collector to the
emitter of a saturated small signal transistor is typically about 0.2 V. The voltage at
In in Fig. 6.30 sets the current I_B via the resistor R2 and influences the current I_C.
When used to boost the output capabilities of an operational amplifier a transistor is
operated in the region between being fully turned off and being saturated. When used
as a switch a transistor is either fully turned off with a large voltage drop between
collector and emitter but no current through the transistor, or it is fully saturated,
conducting appreciable current from collector to emitter but with minimal voltage
drop. In both cases the transistor dissipates little power. PNP transistors work like
NPN transistors with the signs of the voltages and the currents reversed.

The circuit in Fig. 6.31 is a constant current driver for an infrared light emitting
diode. The diode can be switched on and off via the digital input Mod. The transistor
Q1 boosts the output of the amplifier U1. The amplifier U1 controls the base of Q1
in order to keep the voltage drop across the resistor R3 equal to the voltage at the
input I. As long as the input Mod is at ground potential the transistor Q2 is turned off.
When the input is driven to a voltage level of 3.3 V the transistor Q2 saturates and
steers the current out of the emitter of Q1 around the light emitting diode D1. The
current path is never interrupted in this circuit, therefore the control loop can stay in
regulation. The supply voltage of 5 V is necessary to provide enough headroom for
biasing the transistor Q1 into conduction.

A metal oxide semiconductor field effect transistor, a MOSFET, Fig. 6.32, has
three terminals the gate G, the drain D and the source S. MOSFETs come in four

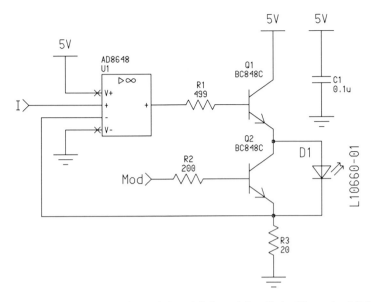

Fig. 6.31 Constant current driver for an infrared light emitting diode. The emitted light can be modulated

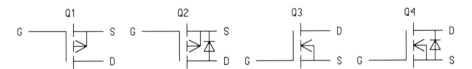

Fig. 6.32 Metal oxide semiconductor field effect transistors, Q1 is a P-channel enhancement mode MOSFET, Q2 is P-channel MOSFET with integrated diode, Q3 is an N-channel MOSFET, and Q4 is an N-channel MOSFET with integrated diode

varieties, the P-channel enhancement mode MOSFETs, the N-channel enhancement mode MOSFETs, the P-channel depletion mode MOSFETs, and the N-channel depletion mode MOSFETs. We do not cover depletion mode MOSFETs.

In a circuit the drain of an N-channel enhancement mode MOSFET usually sits at a more positive potential than the transistor's source. A voltage at the gate of the transistor controls the current from the drain to the source. As long as the voltage between the gate and the source is zero no current can flow from the drain to the source. When the voltage V_{GS} between the gate and the source exceeds the transistor's threshold voltage the drain source connection starts to conduct. Increasing V_{GS} further lowers the resistance R_{DS} between drain and source until the drain source connection is conducting fully. For most N-channel MOSFETs $V_{GS} > 10\,V$ is enough to turn them on fully, for some $V_{GS} > 5\,V$ suffices. The drain of a P-channel enhancement mode MOSFET usually sits at a more negative potential than the transistor's source. A negative voltage between the gate and the source makes the connection between source and drain conduct. But for the inverted signs the behavior of a P-channel enhancement mode MOSFET parallels that of an N-channel one. When used as switches P-channel MOSFETs act as top switches with their sources connected to the positive supply rail while the N-channel MOSFETs act as bottom switches with their sources connected to the ground rail. As N-channel transistors have a lower R_{DS} when conducting fully compared to P-channel transistors of similar construction one uses N-channel transistors for top switches too. The gate drivers then have to supply a gate voltage above the supply voltage for turning the transistors on.

6.9 Bibliographic Notes

The book by Tietze et al. (2008), is a very good reference for analog electronics. The one by Horowitz and Hill (1989) is a somewhat dated but very readable introduction to electronics. The book by Zumbahlen and Analog Devices (2008), discuss analog building blocks and the design of analog systems including construction techniques. The two volumes edited by Dobkin and Williams (2011, 2012), are a collection of application notes written by the staff of Linear Technology. The application notes by the late Jim Williams are among the author's favorites. The book on data converters edited by Kester (2004), with contributions by the technical staff at Analog Devices is a rich source for everything concerning analog-to-digital and digital-to-analog converters, including the history of their development, the architecture of these converts, and the design of actual circuits with these converters. Our treatment of antialiasing filters follows the one found there. The booklet by the staff of Philbrick Researches (1966), is an early reference for operational amplifiers. The components are outdated but the circuits and the theory discussed there are not. The booklet by T.R. Brown, revamped by Carter and published again by Texas Instruments, (Carter and Brown 2001), is another very readable early reference for operational amplifiers. The book edited by Jung (2005) covers operational amplifiers including their historical development, sensor and signal conditioning, filters, and hardware construction. The application note (National Semiconductor 2002) by the staff of then National Semiconductors, now Texas Instruments, is notorious for its brevity. The book by Kitchin and Counts (2006), covers instrumentation amplifiers, a type of amplifier for conditioning sensor signals. A lot of literature is available for free on the semiconductor manufacturers' websites.

6.10 Exercises

Exercise 6.1 Argue (2.20). Consider the signal $w(t)$, (6.3).

Exercise 6.2 You are using a Σ-Δ analog-to-digital converter with a sample rate of 20 MHz and a resolution of 16 bits for sampling a signal with a bandwidth of

50 kHz. The signal's temporal information must be preserved as much as possible. You expect noticeable high-frequency interference. Specify an antialiasing filter.

Exercise 6.3 Derive (6.5).

Exercise 6.4 Derive (6.6). The circuit is essentially an inverting amplifier. Use the complex resistance of the parallel combination of R2 and C1 and solve for the frequency at which the gain is down by 3 dB from the gain at zero frequency.

Exercise 6.5 Design a circuit with three inputs and one output. The input signals range between 0 and 3.3 V, as shall the output signal. The input signals shall each be lowpass filtered with a first-order filter with corner frequency 1.69 kHz. The output shall be the inverted sum of the lowpass filtered inputs.

Exercise 6.6 Derive (6.7).

Exercise 6.7 Apply a sinusoidal voltage $V(t)$ between the inputs L and N of the single-phase rectifier in Fig. 6.26. Sketch the waveform of the voltage $V_0(t)$ between the outputs $+$ and $-$.

Exercise 6.8 Apply the voltage $V_a(t) = V \sin(\omega t)$ between the inputs L1 and L2, $V_b(t) = V \sin(\omega t + 2\pi/3)$ between L2 and L3 and $V_c(t) = V \sin(\omega t + 4\pi/3)$ between L3 and L1 of the three-phase rectifier in Fig. 6.26. Sketch the waveform of the voltage $V_0(t)$ between the outputs $+$ and $-$.

References

Bennett WR (1948) Spectra of quantized signals. Bell Syst Tech J 27:446–472

Carter B, Brown TR (2001) Handbook of operational amplifier applications. Texas Instruments, Application report SBOA092A

Dobkin B, Williams J (eds) (2011) Analog circuit design: a tutorial guide to applications and solutions. Elsevier Science, Amsterdam

Dobkin B, Williams J (eds) (2012) Analog circuit design volume 2: immersion in the black art of analog design, vol 2. Elsevier Science, Amsterdam

Hartley RVL (1928) Transmission of information. Bell Syst Tech J 7:535–563

Horowitz P, Hill W (1989) The art of electronics. Cambridge University Press, Cambridge

Jung W (ed) (2005) Op amp applications handbook. Analog Devices series, Newnes, Amsterdam

Kester W (ed) (2004) The data conversion handbook. Analog Devices series, Newnes, Amsterdam

Kitchin C, Counts L (2006) A designer's guide to instrumentation amplifiers, 3rd edn. Analog Devices

National Semiconductor (2002) Op amp circuit collection, application note 31

Nyquist H (1924a) Certain factors affecting telegraph speed. Bell Syst Tech J 3:412–422

Nyquist H (1924b) Certain factors affecting telegraph speed. Trans AIEE XLIII:412–422. doi:10.1109/T-AIEE.1924.5060996

Nyquist H (1928) Certain topics in telegraph transmission theory. Trans AIEE 47:617–644

Philbrick Researches (1966) Applications manual for computing amplifiers for modelling measuring manipulating and much else. Philbrick Researches Inc

Shannon CE (1948) A mathematical theory of communication. Bell Syst Tech J 27(379–423):623–656

Shannon CE, Weaver W (1949) The mathematical theory of communication. University of Illinois Press, Urbana

Smith P (2004) Little known characteristics of phase noise, application note 741. http://www.analog.com

Tietze U, Schenk C, Gamm E (2008) Electronic circuits: handbook for design and application. Springer, New York

Zumbahlen H (2008) Linear circuit design handbook. Analog Devices, Newnes, Boston

Chapter 7
Energy Conversion—Power Supplies

The power supply circuits in an embedded system convert the electrical energy provided by some source like a battery into electrical energy according to the demand of the rest of the system. Logic and analog circuitry requires a well regulated constant voltage but some applications, such as driving high-power light-emitting diodes, ask for a precisely regulated current. Energy efficient power supplies can often do without extra heat sinks, allowing for the design of small and lightweight devices. In battery operated devices the energy efficiency of the supply circuits translates directly into longer service time between charges. A large part of the success of battery operated devices in the last decade can be attributed mainly to the increased energy efficiency of the electronics and of the supply circuits in particular and to a lesser degree to increases in the capacity of batteries. For the performance of batteries see, e.g., (Reddy 2010).

We restrict ourselves to the conversion of direct current only. Most of the principles, however, apply to conversion of alternating current too. We first study linear regulators, which work by converting excess power into heat. In our quest for energy efficiency we next investigate switchmode converters. The key to their high efficiency is that they switch rapidly between the supply rails and filter the resulting rectangular, pulse-width modulated waveforms using inductors and capacitors. We discuss component selection. To put this theory to use we design a supply for a high-power light-emitting diode. In this supply, software executing on a microcontroller controls the switching action in order to keep the current through the diode constant.

7.1 Linear Regulators

A linear voltage regulator, Fig. 7.1, regulates its output voltage V_o by letting the difference between its larger input voltage V_i and V_o drop across the transistor Q. The input voltage V_i is the voltage between the input rail Vi and ground, the output voltage V_o is the voltage between the output rail Vo and ground. The transistor has to conduct the current I drawn by the load, therefore it has to turn the power $P = (V_i - V_o)I$

© Springer International Publishing Switzerland 2015
P. Hintenaus, *Engineering Embedded Systems*, DOI 10.1007/978-3-319-10680-9_7

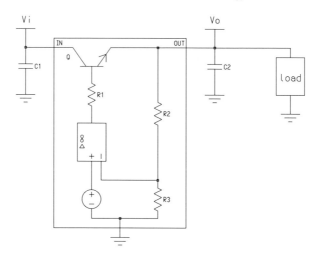

Fig. 7.1 Basic principle of a linear voltage regulator. The operational amplifier controls the voltage drop across the transistor so that the fraction of output voltage produced by the divider R2 and R3 equals the voltage across the reference

into heat, which has to be removed by a heat sink. Disregarding the power required for operating the regulator, the regulator's efficiency is $\eta = V_o/V_i$. For keeping the efficiency high the voltage $V_i - V_o$ must be small. Some regulators allow a minimum difference between V_i and V_o, the so-called dropout voltage, of less than 150 mV. Monolithic regulators contain protection features such as thermal shutdown circuits, which reduce the output current when the regulator becomes too hot. Due to these protection circuits linear regulators are extremely robust. Some linear regulators require an input and an output capacitor for stabilizing the internal control loop, consult the regulator's data sheet. Regardless of the regulator used, these capacitors help with the transient response of the supply.

Their simplicity makes linear regulators attractive. A small chip and two small capacitors most often is all it takes for providing a stable supply. Provided high-frequency content has been sufficiently suppressed by appropriate filtering of the regulator's input voltage, a linear regulator can deliver an exceedingly quiet output as there is no switching action involved in its functional principle. Therefore, linear regulators still are the supplies of choice for delicate analog circuits. Their only drawback is their abysmal efficiency once the difference between the input and the output voltage becomes too large.

The circuit in Fig. 7.2 is a supply for a small battery powered instrument. The instrument can act as a USB device for being configured and for exchanging data. The linear regulator U1 provides five volts for the analog rail. It is powered by four AA batteries connected to the two terminals at the left side of the circuit diagram. The diode D1 protects the circuit from reverse polarity. The ferrite L1 and the lowpass R1 and C1 filter the 5 V rail further in order to provide quiet power for precision circuitry. The diode network D2 and D3 switches the supply of the digital part to USB power as soon as a computer is connected. The linear regulator U2 reduces the 5 V at its

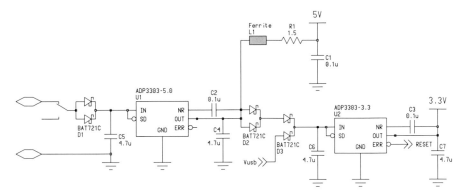

Fig. 7.2 A supply for a small battery powered instrument. The analog circuitry runs off the 5 V rail, the microcontroller off the 3.3 V rail. The diode network D2 and D3 provides power to the controller from the USB bus, as long as the instrument is connected to a computer

input to 3.3 V for running the instrument's microcontroller. The error output of U2 keeps the microcontroller in the reset state until the 3.3 V rail is in regulation.

7.2 Switchmode Converters

Switchmode converters allow to convert electrical power at some input voltage into electrical power at some output voltage above or below the input. These converters consist of capacitors and inductors for short-term energy storage and of semiconductor switches and diodes for steering energy from their input to their output. They work cyclically with cycle time t_c. The cycle time consists of two parts, the first of duration Dt_c, the second of duration $(1 - D)t_c$. The converter changes the state of its switches only at the beginning of the first part and at the beginning of the second part of a switching cycle. The ratio $0 < D < 1$ is called the duty cycle. The input rail in each of the following principal circuits is denoted Vi, the voltage between the input rail and ground is V_i, and the current into the input rail is I_i. We assume that the input voltage is positive although for all converter topologies we discuss equivalent circuits for negative input voltage exist. The output rail is called Vo, the voltage between the output rail and ground is V_o, and the current out of the output rail is I_o. Assuming there are no losses in the converter's capacitors and inductors and that the converter operates under steady-state conditions, that is the input voltage V_i the output voltage V_o and the output current I_o are effectively constant, the converter draws exactly as much energy from its source in each switching cycle as it delivers to its load, in other words

$$V_i I_i = V_o I_o.$$

When computing the change ΔI_L, the so-called current ripple, of the current through an inductor with inductance L during one part of a switching cycle, we assume that the voltage V_L across the inductor stays essentially constant. For the first part of a switching cycle the change of the current through the inductor then is

$$\Delta I_L = D t_c V_L / L. \tag{7.1}$$

For the second part it is

$$\Delta I_L = (1 - D) t_c V_L / L. \tag{7.2}$$

When computing the change of the voltage across a capacitor during a switching cycle, however, we integrate the current through the capacitor. These assumptions allow us to model these converters with high accuracy using systems of linear equations instead of systems of differential equations. When designing a switch-mode converter we start out with the maximum current ripple for each inductor, and the maximum voltage ripple for each capacitor, the peak-to-peak variation during one switching cycle of the voltage across the capacitor. While the load the converter supplies typically sets a maximum allowable voltage ripple for the output capacitor we have to set the current ripples for each inductor ourselves. About 20 % of the average current through an inductor is a good starting point. The arrows next to the inductors in the following figures indicate the positive direction of the current through the inductor.

7.2.1 Step-Down Converters

A step-down (buck) converter, Fig. 7.3 left part, converts a higher voltage at its input to a lower, regulated voltage of the same polarity at its output, while maintaining high efficiency. During the first part of a switching cycle a step-down converter makes the top switch conduct. For the second part of the cycle it makes the bottom

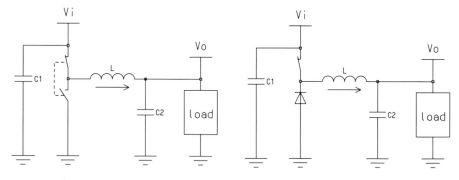

Fig. 7.3 Basic principle of a step-down converter using two controlled switches on the *left*. The two switches operate synchronously in a break before make fashion. Basic principle of a step-down converter using one controlled switch and a diode on the *right*

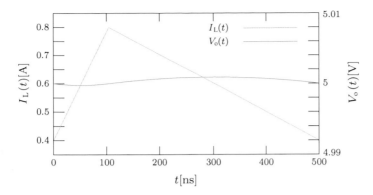

Fig. 7.4 Waveforms of the current $I_L(t)$ through the inductor of a step-down converter and of the output voltage V_o for one switching cycle. The voltage spike that is observable when the top switch opens has been omitted. The cycle time is $t_c = 500$ ns, the inductance of the inductor L is 4.7μH and the capacitance of the output capacitor C2 is 22μF. The converter is stepping down from 24 to 5 V, the load current is 600 mA

switch conduct. The inductor L draws current from the input rail during the first part only, necessitating a large input capacitor, while it delivers current to the output rail during both parts of the switching cycle. Under steady-state conditions, see Fig. 7.4, the current $I_L(t)$ through the inductor L must be periodic with period t_c, $I_L(t) = I_L(t + t_c)$. The voltage across the inductor during the first part of the cycle is $V_i - V_o$, during the second part it is $-V_o$, therefore, according to (7.1) and (7.2)

$$\frac{t_c}{L}(D(V_i - V_o) - (1 - D)V_o) = 0 \tag{7.3}$$

must hold, and the output voltage is

$$V_o = DV_i. \tag{7.4}$$

By commanding the duty cycle D the control unit of a step-down, not shown in Fig. 7.3, controls the output voltage.

The bottom switch can be replaced by a diode, Fig. 7.3 right part. As soon as the top switch stops conducting the inductor instantly reverses the polarity of the voltage across it and increases the magnitude of this voltage until current flows again. This action pulls the one end of the diode below ground and biases the diode into conduction. In order to prevent excessive negative excursions the diode must turn on fast. During the second part of the switching cycle the current $I_L(t)$ through the inductor L can drop to zero in a lightly loaded converter. At this very instance the little energy that is still stored in the circuit will oscillate between the parasitic capacitances of the switches and the inductor. These oscillations may cause electromagnetic interference. This discontinuous mode is not covered by (7.3).

When designing a converter for stepping down from the input voltage V_i to the output voltage V_o we first choose the cycle time t_c and the maximum current ripple $\hat{I}_L$.

The inductor is connected to the Vo rail, therefore the average current through L is I_o, and $\hat{I}_L \approx 0.2 I_o$ is a good starting point. The inductance L of the inductor L then needs to be greater than

$$L > \frac{t_c V_o (V_i - V_o)}{\hat{I}_L V_i}.$$

(7.5)

Choosing too big an inductance, however, is detrimental to the transient response of the converter. For a chosen maximum voltage ripple $\hat{V}_o$ the output capacitor C2 must have a capacitance C_2 greater than

$$C_2 > \frac{t_c^2 V_o (V_i - V_o)}{8 \hat{V}_o V_i L}.$$

(7.6)

The circuit in Fig. 7.5 is a supply for small industrial automation devices such as sensors. It converts the usual 24 V down to 5 V for operating analog circuitry and to 3.3 V for the digital functions. The transistors Q1 and Q2 under the control of U2 disconnect the supply in case of overvoltage or undervoltage and in case of

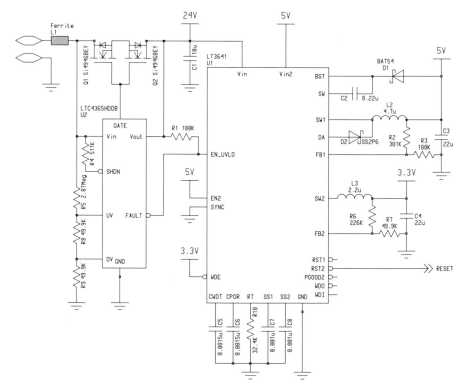

Fig. 7.5 Power supply for an industrial automation device. The 24 V voltage, which is common in industrial automation systems, is first stepped down to 5 V for supplying analog circuitry, and further down to 3.3 V for the digital part. The supply withstands reverse polarity connections

the input voltage being connected in reverse. The undervoltage threshold is 15 V, the overvoltage threshold is 30 V. The dual step-down converter U2 has internal switches. The duration of a switching cycle t_c is 500 ns. The external diode D2 completes the first step-down converter. The diode D1 and the capacitor C2 form a so-called bootstrap circuit. It provides a voltage above the input voltage to keep the top switch of the first step-down converter conducting during the first part of the switching cycle. The reset generator internal to U1 provides a reliable reset signal for the digital circuits in our device. We have taken the converter portion of this circuit from the converter's data sheet.

7.2.2 Step-Up Converters

A step-up (boost) converter, Fig. 7.6, converts a low voltage at its input to a higher, regulated voltage of the same polarity at its output. During the first part of a switching cycle, Fig. 7.7, the bottom switch conducts, therefore the voltage across the inductor is V_i and the current $I_L(t)$ through the inductor ramps up linearly. During the second part of the switching cycle the top switch conducts. The voltage across the inductor is $V_i - V_o$ and the current $I_L(t)$ ramps down linearly. The inductor L draws current from the input rail during both parts of the switching cycle while it delivers current to the output rail during the second part only necessitating a large output capacitance. Under steady-state conditions the current $I_L(t)$ must be periodic with period t_c, $I_L(t) = I_L(t + t_c)$, so according to (7.1) and (7.2)

$$\frac{t_c}{L}(DV_i + (1 - D)(V_i - V_o)) = 0 \qquad (7.7)$$

must hold. Rearranging and solving for V_o we obtain

$$V_o = \frac{V_i}{1 - D}.$$

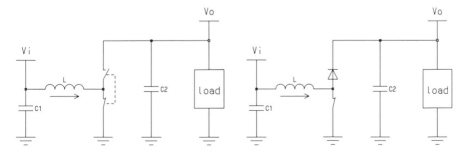

Fig. 7.6 Basic principle of a step-up converter using two controlled switches on the *left*. The two switches operate synchronously in a break before make fashion. Basic principle of a step-up converter using one controlled switch and a diode on the *right*

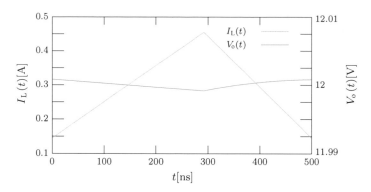

Fig. 7.7 Waveforms of the current I_L through the inductor of a step-up converter and of the output voltage V_o for one switching cycle. The voltage spike that is observable when the bottom switch opens has been omitted. The cycle time is $t_c = 500$ ns, the inductance of the inductor L is $4.7\,\mu$H, and the capacitance of the output capacitor C2 is $22\,\mu$F. The converter is stepping up from 5 to 12 V, the load current is 125 mA

By commanding the duty cycle D the controller, not show in Fig. 7.6, of a step-up converter can control the output voltage.

The top switch can be replaced by a diode. As soon as the bottom switch stops conducting at the end of the first part of a switching cycle the inductor L instantly reverses the polarity of the voltage across it and increases the magnitude of this voltage until current flows again. This action biases the diode into conduction. In this way, a step-up converter with a diode instead of a top switch transfers energy stored in its inductor to its output during the second part of the switching cycle. Note that the diode conducts without any switching action as soon as a positive input voltage V_i is applied to such a step-up. The load of the step-up converter will then see V_i too. Such behavior is undesirable in battery powered devices, as it drains the battery although the load is not powered correctly. Under light load conditions, the current $I_L(t)$ through the inductor can drop to zero and the diode will stop to conduct. What little energy is still stored in the inductor then oscillates between the parasitic capacitances of the diode and the switch, and the inductor. This oscillation may create electromagnetic interference, therefore this discontinuous mode should be avoided; (7.7) does not cover discontinuous mode.

The inductor L is connected to Vi, therefore the average current through L is $I_i = \frac{V_o I_o}{V_i}$. For a given maximum current ripple $\hat{I}_L$, the peak-to-peak variation of the current $I_L(t)$, the inductance L must be greater than

$$L > t_c V_i \frac{V_o - V_i}{\hat{I}_L V_o}, \tag{7.8}$$

and for a given maximum ripple $\hat{V}_o$ of the output voltage V_o the capacitance C_2 must be greater than

$$C_2 > t_c I_o \frac{V_o - V_i}{\hat{V}_o V_o}. \tag{7.9}$$

Fig. 7.8 Basic principle of a
SEPIC converter

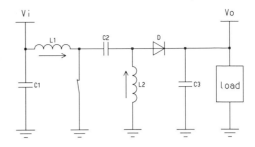

Too big an inductance hinders the converter's transient response, while a capacitance larger than necessary helps it.

7.2.3 SEPIC Converters

A single-ended primary-inductance converter (SEPIC), Fig. 7.8, converts a voltage at its input to a regulated output voltage which is of the same polarity but higher or lower than the input voltage. We consider a switching cycle under steady-state conditions. During the first part of the cycle, while the switch conducts, the voltage across L1 is V_i, during the second part, while the diode conducts, the voltage across L1 is $V_i - V_C - V_o$, where V_C is the voltage across the capacitor C2. The current $I_1(t)$ through the inductor L1 with inductance L_1 must be periodic with period t_c, $I_1(t) = I_1(t + t_c)$, therefore, according to (7.1) and (7.2)

$$\frac{t_c}{L_1}(DV_i + (1 - D)(V_i - V_C - V_o)) = 0 \qquad (7.10)$$

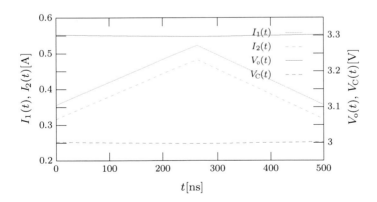

Fig. 7.9 Waveforms of a SEPIC converter for one switching cycle. The current $I_1(t)$ flows through the inductor L1, $I_2(t)$ from ground through L2, V_o is the output voltage and V_C is the voltage across C2. The voltage spike that is observable when the switch opens has been omitted. The cycle time is $t_c = 500$ ns, the inductance of the inductor L1 is $4.7\,\mu$H, the one of L2 is $4.7\,\mu$H, the capacitance of the output capacitor C3 is $22\,\mu$F, and the one of C2 is $22\,\mu$F. The converter is converting a 3 V input into a 3.3 V output, the load current is 400 mA

must hold, Fig. 7.9. During the first part of the cycle the capacitor C2 pulls the one end of L2 below ground and the voltage across L2 is V_C. During the second part the voltage across L2 is $-V_o$. The current $I_2(t)$ through the inductor L2 with inductance L_2 must also be periodic with period t_c, $I_2(t) = I_2(t + t_c)$, therefore, according to (7.1) and (7.2)

$$\frac{t_c}{L_2}(DV_C - (1 - D)V_o) = 0 \tag{7.11}$$

must hold. Solving these two equations for the output voltage V_o and the voltage V_C we obtain

$$V_o = \frac{D}{1 - D}V_i$$
$$V_C = V_i.$$

Under light loads both currents through the inductors can drop to zero during the second part of the switching cycle. The little energy left stored in the inductors then oscillates between the inductors and the capacitor C2. The electromagnetic interference this oscillation may create renders this discontinuous mode of operation problematic; (7.10) and (7.11) do not cover discontinuous mode. The inductor L1 draws current from the input rail during both parts of the switching cycle, while both inductors L1 and L2 deliver current to the output rail during the second part only necessitating a large output capacitance. By commanding the duty cycle D the converter's controller, not shown in Fig. 7.8, can control the output voltage. The inductor L1 is connected to the input rail Vi, therefore the average current through L1 is I_i. The load withdraws a charge of $t_c I_o$ from the output capacitor C3 during a switching cycle, the inductor L1 deposits a charge of $\frac{t_c I_i V_i}{V_o + V_i}$, the inductor L2 must contribute the balance, therefore the average current through L2 is I_o. For a given maximum current ripple $\hat{I}_L$ the inductances L_1 and L_2 must be greater than

$$L_1, L_2 > \frac{t_c V_i V_o}{\hat{I}_L(V_i + V_o)}. \tag{7.12}$$

For a given output voltage ripple $\hat{V}_o$ the output capacitance C_3 must be greater than

$$C_3 > \frac{t_c I_o V_o}{\hat{V}_o(V_i + V_o)}. \tag{7.13}$$

During the first part of the switching cycle the capacitor C2 conducts the current through L2. This current ramps down from $-I_o + \frac{V_i D t_c}{2L_2}$ to $-I_o - \frac{V_i D t_c}{2L_2}$. During the second part of the cycle C2 conducts the current through L1, which ramps down from $\frac{V_o I_o}{V_i} + \frac{V_i D t_c}{2L_1}$ to $\frac{V_o I_o}{V_i} - \frac{V_i D t_c}{2L_1}$. For a given ripple voltage $\hat{V}_C$ the capacitance C_2 of the capacitor C2 must be greater than

$$C_2 > \frac{t_c I_o V_o}{\hat{V}_C(V_o + V_i)}. \tag{7.14}$$

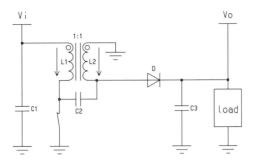

Fig. 7.10 Basic principle of a SEPIC converter with coupled inductors. The two inductors L1 and L2 are wound onto a common core forming a transformer. The field created by a current through the one is seen by the other. The *circles* next to the inductors indicate the winding senses

The capacitor C2 couples all the power the SEPIC converts. This puts high stresses on the capacitor. These stresses can be relieved by coupling some of the energy magnetically. The voltages across inductor L1 and across inductor L2 in Fig. 7.8 are the same during each switching cycle. This and choosing the inductances of both inductors being the same allows for winding both inductors on a common core instead of two separate ones creating a transformer. In an ideal transformer having one winding called the primary with n turns and the other, the secondary, with m turns, an alternating voltage $V(t)$ across the primary appears as $\frac{m}{n}V(t)$ across the secondary and vice versa. The current $I(t)$ through the primary and the current through the secondary $\frac{n}{m}I(t)$ exhibit the inverse relationship. Constant voltages get blocked however. Using a transformer with a 1 : 1 turn ratio instead of two separate inductors, Fig. 7.10, about halves the ripple currents through the inductors when compared with separate inductors of the same inductance.

7.2.4 Buck-Boost Converters

A buck-boost (step-up-down) converter, Fig. 7.11, converts a positive voltage at its input into a negative voltage at its output. The magnitude of the output voltage can be above or below the input voltage. During the first part of the switching cycle the input side switch conducts, therefore the voltage across the inductor L with inductance L is V_i and the current $I_L(t)$ through L ramps up linearly. During the second part of the switching cycle the output-side switch conducts and the voltage across L is V_o. For the current $I_L(t)$ to ramp down V_o must be negative. The inductor L draws current from its input side during the first part of the switching cycle and delivers current to its output during the second part of the switching cycle only. Therefore, both the input capacitance and the output capacitance must be large in order to keep the voltage excursions during a switching cycle small. Under steady-state conditions, the current $I_L(t)$ must be periodic with period t_c, $I_L(t) = I_L(t + t_c)$, therefore, according to

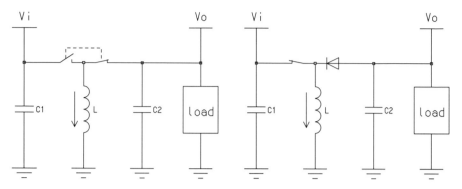

Fig. 7.11 Basic principle of a buck-boost converter using two controlled switches on the *left*. The two switches operate synchronously in a break before make fashion. Basic principle of a buck-boost converter using one controlled switch and a diode on the *right*

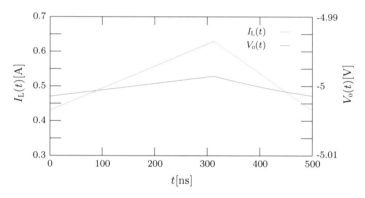

Fig. 7.12 Waveforms of the current I_L through the inductor of a buck-boost converter and of the output voltage V_o for one switching cycle. The voltage spike that is observable when the input side switch opens has been omitted. The cycle time is $t_c = 500\,\mathrm{ns}$, the inductance of the inductor L is $4.7\,\mu\mathrm{H}$, and the capacitance of the output capacitor C2 is $22\,\mu\mathrm{F}$. The converter is converting an input voltage of $3\,\mathrm{V}$ into an output voltage of $-5\,\mathrm{V}$, the load current is $-200\,\mathrm{mA}$

(7.1) and (7.2)

$$\frac{t_c}{L}(DV_i + (1 - D)V_o) = 0 \tag{7.15}$$

must hold, Fig. 7.12. Rearranging and solving for the output voltage V_o we obtain

$$V_o = -V_i \frac{D}{1 - D}.$$

By commanding the duty cycle D the controller of a buck-boost converter, not shown in Fig. 7.11, can control its output voltage. The switch at the output side can be replaced by a diode. As soon as the input side switch stops conducting at the end of the first part of the switching cycle L reverses the polarity of the voltage across it and biases the diode into conduction. Under light load conditions the current through L can drop to zero during the second part of the switching cycle. What little energy is left stored in L then oscillates between the parasitic capacitance of the switch and L. The electromagnetic interference this oscillation may create renders this discontinuous mode of operations problematic; (7.15) does not cover discontinuous mode. During a switching cycle the load withdraws a charge of $t_c I_0$ from the output capacitor C2, which must be replenished by the inductor L. Therefore, the average current through L is $\frac{I_0(V_0 - V_i)}{V_i}$. For a given maximum current ripple $\hat{I}_L$ the inductance L must be greater than

$$ L > \frac{t_c V_i V_0}{\hat{I}_L (V_0 - V_i)}. \tag{7.16} $$

For a given output voltage ripple $\hat{V}_0$ the output capacitance C_2 must be greater than

$$ C_2 > \frac{t_c I_0 V_0}{\hat{V}_0 (V_0 - V_i)}. \tag{7.17} $$

7.2.5 Ćuk Converters

A Ćuk converter, Fig. 7.13, converts a positive voltage at its input into a negative voltage at its output. The magnitude of the output voltage can be above or below the input voltage. We consider a switching cycle under steady state conditions. During the first part of the cycle, while the switch conducts, the voltage across L1 is V_i. At the end of the first part, as soon as the switch turns off, the inductor L1 reverses the polarity of the voltage across it and through C2 biases the diode into conduction. During the second part, while the diode conducts, the voltage across L1 is $V_i - V_C$, where V_C is the voltage across the capacitor C2. The current $I_1(t)$ through the inductor L1 with inductance L_1 must be periodic with period t_c, $I_1(t) = I_1(t + t_c)$, therefore, according to (7.1) and (7.2)

$$ \frac{t_c}{L_1} (DV_i + (1 - D)(V_i - V_C)) = 0 \tag{7.18} $$

Fig. 7.13 Basic principle of a Ćuk converter

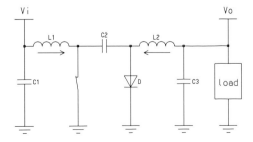

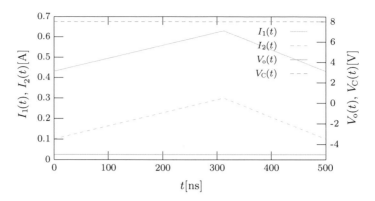

Fig. 7.14 Waveforms of a Ćuk converter for one switching cycle. The current $I_1(t)$ flows through the inductor L1, $I_2(t)$ through L2, V_o is the output voltage and V_C is the voltage across C2. The voltage spikes that is observable when the switch opens has been omitted. The cycle time is $t_c = 500\,\text{ns}$, the inductance of the inductor L1 is $4.7\,\mu\text{H}$, the one of L2 is $4.7\,\mu\text{H}$, the capacitance of the output capacitor C3 is $22\,\mu\text{F}$ and the one of C2 is $22\,\mu\text{F}$. The converter is converting a 3 V input into a $-5\,\text{V}$ output, the load current is $-200\,\text{mA}$

must hold, Fig. 7.14. During the first part of the cycle the capacitor C2 pulls the one end of L2 below ground and the voltage across L2 is $V_o + V_C$. During the second part the voltage across L2 is V_o. The current $I_2(t)$ through the inductor L2 with inductance L_2 too must be periodic with period t_c, $I_2(t) = I_2(t + t_c)$, therefore, according to (7.1) and (7.2)

$$\frac{t_c}{L_2}\left(D(V_o + V_C) + (1 - D)V_o\right) = 0 \qquad (7.19)$$

must hold. Solving these two equations for the output voltage V_o and the voltage V_C we obtain

$$V_o = -V_i\frac{D}{1 - D}$$
$$V_C = V_i - V_o.$$

Under light loads both currents through the inductors can drop to zero during the second part of the switching cycle. The little energy left stored in the inductors then oscillates between the inductors and the capacitor C2. The electromagnetic interference this oscillation may create renders this discontinuous mode of operation problematic; (7.18) and (7.19) do not cover discontinuous mode. The inductor L1 draws current from the input rail during both parts of the switching cycle, and the inductor L2 delivers current to the output rail during both parts of a switching cycle. By commanding the duty cycle D the converter's controller, not shown in Fig. 7.13, can control the output voltage. The inductor L1 is connected to the input rail Vi, therefore the average current through it is $I_i = \frac{V_o I_o}{V_i}$. The inductor L2 is connected to the output rail, therefore the average current through it is $-I_o$. For a given maximum

Fig. 7.15 Basic principle of a Ćuk converter with coupled inductors. The two inductors L1 and L2 are wound onto a common core forming a transformer. The circles next to the inductors indicate the winding senses

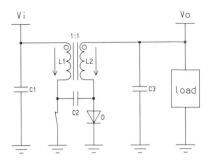

current ripple $\hat{I}_L$ the inductances L_1 and L_2 must be greater than

$$L_1, L_2 > \frac{t_c V_i V_o}{\hat{I}_L (V_o - V_i)}. \tag{7.20}$$

For a given output voltage ripple $\hat{V}_o$ the output capacitance C_3 must be greater than

$$C_3 > \frac{t_c^2 V_i V_o}{8 \hat{V}_o L_2 (V_o - V_i)}. \tag{7.21}$$

During the first part of the switching cycle the capacitor C2 conducts the current through L2. This current ramps up from $-I_o - \frac{V_i D t_c}{2L_2}$ to $-I_o + \frac{V_i D t_c}{2L_2}$. During the second part of the cycle C2 conducts the current through L1, which ramps down from $\frac{V_o I_o}{V_i} + \frac{V_i D t_c}{2L_1}$ to $\frac{V_o I_o}{V_i} - \frac{V_i D t_c}{2L_1}$. For a given ripple voltage $\hat{V}_C$ the capacitance C_2 of the capacitor C2 must be greater than

$$C_2 > \frac{t_c I_o V_o}{\hat{V}_C (V_o - V_i)}. \tag{7.22}$$

The capacitor C2 couples all the power the converter converts. This puts high stresses on the capacitor. These stresses can be relieved by coupling some of the energy magnetically, Fig. 7.15. The voltages across L1 and L2 are the same during the first and the second part of a switching cycle but for the sign, therefore a transformer with a winding ratio of 1 : 1 will do.

7.2.6 Flyback Converters

A flyback converter, Fig. 7.16, converts a positive voltage at its input into one or several voltages of arbitrary polarity, which moreover can be isolated from the input. We consider a switching cycle under steady state conditions, Fig. 7.17. During the first part of the switching cycle of duration $t_c D$, while the switch conducts, the voltage across L1 is the input voltage V_i and the voltage across L2 is $-\frac{m}{n} V_i$. The diode is

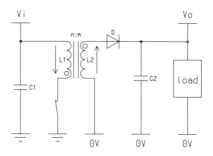

Fig. 7.16 Basic principle of a flyback converter. Provided the feedback signal does not break the isolation barrier the output is isolated from the input

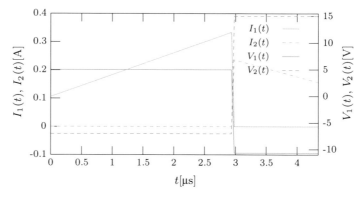

Fig. 7.17 Waveforms of a flyback converter for one switching cycle. The current $I_1(t)$ flows through the inductor L1, $I_2(t)$ through L2, V_1 is the voltage across L1 and V_2 is the voltage across L2. The voltage spikes that are observable when the switch opens have been omitted. The cycle time is $t_c = 4.34\,\mu s$, the inductance of the inductor L1 is $64\,\mu H$, the one of L2 is $127.5\,\mu H$. The converter is converting a 5 V input into a 15 V output, the load current is 50 mA

reverse biased and no current can flow through L2. At the end of the first half, as soon as the switch opens, the starting breakdown of the magnetic flux through L1 and L2 reverses the polarities of the two voltages across the two inductors. The voltage across L2 becomes large enough to bias the diode D forward and current now flows through L2, so in essence the current jumps from flowing through L1 to flowing through L2. The current through L2 then maintains the magnetic flux during the second part of the switching cycle. As coupled inductors cannot transfer constant voltages the average voltage across L2 during one switching cycle must be zero, therefore

$$- t_c D \frac{m}{n} V_i + t_c (1 - D) V_o = 0 \tag{7.23}$$

must hold. By rearranging and solving for V_o we obtain

$$V_o = \frac{m V_i D}{n(1 - D)}.$$

By adjusting the turn ratio $n : m$ between the two inductors we can bridge a large voltage difference between the input and the output without having to resort to impractical duty cycles. Under light loads the current through L2 can drop to zero. What little energy is still stored in the inductor will oscillate between the inductor L2 and the parasitic capacitance of the diode and between the inductor L1 and the parasitic capacitance of the switch. The possible problems with electromagnetic interference make this discontinuous mode of operations undesirable; (7.23) does not cover discontinuous mode. The inductor L1 draws current from the input rail during the first part of the switching cycle only, while the inductor L2 delivers current to the output rail during the second part only, calling for a large input and a large output capacitance. By commanding the duty cycle D the converter's controller, not shown in Fig. 7.16, can control the output voltage. No electrical connection exists between the input side and the output side as a flyback converter transfers energy from the input to the output magnetically. As long as the feedback signal, derived from the output voltage V_o, is delivered to the converter's controller in a nonelectrical way, this isolation barrier is maintained, making flyback converters suitable for being powered directly by the mains. We can produce additional output voltages by winding additional inductors onto the common magnetic core.

For a given maximum increase of the current $\hat{I}_1$ through L1 during the first part of the switching cycle, which translates into a maximum decrease $\hat{I}_2$ of the current through L2 during the second part of the switching cycle, $\hat{I}_2 = n/m\,\hat{I}_1$, the inductance L_1 of L1 must be greater than

$$L_1 > \frac{t_c n\, V_i V_o}{\hat{I}_1 (n V_o + m V_i)}. \tag{7.24}$$

For a given maximum output voltage ripple $\hat{V}_o$ the capacitance C_2 of the output capacitor C2 must be greater than

$$C_2 > \frac{t_c m\, V_i I_o}{\hat{V}_o (n V_o + m V_i)}. \tag{7.25}$$

The circuit in Fig. 7.18 is a flyback converter converting 5 V at the input into 15 V and -15 V at the outputs. It provides the supplies for precision analog circuitry, therefore the outputs are heavily filtered. The converter's controller contains the switch. It allows for controlling both the voltage slope and the current slope during switching actions. While slow switching increases the losses it reduces the high-frequency content of the switching waveforms. Reducing this high-frequency content enables the supply to produce very quiet supplies, a must for precision analog circuitry. In case the outputs are unloaded the diodes D2 and D3 provide a path for the current to flow during the second part of the switching cycle. The controller supervises both output voltages and takes the one which deviates from its nominal voltage most as feedback. No insulation is needed for this application, therefore the feedback path is electrical. We have taken this circuit from the controller's data sheet.

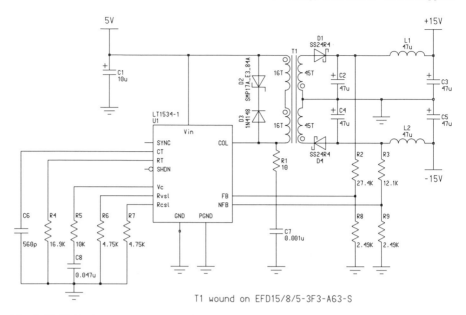

Fig. 7.18 Flyback converter for supplying precision analog circuitry. Note the heavy filtering of the outputs!

7.2.7 Control of Switchmode Converters

Voltage mode control, Fig. 7.19, commands the duty cycle of the switches in order to control the output voltage. It applies to all converters covered. For simplicity we demonstrate voltage mode control for a step-down converter. The oscillator U1 sets the converter's switching cycle. It has two outputs, a logic output for indicating the beginning of a switching cycle and a sawtooth signal for supporting the generation of the pulse width modulated switching pattern. At the beginning of a switching cycle it sets the flip-flop U3 and makes the top switch conduct. The comparator U2 compares the command signal provided by the PI controller U4 with the sawtooth signal provided by the oscillator. As soon as the sawtooth signal exceeds the command signal the falling edge at the comparator's output resets the flip-flop, which in turn disables the top switch and makes the bottom switch conduct thus starting the second part of the switching cycle. The error amplifier U5 provides the difference between the output of the voltage reference U6 and the feedback signal to the PI controller. The resistive divider R1 and R2 derives the feedback signal from the output voltage. Switchmode converters are sampled systems with sampling frequency $f_s = 1/t_c$. We cannot expect them to follow control inputs with frequencies beyond $f_s/2$ in any meaningful way. The process of the control system consists of the converter's inductors and capacitors and the load. Having at least two interacting energy storage devices in this system, and at least four for the SEPIC converter and the Ćuk converter, poses a challenge when we try to tune the PI controller for fast transient response.

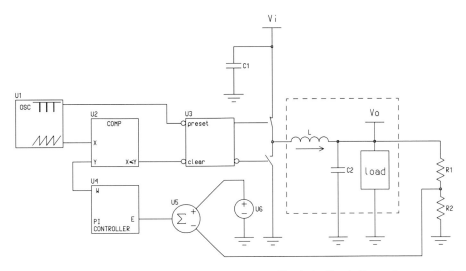

Fig. 7.19 Voltage mode control for a step-down converter. The *dashed box* indicates the controller's process

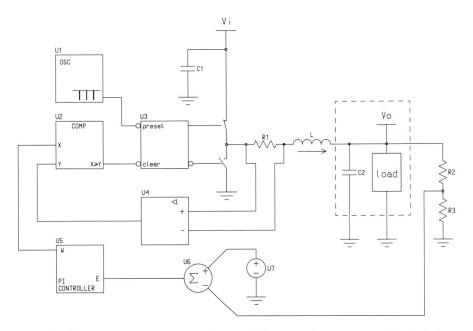

Fig. 7.20 Constant frequency current mode control for a step-down converter. The *dashed box* indicates the controller's process

Current mode control, Fig. 7.20, commands the sum of the current through the load and the current through the output capacitor. This current can be measured right before the output capacitor. Depending on the type of converter this current may be available at other more appropriate locations in the circuit during part of the switching cycle too. A cascaded control scheme consisting of two control loops is used. The inner loop controls the average sum of the currents through the load and through the output capacitor while the outer loop provides the desired current from measurements of the output voltage. We describe current mode control for a step-down converter but the principles apply to all converters covered. The oscillator U1 sets the converter's switching cycle. At the beginning of a switching cycle it sets the flip-flop U3, which makes the top switch conduct. The current through L ramps up linearly, as does the voltage across the shunt R1, which is amplified by the current sense amplifier U4. Ideally the comparator U2 resets the flip-flop U3 as soon as the average current for the switching cycle, as expectable from the output of U4, equals the desired current provided by the PI controller U5. In most implementations either the tips or the valleys of the current waveform are used instead of the averages. A slope compensation scheme, which is beyond our scope, is needed then. The error amplifier U6 provides the difference between the feedback signal and the output of the voltage reference U7 to the PI controller. The voltage divider R2 and R3 derives the feedback signal from the output voltage. Current mode control reduces the process to the output capacitor C2 and the load. The lower number of interacting energy storage components in this system compared to the system when voltage mode control is employed makes tuning the PI controller for fast transient response much easier. Furthermore, current mode control simplifies building over-current protection, limiting the output of the PI controller to acceptable values is all that is needed.

7.2.8 Components for Switchmode Converters

While the losses in the real components cannot be neglected in switchmode converters for high-power applications we opted to exclude them from our presentation until now. Instead of generalizing the equations to include the non-ideal characteristics of real components we mention the most important loss mechanisms in an informal way.

The semiconductor switches in a switchmode converter introduce conduction losses, due to the voltage drop across a switch while the switch is fully on, switching losses, due to the nonzero time it takes the switch to turn fully on or fully off and losses incurred from driving the control terminal of the switch. The types of semiconductor switches available for the voltage and current range a converter has to be designed for limit the switching frequency at which the converter can operate. High switching frequencies enable the use of physically small inductors and capacitors. For maximum voltages below 100 V bipolar NPN power transistors and low voltage N-channel power MOSFET transistors are the switches of choice allowing switching frequencies of up to about 2 MHz. Even for the top switch NPN and N-channel transistors are preferred over their PNP and P-channel counterparts for their superior performance. A voltage above the positive supply rail is then necessary for driving their base or gate respectively, which must be generated within the converter. For output currents below 15 A these transistors may be integrated together with the controlling functions into a switching regulator chip, as in the circuits in Figs. 7.5 and 7.18. For converters operating directly off the mains voltage high voltage N-channel power MOSFET transistors allow currents below 10 A and switching frequencies below 300 kHz. For higher voltages or higher currents isolated gate bipolar transistors operating at switching frequencies between 20 kHz and 100 kHz dominate.

The diodes in switchmode converters must be fast. For voltages below roughly 40 V and currents below 10 A Schottky barrier diodes are the best choice. They exhibit a voltage drop of about 0.5 V when fully conducting, therefore keeping losses low. For higher voltages or higher currents fast recovery diodes with voltage drops around 1 V must be chosen.

The capacitors in a switchmode converter must conduct high-frequency currents. A real capacitor not only exhibits a capacitance but has to be considered as an ideal capacitor with a resistor and an inductor in series. More than the sheer capacitance the equivalent series resistance (ESR) and the equivalent series inductance (ESL) determines its suitability. Aluminum electrolytic capacitors with liquid electrolyte offer high capacitances but suffer from high equivalent series resistance and high equivalent series inductance which together dominate their behavior at high frequencies. When used in switchmode converters aluminum electrolyte capacitors should be paralleled with ceramic capacitors. The loss of electrolyte to evaporation at elevated temperatures limits their lifespan. Solid electrolyte tantalum capacitors specially designed for low equivalent series resistance are available. These capacitors are sensitive to the stress created by pulsating currents. To prevent their early rather violent demise this stress should be distributed over several paralleled capacitors. At high switching frequencies, where their limited capacitance is acceptable, ceramic capacitors are the best choice. They have both very low equivalent series resistance and very low equivalent series inductance.

For achieving the inductance required the inductors used in switchmode converters are wound on cores made out of appropriate ferromagnetic materials. Their inductance depends on temperature and more importantly on the current through the inductor. Above a certain current the inductance breaks down rapidly, the inductor is said to saturate. Saturation has to be avoided by all means as the resulting current peaks will most probably destroy the converter's switches and diodes.

For moderate power ranges the semiconductor industry offers a wide variety of integrated circuits for building switchmode converters, both with integrated switching transistors and for the use together with external switches. In order to achieve the desired switching frequencies to allow the use of tiny capacitors and inductors these circuits are analog devices. Digital control is possible but generating the pulse width modulation digitally limits the switching frequency attainable with acceptable effort to about 100 kHz, see Sect. 5.6.4. At high-power levels the switching elements limit the maximum switching frequency anyway. The flexibility offered by programmable devices in implementing complicated control laws then is most welcome.

7.3 Constant Current Supply for LED Illumination

Light emitting diodes for illumination are low-power devices. As such, power
supplies for them lend themselves more to analog control. Designing a digitally
controlled power supply for light-emitting diodes, however, allows us to study such
supplies without having to cope with the dangers introduced by high voltages and
high currents.

The supply's power stage in Fig. 7.21 operates off an input voltage between 6 V
and 12 V. It powers a single white LED, which draws a power of approximately
1 W. A voltage of about 3.4 V drops across the LED when it operates at maximum
intensity, therefore a step-down converter is appropriate. The microcontroller for
operating the power stage must feature a pulse width modulation unit with at least
one channel, an analog-to-digital converter with at least three multiplexed inputs
and a timer for periodically triggering analog-to-digital conversions. The output of
the pulse width unit operates the switching transistor, while the analog inputs are
used for measuring the current through the LED, for measuring the temperature
of the LED and for reading the position of a potentiometer, which allows to set

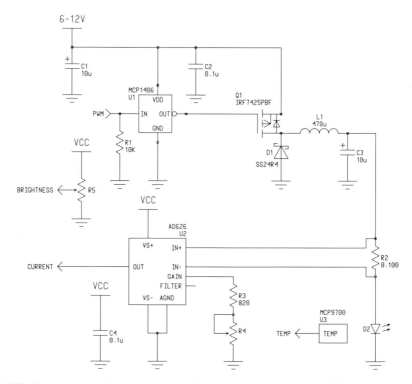

Fig. 7.21 Power stage and current feedback for a step-down converter providing constant current
for LED illumination. The microcontroller running the control algorithms is not shown

a desired brightness. When designing the software we first set up the pulse width modulation unit. The power stage requires a switching frequency of 40 kHz, which can be produced digitally while still maintaining adequate resolution. We set the duty cycle in such a way that the power stage applies about 2 V across the LED. When we operate the supply with constant duty cycle while we measure the current through the LED with a multimeter we observe that this current increases while the LED heats up. Operating an illumination LED with constant voltage, therefore, creates a runaway condition. The losses heat the LED, which make the LED consume more power, which increases the LED's temperature, which in turn increases the power, and so on until excessive temperature destroys the LED. Keeping the current through the LED constant, however, counteracts any increase in temperature by decreasing the power the LED consumes when it heats up. For closing the control loop we next have to set up the analog-to-digital converter and the timer in order to sample the current signal provided by the power stage periodically. The time between analog-to-digital conversions should be somewhat shorter than the time for a pulse width modulation cycle. Having the measurement times wandering with respect to the pulse width modulation cycles eliminates the influence the current ripple has on the measurements. Once we have verified that sampling the current works we can proceed with writing the interrupt service routine, which handles the result of every single conversion. This interrupt service routine contains a PI controller. Fixed point arithmetic suffices for writing the controller, moreover it introduces automatic anti-windup for the controller's integral term, see Sect. 2.10. The LED is relatively simple to control. The controller can be tuned empirically until no flicker can be seen in response to a step of the supply voltage. Protecting the LED from excessive heat using the temperature sensor and evaluating the potentiometer's position finishes the program.

The P-channel MOSFET Q1, the Schottky diode D1, the inductor L1, and the capacitors C1 and C3 in Fig. 7.21 form the step-down converter. The gate driver U1 level-shifts the pulse width modulation signal and drives the gate of Q1. The current sense amplifier U2 amplifies the current signal provided by the shunt R2. In order to faithfully reproduce the LED's temperature the sensor U3 has to be in close thermal contact with the LED D2.

7.4 Bibliographic Notes

Robert Widlar, who laid the groundwork for the design of analog integrated circuits as we know it today, in (Widlar 1971) introduced the first viable integrated linear regulator. David R. Middlebrook and Slobodan Ćuk introduced the Ćuk converter in (Middlebrook and Ćuk 1977). The book (Mohan et al. 2003) is a rich resource when it comes to power electronics applications. The books (Dobkin and Williams 2011, 2012) are a collection of application notes written by the technical staff of Linear Technology. Among these are a number of excellent essays on power supply circuits by the late Jim Williams. The book (Kester and Analog Devices 1998) contains a good introductory chapter to switchmode converters. Many semiconductor manufacturers have published application notes on power supply design. These are available on the manufacturers' web sites.

7.5 Exercises

Exercise 7.1 Derive inequality (7.5). Use the inductor law (3.7).

Exercise 7.2 Derive inequality (7.6). Use the capacitor law (3.6).

Exercise 7.3 Derive inequality (7.8). Use the inductor law (3.7).

Exercise 7.4 Derive inequality (7.9). Use the capacitor law (3.6).

Exercise 7.5 Derive inequality (7.12). Use the inductor law (3.7).

Exercise 7.6 Derive inequality (7.13). Use the capacitor law (3.6).

Exercise 7.7 Derive inequality (7.14). Use the capacitor law (3.6).

Exercise 7.8 Derive inequality (7.16). Use the inductor law (3.7).

Exercise 7.9 Derive inequality (7.17). Use the capacitor law (3.6).

Exercise 7.10 Derive inequality (7.20). Use the inductor law (3.7).

Exercise 7.11 Derive inequality (7.21). Use the capacitor law (3.6).

Exercise 7.12 Derive inequality (7.22). Use the capacitor law (3.6).

Exercise 7.13 Derive inequality (7.24). Use the inductor law (3.7).

Exercise 7.14 Derive inequality (7.25). Use the capacitor law (3.6).

Exercise 7.15 The circuit in Fig. 7.22 can be used to step up from left to right or to step up from right to left or to step down from left to right or to step down from right to left. Find the switching pattern for each of the four cases.

Fig. 7.22 Circuit for stepping up or down in both directions

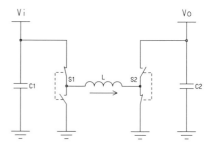

7.6 Lab Exercises

Exercise 7.16 Implement the control algorithms for the LED supply. Tune your controller's response to line-side disturbances.

Exercise 7.17 Design, implement and evaluate a control law for controlling the LED's temperature.

References

Dobkin B, Williams J (2011) Analog circuit design: a tutorial guide to applications and solutions. Elsevier, USA

Dobkin B, Williams J (2012) Analog circuit design: immersion in the black art of analog design. Elsevier, USA

Kester W, Analog Devices (1998) Practical design techniques for power and thermal management. Analog Devices, USA

Middlebrook RD, Ćuk S (1977) A general unified approach to modeling switching-converter power stages. Int J Electron 42(6):521–550. doi:10.1080/00207217708900678

Mohan N, Undeland T, Robbins W (2003) Power electronics: converters applications and design. Wiley, Australia

Reddy T (2010) Linden's Handbook of Batteries, 4th edn. Mcgraw-hill, New York

Widlar R (1971) New developments in IC voltage regulators. IEEE J. Solid-State Circ 6(1):2–7. doi:10.1109/JSSC.1971.1050151

Chapter 8
Energy Conversion—Motor Control

Electrical drives are ubiquitous. In a modern car, for example, more than 50 electrical motors, from the starter to the actuators in power windows, provide auxiliary functions. Most of these are permanent magnet DC motors. While these motors are very cheap they suffer from limited reliability by their very operational principle. The commutator and the brushes necessary for switching between the motor's windings at precise moments in order to keep the motor running are bound to wear out until the motor fails. Modern power electronics paired with fast microcontrollers allow us to replace these motors with ones having no commutators and brushes, the permanent magnet synchronous machines. Doing so allows us to operate these auxiliary drives at exactly the power needed at any point in time at no additional cost, thereby reducing fuel consumption. For electric cars the permanent magnet synchronous motor is also one of the candidates for providing propulsion, see, e.g., Braess and Seiffert (2005).

Electric machines are studied very well. This allows us to derive a mathematical model of such a motor from first principles. By subjecting this model to different operational strategies in a simulation environment we can learn about the behavior of these machines before we start designing hard- and software in detail.

8.1 Magnetism

Before looking at electric motors we state some facts about electromagnetism and ferromagnetic materials without discussing these in detail. A magnetic field $\vec{B}$ exerts a force on the individual moving charges in a current carrying conductor, the Lorentz force (3.2). By considering the product of charge and velocity necessary to maintain a current I through a straight homogeneous wire with direction and length $\vec{l}$ and assuming that the field $\vec{B}$ is constant along the wire we can compute the Lorentz force $\vec{F}$ acting on the center of gravity of the wire,

$$\vec{F} = I\vec{l} \times \vec{B}. \tag{8.1}$$

© Springer International Publishing Switzerland 2015
P. Hintenaus, *Engineering Embedded Systems*, DOI 10.1007/978-3-319-10680-9_8

The law of Biot–Savart (3.1) allows us to compute the magnetic field $\vec{B}$ at any point in space. It requires us, however, to know every single current. This sounds doable naively considered, but the properties of matter complicate the situation to such an extent, that the task becomes insurmountable. The movement and the spin of every single electron, for example, forms a current along a microscopic closed path. These currents within matter are called bound in contrast to the free currents in our circuits. Besides the free currents, which we know, we have to consider every single bound current when computing the total magnetic field. We may hope that these currents are not coordinated much at all, and therefore almost cancel. This is true for almost all materials. The notable exception are the ferromagnetic materials.

In ferromagnetic materials so-called magnetic domains form spontaneously. In such a magnetic domain the axes around which the microscopic currents circulate all have the same direction. Inside such a domain the microscopic currents cancel but on the surface of the domain they form a current that flows around the whole domain. Thermal movements of the atoms within the material disturb the order in the magnetic domains. Above a certain temperature, the material's Curie temperature, these movements destroy the order and the material looses its ferromagnetic properties.

In order to manage this complicated situation one introduces another vector field $\vec{H}$ which is created by the free currents only. The unit of $\vec{H}$ is $[\mathrm{Am}^{-1}]$. The terminology has changed over the years. For avoiding confusion we call the field $\vec{B}$ that effects the Lorentz force (3.2) on moving charges the magnetic field or B-field and the field $\vec{H}$ the H-field. In vacuum the B-field and H-field are related by

$$\vec{B} = \mu_0 \vec{H}, \tag{8.2}$$

where $\mu_0 = 4\pi \times 10^{-7} \mathrm{VsA}^{-1}\mathrm{m}^{-1}$ is the magnetic constant. A current I through a long cylindrical coil which has n turns per meter of its length produces an H-field within the coil of magnitude

$$H = nI.$$

The direction of the H-field within the coil is parallel to the coil's axis and obeys the right-hand rule. When the thumb of the right hand points in the direction of $\vec{H}$ then the fingers indicate the direction of the current. Related with this H-field there is alway a magnetic field $\vec{B}$. When this magnetic field interacts with a piece of ferromagnetic matter some magnetic domains within the piece will align with the external field amplifying it. With increasing strength of the field $\vec{B}$ the aligned domains grow and more domains align themselves, the piece becomes magnetized. Once all domains are aligned the ferromagnetic piece does not amplify further increases of the strength of both the field $\vec{B}$ and of the related H-field; then the piece is called saturated. More precisely, once the piece is saturated, the relationship between a further increase of the H-field and the increase of the B-field follows (8.2). The end of a magnetized piece of matter in the direction of the piece's magnetic field is called the north pole of the piece, the one in the direction opposing the field the south pole. Magnetic poles always come in north-south pairs, a single pole has never been observed.

8.1.1 Magnetically Soft Materials

For some ferromagnetic materials little work is needed for aligning the magnetic domains. These materials are called magnetically soft. As long as a magnetically soft material is not close to being saturated the magnitude of the B-field, the superposition of the external field, and the field created by the bound currents, is approximately proportional to the strength of the external field. The constant of proportionality μ_r is called the permeability of the material. For common magnetically soft materials $\mu_r \gg 1,000$. The relationship between the magnitude of an H-field generated by a current flowing through a coil in which a piece of ferromagnetic matter has been placed and the magnitude of the resulting B-field is described in magnetization diagrams, see Fig. 8.1. Magnetically soft materials have a diagram like that on the left. When an H-field is applied to an unmagnetized piece of magnetically soft material and later removed, the material stays slightly magnetized. When the H-field later is applied in the reverse direction the piece will become magnetized in the reversed direction. In this process it dissipates magnetic energy and turns it into heat. The area in the magnetization diagram is the work it takes to change the magnetization of $1\,m^3$ from being fully saturated in the one direction to being fully saturated in the opposite direction and back.

8.1.2 Magnetic Circuits

When a magnetic field $\vec{B}$ meets the surface of a piece of unsaturated soft magnetic matter at an angle the piece changes the field's direction to being practically parallel to the surface of the piece within the piece. Whenever a magnetic field exits from the piece into free space, the direction of the field in free space, close to the surface of the piece, is quite perpendicular to the surface. These properties allow us to steer magnetic

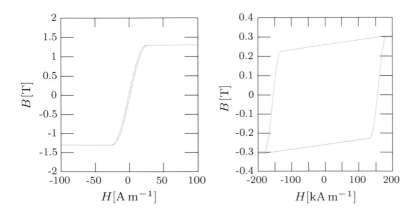

Fig. 8.1 Sketches of magnetization diagrams, for a soft material to the *left*, for a hard one to the *right*

Fig. 8.2 A magnetic circuit
with a winding and an air gap
of length d. The magnetic
flux through the plane surface
S_1, cutting through the core
completely, equals the
magnetic flux though the
surface S_2 cutting completely
through the air gap

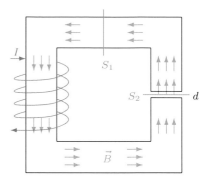

fields in so-called magnetic circuits using pieces of magnetically soft materials as
core. In a magnetic circuit the field $\vec{B}$ stays within the core as long as possible. It
bridges small air gaps without spreading much. As the magnetic field has neither
sources nor sinks, the magnetic flux Φ_S through any cross section S of a shank of the
core must equal the flux Φ_G through the air gap, see Fig. 8.2. In this situation the flux
through each turn of the winding is Φ_G, therefore the flux Ψ through the winding with
N turns wound tightly around the core is $N\Phi_G$. In more complicated windings we
have to consider the flux through each turn of the winding separately and add all these
fluxes. When the length of the air gap is large enough it dominates the inductance of
the winding in Fig. 8.2 over the influence of the core. The inductance L then is

$$L = \frac{\mu_0 N^2 A}{d},$$

where A is the area of the cross section of the winding.

When a nonstationary magnetic field interacts with a piece of electrically conducting matter Faraday's law comes into play. In each conducting loop within the piece,
and the piece is full of these, a voltage is induced. These loops form short circuits,
therefore currents, the so-called eddy currents, flow. The piece dissipates magnetic
power and converts it into heat. Many magnetic materials are electric conductors. In
order to minimize eddy current losses cores built from such a material are laminated,
that is, the cores are constructed from thin sheets of this material, which are insulated
from each other.

8.1.3 Magnetically Hard Materials

Some ferromagnetic materials require a considerable amount of work to become
magnetized. These materials retain their magnetization once the external field is
removed. They are called magnetically hard and lend themselves to use as permanent magnets, see Fig. 8.1. An unmagnetized piece of magnetically hard material,
when subjected to a strong H-field becomes magnetized, but it retains most of its

magnetization when the H-field is removed. Even a rather intense H-field in the opposite direction does not change the magnetization of the piece. This is the typical area for a permanent magnet to work in. Only when the intensity of the opposing H-field exceeds a certain, temperature dependent value, magnetic domains in the material start to realign themselves, the magnet becomes (partially) demagnetized and, therefore, unfit for its intended use. The dependence of the behavior of a piece of magnetic material on its magnetic history is called hysteresis.

8.2 The Permanent Magnet Synchronous Machine

In an interior rotor permanent magnet synchronous machine the stator has a cylindrical bore, in which the (approximately) cylindrical rotor is allowed to rotate. The air gap separates the stator and the rotor, Fig. 8.3. The windings sit in slots, which are cut outwards from the stator's bore into the material of the stator's core. Mounting the windings in the slots protects the delicate insulation of the windings' individual wires from mechanical damage and provides good thermal contact for removing the waste heat generated in the windings. Appropriate currents through the stator's windings generate a magnetic field in the air gap, which is directed radially and rotates around the axis of the bore. In order to amplify and shape the magnetic field generated by the windings the stator's core is made of magnetically soft steel. As the stator's core is subjected to a rotating magnetic field, it is laminated from insulated sheets, in order to minimize eddy current losses. The stator carries at least two windings, the

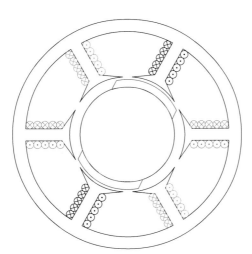

Fig. 8.3 Cut through an interior rotor permanent magnet synchronous machine. The rotor in the *middle* carries the magnets, the stator on the outside the windings. The windings for phase *a* are drawn in *brown*, the ones for phase *b* in *black*, and the ones for the phase *c* in *gray*. A wire decorated with a dot, ⊙, emerges from the drawing, while one decorated with a cross, ⊗, dips into the drawing. The north poles of the permanent magnets are drawn in *red*, the south poles in *blue*

phase windings. Three phase machines with three windings are the most common. The cylindrical rotor core consists of steel. The rotor core carries an even number, $2Z_p$, of permanent magnets, whose orientation alternates between south pole toward the core and north pole toward the core. The number Z_p is called the number of the machine's pole-pairs, it is one in Fig. 8.3. The interaction of the rotor's magnetic field with the currents in the stator's winding produces torque at the rotor's shaft. For this torque to be useful the stator's field and the rotor must turn in sync. For a machine with one pole-pair the angle ϑ between the direction of the two fields must satisfy $-\pi < \vartheta < \pi$. When the machine is operated under these constraints the magnetic field in the rotor changes only when the angular velocity or the applied torque changes. Therefore, the rotor's core need not be laminated. The machines we consider in this chapter have their magnets glued to the surface of the rotor core.

The rotor's permanent magnets cling to the teeth formed in the stator's core between two subsequent slots. Therefore, the machine has preferred positions when not powered. This effect is called cogging. Cogging introduces a small torque ripple, which can be minimized by constructive means.

In the other common form, the exterior rotor permanent magnet synchronous machine, the stator core has cylindrical shape. The slots for the windings are cut from the perimeter of the cylinder toward the center. The stator again exhibits a laminated construction. The rotor, made from steel, is a cylindrical shell rotating around the stator. It carries the machine's permanent magnets on its inside, see Fig. 8.4 for a three phase machine with 7 pole-pairs.

8.2.1 Sine-Distributed Windings

In the following, we consider a hypothetical permanent magnet machine with a cylindrical rotor core of length l and a stator with a cylindrical bore of the same length.

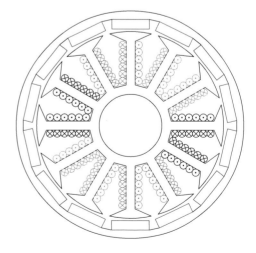

Fig. 8.4 Cut through an exterior rotor permanent magnet synchronous machine. The stator in the *middle* carries the windings, the rotor on the outside the magnets

The stator shall have no slots for the windings. Two permanent magnets sit at the cylindrical surface of the rotor and form a single pole-pair. The stator and the rotor are separated by the air gap with effective radius r and effective width g in the radial direction. We analyze the field and the magnetic flux of as single phase winding of a specific shape. Every phase winding under consideration consists of conductors, which run on the cylindrical inside of the stator bore, parallel to the bores axis, and of connecting conductors, which run on the front and back of the stator. The first group of conductors, the one in the bore, participates in the production of the machine's torque, the second, the end turns, do not. In Figs. 8.5 and 8.6 we show cuts through the machine, which are perpendicular to the stator's bore. The torque-producing conductors show as circles, the end turns do not show at all. A conductor decorated with a $\odot$ carries a current which emerges from the drawing, while one decorated with a

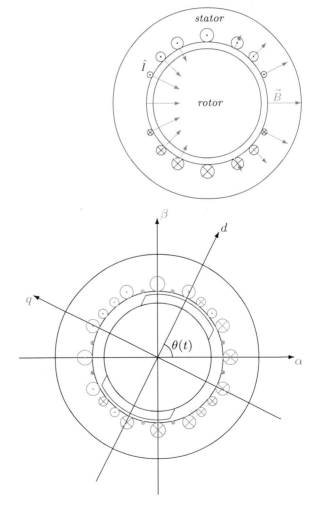

Fig. 8.5 A stator-winding with sinusoidal conductor distribution. A current $\hat{I}$ flowing through such a winding produces the magnetic field $\vec{B}$ in the air gap. The permanent magnets are omitted

Fig. 8.6 Equivalent two phase machine. A current through the winding α produces a *horizontal* field, a current through the winding β a *vertical* one. The axis of the rotor coordinate system in the direction of the rotor magnetic field is called the direct axis d, the one perpendicular to the field is called the quadrature axis q

$\otimes$ carries a current which dips into the drawing. The end turns connect the torque producing conductors in a series connection. Any such winding, consisting of N turns, shall have a sinusoidal spatial conductor density $C(\gamma)$,

$$C(\gamma) = \frac{N}{2} \sin \gamma.$$

A current $\hat{I}$ flowing through this winding has the spatial distribution

$$I(\gamma) = \hat{I} C(\gamma) = \frac{N\hat{I}}{2} \sin \gamma.$$

Assuming that the magnetically soft iron in the rotor and the stator is linear and therefore, does not saturate, this current distribution produces a magnetic field $\vec{B}$ in the air gap. The field $\vec{B}$ has radial direction; its and magnitude is sinusoidally distributed,

$$B(\gamma) = \hat{B} \cos \gamma,$$

with $\hat{B} = \frac{\mu_0 \hat{I} N}{2g}$ and $\mu_0 = 4\pi \times 10^{-7} \mathrm{VsA^{-1}m^{-1}}$ being the magnetic constant, see Fig. 8.5. Note that the spatial amplitude $\hat{B}$ is proportional to the current $\hat{I}$. For deriving this result one has to argue about the interaction between the magnetic field of the current distribution $I(\gamma)$ and the magnetic material in the rotor and the core. This is beyond our scope. We define the overall direction of $\vec{B}$ to be the direction of the average field vector of $\vec{B}$ in the air gap.

Next, we study a winding consisting of N turns with sinusoidal conductor density and a sinusoidally distributed magnetic field $\vec{B}$ where the axis of the winding and the overall direction of the field enclose an angle δ. The magnetic flux $\Phi(\gamma, \delta)$ of $\vec{B}$ through a single turn of the winding, whose one torque producing conductor sits at an angle γ and the other at $-\gamma$, is

$$\Phi(\gamma, \delta) = \int_{-\gamma}^{\gamma} \hat{B} r l \cos(\rho + \delta) \, d\rho$$

$$= 2\hat{B} r l \sin \gamma \cos \delta.$$

The magnetic flux $\Psi(\delta)$ of $\vec{B}$ through the whole winding then is

$$\Psi(\delta) = \int_{0}^{\pi} \frac{N}{2} \Phi(\rho, \delta) \sin \rho \, d\rho$$

$$= N\hat{B} r l \int_{0}^{\pi} \cos \delta \sin^2 \rho \, d\rho$$

$$= \frac{\pi}{2} N \hat{B} r l \cos \delta. \tag{8.3}$$

Substituting 0 for δ in (8.3) we get the flux Ψ when both the field $\vec{B}$ and the axis of the winding are aligned.

$$\Psi = \frac{\pi}{2} N \hat{B} r l. \tag{8.4}$$

When the direction of the field and the axis of the winding are perpendicular the flux of $\vec{B}$ through the winding is zero. A magnetic field produced in one winding does not create a magnetic flux in another winding as long as the axes of the two windings are perpendicular. Moreover, the analysis of the superposition of the magnetic fields produced in any number of stator windings can be reduced to the analysis of the superposition of the fields of just two windings whose axes are perpendicular.

8.2.2 Space Vectors

We start out with a machine with three stator windings a, b, and c. Each of these windings has N turns and a sinusoidally distributed conductor density. The axis of winding a is horizontal, the axis of b encloses an angle of $2\pi/3$ with the horizontal, and the axis of c encloses an angle of $4\pi/3$ with the horizontal. We operate the machine with balanced voltages V_a, V_b, and V_c across the windings a, b, and c, i.e. $V_a + V_b + V_c = 0$, and, because of symmetry, balanced currents I_a, I_b, and I_c through the windings a, b, and c, i.e. $I_a + I_b + I_c = 0$. In order to facilitate computation of rotations around the axis of the machine we represent a vector not in its Cartesian form, (x, y), but as the complex number $x + iy$. A rotation by an angle ϕ in the counterclockwise direction then amounts to the complex multiplication $(x + iy)e^{i\phi}$.

We consider currents $I_a(t)$, $I_b(t)$, and $I_c(t)$ through the three windings such that the resulting magnetic fields have the magnitudes $\hat{B}_a(t) = \cos \omega t$, $\hat{B}_b(t) = \cos(\omega t - 2\pi/3)$, and $\hat{B}_c(t) = \cos(\omega t - 4\pi/3)$ and are sinusoidally distributed in the air gap. The superposition of these three fields is a rotating magnetic field in the air gap whose direction and magnitude is represented by $\tilde{\mathbf{B}}(t)$:

$$\begin{aligned}
\tilde{\mathbf{B}}(t) &= \hat{B}_a(t) + \hat{B}_b(t)e^{i2\pi/3} + \hat{B}_c(t)e^{i4\pi/3} \\
&= \cos \omega t + \cos(\omega t - 2\pi/3)\, e^{i2\pi/3} + \cos(\omega t - 4\pi/3)\, e^{i4\pi/3} \\
&= \frac{3}{2} (\cos \omega t + i \sin \omega t) \\
&= \frac{3}{2} e^{i\omega t}.
\end{aligned}$$

Two obstacles keep us from creating a model for the three phase machine directly: first the redundancy inherent in the balanced three phase system and second the magnetic flux created in two windings by the magnetic field generated in the third.

Instead we imagine a hypothetical two phase machine with windings α and β whose axes are perpendicular, see, Fig. 8.6. We choose the real axis of the stator coordinate system to coincide with the axis of the winding α and the imaginary axis with the axis of the winding β. We transform the three voltages V_a, V_b, and V_c into two voltages V_α across the hypothetical winding α and V_β across β and combine these into one complex quantity, the voltage space vector $\tilde{\mathbf{V}}$,

$$
\begin{aligned}
\tilde{\mathbf{V}} &= V_\alpha + \mathrm{i}V_\beta \\
&= \sqrt{\frac{2}{3}}\left(V_a + \mathrm{e}^{\mathrm{i}2\pi/3}V_b + \mathrm{e}^{\mathrm{i}4\pi/3}V_c\right) \\
&= \sqrt{\frac{2}{3}}\left(V_a - \frac{V_b}{2} - \frac{V_c}{2} + \mathrm{i}\sqrt{3}(V_b - V_c)\right).
\end{aligned}
\tag{8.5}
$$

This transformation is called the unitary Clarke[1] transformation. Throughout this chapter we indicate space vectors in bold face. Moreover, space vectors in the stator coordinate system are marked with a tilde while the ones in rotor coordinates come unmarked. By the same token we compute the current space vector $\tilde{\mathbf{I}}$ by

$$
\begin{aligned}
\tilde{\mathbf{I}} &= I_\alpha + \mathrm{i}I_\beta \\
&= \sqrt{\frac{2}{3}}\left(I_a - \frac{I_b}{2} - \frac{I_c}{2} + \mathrm{i}\sqrt{3}(I_b - I_c)\right).
\end{aligned}
\tag{8.6}
$$

The factor $\sqrt{\frac{2}{3}}$ in both equations ensures that the three phase machine and the equivalent two phase machine convert the same power. The transform of a current space vector back into balanced three phase currents is given by

$$
\begin{pmatrix} I_a \\ I_b \\ I_c \end{pmatrix} = \sqrt{\frac{2}{3}}\begin{pmatrix} 1 & 0 \\ -\frac{1}{2} & \frac{\sqrt{3}}{2} \\ -\frac{1}{2} & -\frac{\sqrt{3}}{2} \end{pmatrix}\begin{pmatrix} I_\alpha \\ I_\beta \end{pmatrix}.
\tag{8.7}
$$

A similar transform translates the voltages V_α and V_β into the phase voltages V_a, V_b, and V_c.

The flux space vector $\tilde{\mathbf{\Psi}}$ is composed of the magnetic flux Ψ_α through the virtual winding α and the flux Ψ_β through the virtual winding β,

$$
\tilde{\mathbf{\Psi}} = \Psi_\alpha + \mathrm{i}\Psi_\beta.
$$

---

[1] Named after Edith Clarke, the first female University professor in Electrical Engineering in the United States.

8.2.3 A Model of the Permanent Magnet Synchronous Machine

The voltage across each winding in Fig. 8.6 equals the sum of the resistive drop across
the winding and the voltage induced in the winding by the change of the magnetic
flux through it, in space vector form:

$$\tilde{\mathbf{V}}(t) = R\tilde{\mathbf{I}}(t) + \frac{d\tilde{\mathbf{\Psi}}(t)}{dt}. \tag{8.8}$$

The space vector for the magnetic flux $\tilde{\mathbf{\Psi}}(t)$ depends on the space vector for the
current $\tilde{\mathbf{I}}(t)$ and on the space vector describing the magnetic flux of the permanent
magnets in the rotor. Both these space vectors rotate in α-β coordinates, but the latter
is stationary in d-q coordinates. We transform (8.8) by substituting $\tilde{\mathbf{V}}(t) = e^{i\theta(t)}\mathbf{V}(t)$,
$\tilde{\mathbf{\Psi}}(t) = e^{i\theta(t)}\mathbf{\Psi}(t)$ and $\tilde{\mathbf{I}}(t) = e^{i\theta(t)}\mathbf{I}(t)$, where $\theta(t)$ is the angular position of the
rotor with respect to the stator and $\omega(t) = \frac{d\theta(t)}{dt}$ is the angular velocity of the rotor
both at time t,

$$e^{i\theta(t)}\mathbf{V}(t) = Re^{i\theta(t)}\mathbf{I}(t) + \frac{d}{dt}\left(e^{i\theta(t)}\mathbf{\Psi}(t)\right)$$

$$= e^{i\theta(t)}\left(R\mathbf{I}(t) + \frac{d\mathbf{\Psi}(t)}{dt} + i\omega(t)\mathbf{\Psi}(t)\right),$$

to obtain the voltage equation in d-q coordinates,

$$\mathbf{V}(t) = R\mathbf{I}(t) + \frac{d\mathbf{\Psi}(t)}{dt} + i\omega(t)\mathbf{\Psi}(t). \tag{8.9}$$

The transform from the stator system into the rotor system is called the Park
transform.[2]

The permeability μ_r of the permanent magnets is close to one, therefore the air
gap is practically uniform, and the inductances of the two windings in Fig. 8.6 do
not depend on the position of the rotor[3] and are both L. Therefore, the component
of the space vector $\mathbf{\Psi}(t)$ for the magnetic flux in the d direction is

$$\Psi_d(t) = LI_d(t) + \Psi_P, \tag{8.10}$$

where Ψ_P is the magnetic flux produced by the rotor's permanent magnets, and the
component in the q direction is

$$\Psi_q(t) = LI_q(t). \tag{8.11}$$

---

[2] Named after Robert H. Park, who first formulated a model of a synchronous machine in rotor
coordinates.

[3] The windings of permanent magnet machines which have a rotor with a core that is not cylindrical
exhibit inductances which depend on the position of the rotor.

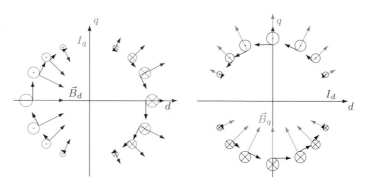

Fig. 8.7 Lorentz forces, in *black*, acting on the stator due to the magnetic field $\vec{B}_d$ and the current distribution I_q on the *left*, due to the magnetic field $\vec{B}_q$ and the current distribution I_d on the *right*. Note that the forces in the *left* diagram run clockwise, while in the *right* diagram the forces run counterclockwise

The magnetic field which creates the flux $\boldsymbol{\Psi}(t)$ is the superposition of the two fields $\vec{B}_d(t)$ and $\vec{B}_q(t)$. The field $\vec{B}_d(t)$ has the magnitude distribution $\hat{B}_d(\gamma, t) = B_d(t)\cos\gamma$, the field $\vec{B}_q(t)$ has the magnitude distribution $\hat{B}_q(\gamma, t) = B_q(t)\sin\gamma$. Both fields are directed radially at any point in the air gap. For computing the torque $T(t)$ produced by the machine we consider the interaction between these magnetic fields and the current distribution related to the space vector $\mathbf{I}(t)$. Considering the Lorentz forces (8.1) and Fig. 8.7, the magnitude of the torque is

$$
\begin{aligned}
T(t) &= \int_0^{2\pi} rl\hat{I}_q(\gamma, t)\hat{B}_d(\gamma, t)\,\mathrm{d}\gamma - \int_0^{2\pi} rl\hat{I}_d(\gamma, t)\hat{B}_q(\gamma, t)\,\mathrm{d}\gamma \\
&= rl\int_0^{2\pi} \frac{N}{2}I_q(t)\cos\gamma\, B_d(t)\cos\gamma\,\mathrm{d}\gamma - rl\int_0^{2\pi} \frac{N}{2}I_d(t)\sin\gamma\, B_q(t)\sin\gamma\,\mathrm{d}\gamma \\
&= \frac{r\pi l B_d(t)}{2}I_q(t) - \frac{r\pi l B_q(t)}{2}I_d(t).
\end{aligned}
$$

The minus sign in the first equation accounts for the forces running counterclockwise and clockwise, respectively. For computing the torque acting on the rotor instead of the stator the direction of the Lorentz forces have to be reversed. By using (8.4) twice and by substituting for the fluxes $\Psi_d(t)$ and $\Psi_q(t)$ we can simplify further,

$$
\begin{aligned}
T(t) &= \Psi_d(t)I_q(t) - \Psi_q(t)I_d(t) \\
&= \Psi_{\mathrm{P}}I_q(t).
\end{aligned} \tag{8.12}
$$

In order to check whether our model is plausible we consider the machine operating at a constant velocity ω,

$$V_d = RI_d - \omega\Psi_q$$

$$V_q = RI_q + \omega\Psi_d,$$

and delivering constant torque T. We compute both the electrical power P_e,

$$P_e = V_d I_d + V_q I_q$$
$$= (RI_d - \omega\Psi_q)I_d + (RI_q + \omega\Psi_d)I_q$$
$$= RI_d^2 + RI_q^2 + \omega\Psi_P I_q,$$

and the mechanical power P_m,

$$P_m = \omega T = \omega\Psi_P I_q.$$

But for the resistive losses included in the electrical power, P_e equals the mechanical power P_m, therefore energy is preserved.

When compared to a machine with just one pole-pair operating at an angular velocity of $\omega(t)$ and producing a torque $T(t)$ a machine with Z_p pole-pairs but with identical flux of the permanent magnets operates at an angular velocity of $\frac{\omega(t)}{Z_p}$ producing a torque of $Z_p T(t)$. To model this situation we introduce the electrical angular velocity $\omega_e(t)$, the angular velocity of the equivalent machine with a single pole-pair, and the electrical torque $T_e(t)$, the torque produced by the equivalent machine,

$$\frac{d\Psi_d(t)}{dt} = V_d(t) - RI_d(t) + \omega_e(t)\Psi_q(t) \tag{8.13}$$

$$\frac{d\Psi_q(t)}{dt} = V_q(t) - RI_q(t) - \omega_e(t)\Psi_d(t) \tag{8.14}$$

$$\Psi_d(t) = \Psi_P + LI_d(t) \tag{8.15}$$

$$\Psi_q(t) = LI_q(t) \tag{8.16}$$

$$T_e(t) = \Psi_P I_q(t). \tag{8.17}$$

The mechanical angular velocity $\omega_m(t)$ and thus the electrical angular velocity depend on the mechanical torque $T_m(t)$ produced by the machine, on the torque required for accelerating or decelerating the rotor and on the external load,

$$T_m(t) = Z_p T_e(t) \tag{8.18}$$

$$I_R \frac{d\omega_m(t)}{dt} = T_m(t) - T_L(t) \tag{8.19}$$

$$\omega_e(t) = Z_p\omega_m(t), \tag{8.20}$$

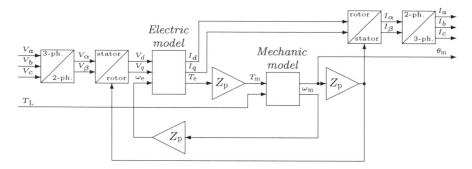

Fig. 8.8 Model of a three phase machine. The electric model of the equivalent two phase machine consists of (8.13)–(8.17), the mechanic model of (8.19)

where I_R is the moment of inertia of the rotor and $T_L(t)$ is the torque required by the external load. Equations (8.13)–(8.20) constitute the classical d-q model of a permanent magnet synchronous machine. In order to turn the d-q model into the model of a three phase machine we have to augment it with the necessary coordinate transforms, Fig. 8.8.

For analyzing (8.17) we pick a current space vector **I** of constant magnitude. As long as **I** points in the positive d direction the machine does not produce any torque. When I_q is positive the machine produces torque, it acts as motor. For a given magnitude of **I** the torque the machine produces is maximal for $I_d = 0$. For I_q negative the machine consumes torque, it converts mechanical power into electrical power, it operates as generator. Problems arise when I_q changes sign while $I_d \leq 0$. Let us assume the machine operates as a generator being driven by an external torque. If this torque is large enough to accelerate the rotor with respect to the stator field against the torque consumed by the machine, I_q will become less and less negative. Eventually I_q will become positive and now the torque produced by the machine adds to the external torque. In this very moment the machine looses synchronization and the rotor spins out of control. The machine then switches rapidly between acting as motor and acting as generator and the torque produced by the machine will show large and rapid oscillations, both into the positive and the negative. Similar things happen when an external torque decelerates the machine so hard while the machine is acting as motor that I_q becomes negative. The rotor will coast to a stop while being subjected to oscillatory torque. If the external moment persists the rotor will even start to turn against the stator field.

When the machine operates with $I_d(t) < 0$ the field generated by $I_d(t)$ opposes the field of the permanent magnets. This mode, called flux weakening, allows to operate the machine at high angular velocities without exceeding the machine's rated power. The large effective air gap in the machines we consider, however, effects small winding inductances, which diminishes the effectivity of flux weakening (El-Refaie et al. 2006).

8.3 Open Loop Operation

In the following, we simulate a permanent magnet synchronous motor with a rated power of 70 W driving a small centrifugal fan.[4] The torque $T(\omega)$ such a fan requires is proportional to the square of its angular velocity ω. After having translated the differential equations, which constitute the model, into Simulink® block diagrams we are ready to evaluate strategies for operating a permanent magnet synchronous machine. For an application like this, which requires little dynamics, we might get away with considering the steady state only.

In the steady state, where the voltages V_d, V_q, the currents I_d, I_q, and the torque T_L required by the load depend on the angular velocity ω_e but not on the time t, the two voltage equations (8.13) and (8.14) reduce to

$$V_d(\omega_e) = RI_d(\omega_e) - \omega_e L I_q(\omega_e)$$
$$V_q(\omega_e) = RI_q(\omega_e) + \omega_e(\Psi_P + LI_d(\omega_e)),$$

and the torque equation (8.19) reduces to

$$T_L(\omega_e) = Z_p \Psi_P I_q(\omega_e).$$

In order to minimize resistive losses we want to operate the machine with $I_d(\omega_e) = 0$. Then the voltage equations reduce to

$$V_d(\omega_e) = -\omega_e L \frac{T_L(\omega_e)}{Z_p \Psi_P}$$
$$V_q(\omega_e) = R \frac{T_L(\omega_e)}{Z_p \Psi_P} + \omega_e \Psi_P.$$

The magnitude $A(\omega_e)$ of this space vector is

$$A(\omega_e) = \frac{1}{Z_p \Psi_P} \sqrt{\left((T_L(\omega_e)R + \omega_e Z_p \Psi_P^2)^2 + \omega_e^2 L^2 T_L(\omega_e)^2\right)},$$

therefore three phase voltages with minimal amplitude for the required angular velocity and the required torque are

$$\begin{pmatrix} V_a(\omega_e, t) \\ V_b(\omega_e, t) \\ V_c(\omega_e, t) \end{pmatrix} = \sqrt{\frac{2}{3}} A(\omega_e) \begin{pmatrix} \cos(\omega_e t) \\ \cos\left(\omega_e t - \frac{2\pi}{3}\right) \\ \cos\left(\omega_e t - \frac{4\pi}{3}\right) \end{pmatrix}. \tag{8.21}$$

---

[4] Also known as squirrel cage fan.

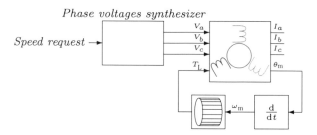

Fig. 8.9 Scheme for open loop operation. The synthesizer for the phase voltages reacts to a request for a change of speed by gently ramping up or down both the frequency and the amplitude of the three phase voltages

Our first attempt, Fig. 8.9, at operating our motor is to follow (8.21). We estimate the torque T_L consumed by the load. In order to cope with the temperature dependence of the winding resistance, with uncertainties when measuring the inductance of the windings and the magnetic flux of the permanent magnets, and to provide some headroom for accelerating, we increase the theoretical amplitude by 10 %. Furthermore, we add a small offset to the amplitude for facilitating startup from standstill. All we have to build is a synthesizer for converting a speed request into three phase voltages of appropriate frequency and amplitude. This synthesizer has to react gently to requests for large changes in speed so that it does not force the machine to deviate from the steady state too much.

In our simulation we let the motor accelerate from standstill to 9,000 revolutions per minute during the first 14 s and let it then run with constant speed. The small offset we added to the amplitude of the three phase voltages introduces a large relative amplitude error at standstill, Fig. 8.10. Therefore, the phase currents are relatively large while the rotor is not turning but they drop off once the machine is running, see Fig. 8.11. So far the behavior of the machine when starting from standstill is fine.

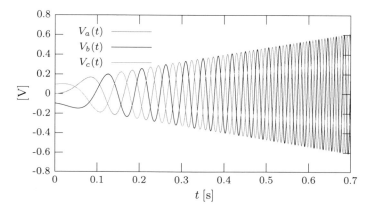

Fig. 8.10 Simulated phase voltages during startup. Starting with a nonzero amplitude from standstill helps to get the motor turning

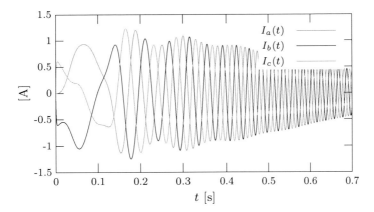

Fig. 8.11 Simulated phase currents during startup. After 0.2 s the rotor starts to turn and from then on stays synchronized with the stator field

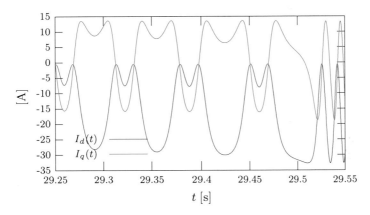

Fig. 8.12 Simulated currents in rotor coordinates. The rotor oscillates between lagging behind the stator field, while I_q is positive, and leading it, while I_q is negative. At 29.5 s the machine looses synchronization and coasts to a stop

Once the constant speed mode of operation has been reached, however, the machine does not behave steady at all. The currents I_d and I_q start to oscillate and, after some time, the rotor coasts to a stop despite the machine still being supplied with the appropriate voltages, see, Figs. 8.12 and 8.13. The machine oscillates rapidly between acting as a motor and acting as a generator, until it slips a turn and looses synchronization.

The angular velocity $\omega_e(t)$ of the rotor and the torque $T_e(t)$ produced by the machine oscillate rapidly too, see, Fig. 8.14. Once the machine has lost synchronization the interaction between the stator currents and the rotor field produces rapid torque excursions in both directions, but the average torque is close to zero.

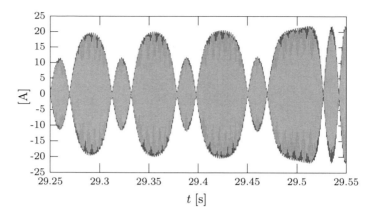

Fig. 8.13 The tragedy unfolds again, this time showing the three phase currents. Note that these current waveforms do not make much sense at all, while the same currents in rotor coordinates present a clear picture of the state of the machine

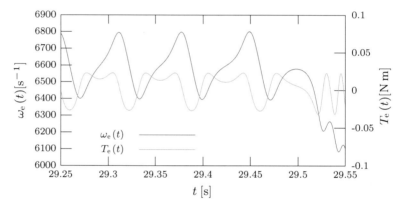

Fig. 8.14 A third view of the disaster, from a mechanical vantage point. Note how the angular velocity ω_e drops at 29.5 s instead of picking up again. The wild torque fluctuations render such a mode of operations impractical

Reconsidering the whole situation this oscillatory behavior does not come as a surprise. Let us assume that the current space vector $\tilde{\mathbf{I}}(t)$ in the stator system has constant magnitude and rotates with constant angular velocity ω_e. This assumption is definitely false when excursions are large, but for oscillations with small amplitude the errors are small. Moreover, this assumption allows for a qualitative analysis of the situation. Let $\vartheta(t)$ be the angle between the current space vector $\tilde{\mathbf{I}}(t)$ and the space vector of the rotor flux $\tilde{\mathbf{\Psi}}_P(t)$. Furthermore we assume that no load is present, that is $T_L = 0$. We can then rewrite the torque equation (8.19) into an equation of the form

$$\frac{\mathrm{d}^2 \vartheta(t)}{\mathrm{d}t^2} + C \sin \vartheta(t) = 0,$$

for some constant C. This equation describes the behavior of an undamped pendulum. A pendulum can oscillate, so can the rotor in a permanent magnet synchronous machine. A pendulum can also go over the top. When the rotor behaves this way the machine looses synchronization.

8.4 Torque Control

In order to prevent the oscillatory behavior we have to adopt a closed-loop control scheme. According to (8.17) the torque produced by the machine is proportional to the current I_q. This current however is a virtual one and therefore, cannot be measured directly anywhere. What we can measure however are the phase currents. When we assume that we know the rotor position θ_m, provided for example by an optical sensor attached to the rotor's shaft, we can convert the phase currents I_a, I_b, and I_c into the virtual currents I_d and I_q using the Clarke transform followed by the Park transform. We use two PI controllers, one manipulating the virtual voltage V_d in order to achieve $I_d = 0$, the other manipulating V_q in order to achieve the desired torque. By transforming these two virtual voltages back into the three phase system using an inverse Park transform and an inverse Clarke transform we arrive at the voltages for driving the phase windings, see Fig. 8.15. This scheme is called field oriented control or vector control.

In order to evaluate this control scheme we requested a torque step of 0.006 Nm. The simulated machine accelerates to an angular velocity of about $6{,}600\,\mathrm{s}^{-1}$ within about 3 s, Figs. 8.16 and 8.17. The oscillatory behavior of the machine is tamed perfectly.

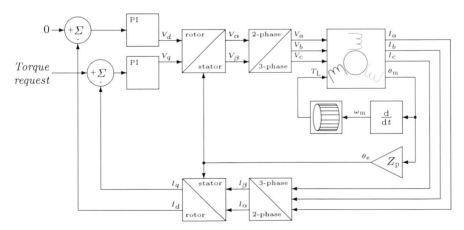

Fig. 8.15 Field-oriented torque control. The three phase currents I_a, I_b, and I_c are converted into the virtual currents I_d and I_q using the measured rotor position θ_m. Two PI controllers produce appropriate virtual voltages V_d and V_q. These voltages are transformed into the three phase voltages V_a, V_b, and V_c, in order to drive the machine's phase windings

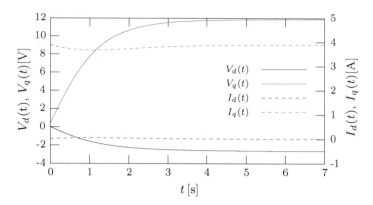

Fig. 8.16 Simulated virtual voltages V_d and V_q and virtual currents I_d and I_q for the field-oriented torque control in response to a requested step at time zero. No oscillations can be observed. Due to the limited torque available for speeding up the rotor and the fan it takes about 3 s to reach full speed

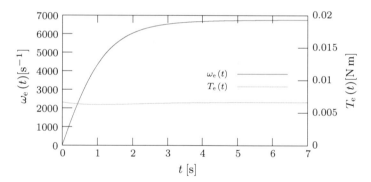

Fig. 8.17 Simulated angular velocity and torque for the field-oriented torque control in response to a requested step at time zero

8.5 Speed Control

By adding a PI controller which produces a torque request from a speed input we can convert the field-oriented torque control into a field-oriented speed control, Fig. 8.18.

We evaluate the control scheme with a speed step of $6{,}600\,\mathrm{s}^{-1}$, Figs. 8.19 and 8.20. Again no oscillations can be observed. In order to accelerate the rotor and the fan to the requested speed within one second the machine initially draws a current of about 14 A. Extending the speed control scheme to position control requires a sensor for

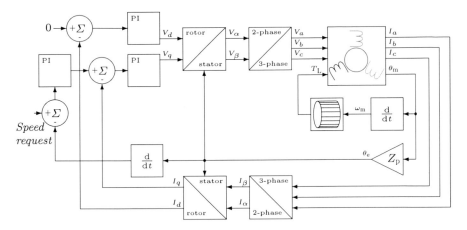

Fig. 8.18 Field-oriented speed control. An additional PI controller produces a torque request in response to a requested speed

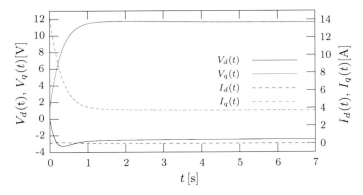

Fig. 8.19 Simulated virtual voltages V_d and V_q and virtual currents I_d and I_q for the field-oriented speed control in response to a speed step

the absolute position of the rotor's shaft and another controller, which produces a reference speed from the position error.

8.6 Sensorless Operation

The field-oriented control schemes require a means to sense the position of the rotor's shaft. This additional sensor adds cost and complexity. For applications which do not require precise knowledge of the rotor position one can estimate it from the voltages V_α and V_β and from the currents I_α and I_β alone. We have to compute these parameters for the field-oriented control anyway.

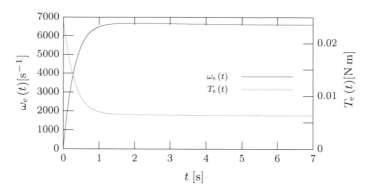

Fig. 8.20 Simulated angular speed and torque for the field-oriented speed control in response to a speed step

Using $\tilde{\boldsymbol{\Psi}}(t) = e^{i\theta(t)}\boldsymbol{\Psi}(t)$ and $\tilde{\mathbf{I}}(t) = e^{i\theta(t)}\mathbf{I}(t)$ and using (8.10) and (8.11) we can rewrite (8.8),

$$L\frac{d\tilde{\mathbf{I}}(t)}{dt} + \frac{d}{dt}\left(\Psi_P e^{i\theta(t)}\right) = \tilde{\mathbf{V}}(t) - R\tilde{\mathbf{I}}(t).$$

Integrating both sides allows us to compute the position $\theta(t)$ of the rotor from electrical quantities only,

$$\Psi_P \cos\theta(t) = LI_\alpha(t) - \int_0^t (V_\alpha(\tau) - RI_\alpha(\tau))\,d\tau$$

$$\Psi_P \sin\theta(t) = LI_\beta(t) - \int_0^t (V_\beta(\tau) - RI_\beta(\tau))\,d\tau.$$

The two integrals in these equations tend to accumulate errors until the estimates are not useful any more. Therefore, it is better in practice to replace the integrals with first-order lowpass filters. For accurate approximations the phase frequency of the smallest angular velocity for which the estimate will be used has to fall into the stop band of the filters by at least a decade. For small angular velocities this estimate delivers no useful result. As long as little torque is required at standstill, we may use an open-loop scheme, Sect. 8.3, and switch to field-oriented control once the angular velocity of the rotor is sufficient for the estimate to work.

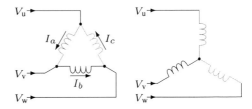

Fig. 8.21 Delta connection of the phase windings on the *left*, wye connection on the *right*

8.7 Delta and Wye Connection

According to our deliberations so far it seems that we need six wires for connecting to such a machine. Actually, there are two ways to connect the windings at the machine which allow us to use only three wires, see Fig. 8.21. In the delta scheme the end of a winding is connected to the start of the next, resulting in a triangular wiring arrangement. In order to achieve the voltages V_a, V_b, and V_c across the windings a, b, and c respectively the supply voltages V_u, V_v, and V_w have to be driven to

$$V_u = (V_a - V_c)/3$$
$$V_v = (V_b - V_a)/3$$
$$V_w = (V_c - V_b)/3.$$

The currents I_a, I_b, and I_c are related to the currents I_u, I_v and I_w by $I_a = -I_w$, $I_b = -I_u$, and $I_c = -I_v$. Creating a delta configuration is particularly easy, it requires a single wire only for creating the windings.

In the wye configuration the ends of the three windings are connected in a common star point. The supply voltages have to be set to $V_u = V_a$, $V_v = V_b$, and $V_w = V_c$ in order to achieve the desired voltages across the phase windings. For producing the same mechanical power a machine wired in the wye configuration draws less current at higher voltages than the same machine wired in the delta configuration, reducing resistive losses in the supply wires.

8.8 Implementation of Field-Oriented Motor Control

The high power that has to be managed for controlling an electrical machine pre-cludes linear approaches for shaping the phase voltages because of their abysmal energy efficiency. Instead one resorts to a switching approach. A three phase power stage, Fig. 8.22, consists of six solid-state power switches, two for each phase. The top switch in a phase circuit connects the phase line directly to the positive supply rail, the bottom switch to the negative rail. The rails are connected to a supply capable of sourcing and sinking the electrical power necessary for operating the machine. Note that at no time both switches may be on simultaneously. The resulting short will

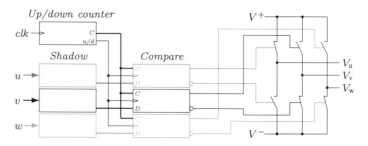

Fig. 8.22 Three phase power stage with pulse width modulation unit

destroy the power stage before any protective measure can become effective! A pulse width modulation unit provides the control signals for the switches under software control. It consists of a counter, which repeatedly counts up to a threshold and down to zero. The time it takes the counter to count from zero to the threshold and back down again is the cycle time $\hat{t}$, its inverse is the pulse width modulation frequency $\hat{f}$. Each phase has a separate compare block. Such a block contains a compare register whose content P it compares with the counter value C. It drives the output for the control of the bottom switch active as long as $C \leq P$. It drives the output for the control of the top switch active as long as $C \geq P+I$ where I, the so-called interlock delay, is stored in a register, not shown, common to all three compare units. The interlock delay allows to compensate for the time it takes a switch to stop conducting. In short the counter, one compare block, and the associated switches produce a pulse width modulated waveform at the phase output, whose duty cycle is set by the value P. The compare registers cannot be updated directly, instead the pulse width modulation unit loads them from three shadow registers whenever the counter reaches zero. As long as the controlling program writes all shadow registers before the next update event, it is guaranteed that consistent data is loaded simultaneously into all three compare blocks.

A gate driver, not shown in Fig. 8.22, sits between the compare blocks and the switches. It shifts the logic levels to the voltages necessary for operating the switches. In case of error conditions such as illegal switching patterns on overload it deactivates all switches in order to protect the power stage and the electrical machine. These protection features form the last line of defense during development when the controlling software cannot be expected to work correctly. The electrical power that has to be managed when controlling a machine capable of producing several kilowatts of mechanical power definitely suffices to cause explosive destruction when being misguided! The level shifter might also provide electrical insulation between the low voltage logic controlling the power stage and the high voltage power circuits.

When choosing the pulse width modulation frequency $\hat{f}$ we shall strive for a frequency outside the audible range. Moreover, $\hat{f}$ has to be high enough so that the required waveforms can be reproduced with sufficient accuracy and that the ripple of the currents I_a, I_b, and I_c stays within limits. An input clock of 40 MHz allows for a switching frequency of about 20 kHz and a resolution of 10 bits. A machine with seven pole-pairs operated at 8,500 revolutions per minute requires an electrical frequency of about 1 kHz, therefore about 20 pulse width modulation cycles produce one cycle of the output signal. This is sufficient for reproducing the phase voltages, see, e.g. Schröder (2009). The software implementing the torque control loop in Figs. 8.15 and 8.18 best runs synchronously with the switching cycle. It can start working once the phase currents for the previous switching cycle are available. We can measure the current of a phase at three locations in Fig. 8.22, between the phase's bottom switch and the negative supply rail V^-, between the phase's top switch and the positive supply rail V^+ and between the phase's output and the machine. A measurement at the bottom location is only possible while the bottom switch conducts, one at the top location while the top switch conducts. In order to dodge the current ripple, current measurements should only be undertaken while all three phases are at the same potential. Therefore, we have to synchronize current measurements with the pulse width modulation cycle, even when measuring at the output location. One good time for measuring at the bottom or at the output position is at the beginning of a cycle, when the bottom switches are conducting in all phase circuits. At this time the three phase system is balanced, therefore it suffices to measure I_u and I_v only and compute I_w from $I_w = -I_u - I_v$. This measurement strategy samples two signals simultaneously, therefore it requires two independent analog-to-digital converters. As long as the supply rail V^- is close to the ground potential of the logic, we can base measurements at the bottom location on shunts. At the other locations current sensors, which rely on measuring the conductor's magnetic field, are more appropriate. In order to faithfully reproduce the current in the short time available for measurement the current sense circuits must have enough bandwidth.

Many microcontroller chips targeted for the control of electrical machines allow for triggering analog-to-digital conversions automatically whenever the pulse width modulation counter hits zero. Relying on such a function we implement the torque control loop in the interrupt service routine handling the completion of the analog-to-digital conversions. The execution of the torque control loop must be finished within the pulse width modulation cycle it was started in, or else the machine's windings will be supplied with erroneous voltages. This will most probably trigger an over current event and cause the power stage to switch off.

8.8.1 Synthesizing Three Phase Voltages

For converting the two voltages V_α and V_β into the three phase voltages V_a, V_b, and V_c we may just apply the inverse unitary Clarke transform to V_α and V_β. Doing so limits the maximal amplitude for the voltages V_a, V_b, and V_c to $V = \frac{V^+ - V^-}{2}$ for a wye connected machine and to $\sqrt{3}\,V$ for a delta connected one. We can improve on that by deviating from purely sinusoidal voltages for V_u, V_v, and V_w and allow some third harmonic content, which does not effect the machine, see, e.g. Schröder (2012). The procedure THREEPHASEPWM, Fig. 8.23, achieves a maximum amplitude of $1.15\,V$ for a wye connected machine and of $1.15\sqrt{3}\,V$ for a delta connected one.

1: **procedure** THREEPHASEPWM(V_α, V_β)

2:      $\theta \leftarrow \arctan \frac{V_\beta}{V_\alpha}$

3:      $A \leftarrow \sqrt{V_\alpha^2 + V_\beta^2}$

4:      $t_u \leftarrow \hat{t}A/V \left(\cos\theta - \frac{1}{6}\cos 3\theta \right)$

5:      $t_v \leftarrow \hat{t}A/V \left(\cos\left(\theta + \frac{2\pi}{3}\right) - \frac{1}{6}\cos 3\theta \right)$

6:      $t_w \leftarrow \hat{t}A/V \left(\cos\left(\theta + \frac{4\pi}{3}\right) - \frac{1}{6}\cos 3\theta \right)$

7: **end procedure**

Fig. 8.23 Procedure THREEPHASEPWM. By adding some third harmonic content to the voltages produced by the power stage the maximum voltage amplitudes delivered to the machine's windings is increased

Table 8.1 Space vectors reachable with simple switching patterns

U	V	W	Space vector	U	V	W	Space vector
T	B	B	V	B	T	T	$-V$
T	T	B	$V/2(1+i\sqrt{3})$	B	B	T	$V/2(-1-i\sqrt{3})$
B	T	B	$V/2(-1+i\sqrt{3})$	T	B	T	$V/2(1-i\sqrt{3})$
B	B	B	0	T	T	T	0

In each phase circuit the letter T indicates a conducting top switch, the letter B a conducting bottom switch

The procedure THREEPHASEPWM will be called by the torque control loop. In order to meet the hard real-time requirements the computation of the mathematical functions called inside THREEPHASEPWM must be fast. This we achieve by tabulating these functions. The arc tangent function, also necessary for sensorless operation, has to work for all four quadrants, therefore it has to perform a case analysis on the signs of the numerator and the denominator of its argument. Exploiting the symmetry of the voltage waveform we have to tabulate it for a quarter-period only.

When we consider all three phases at once we can arrive at a procedure, called space vector modulation, which delivers the same maximum amplitude, but with reduced memory requirements for the tabulated waveform for the phases, see, e.g. Schröder (2012). In Table 8.1 we consider all possible switching patterns in the power stage in Fig. 8.22. We recognize that the power stage can produce six nonzero space vectors, all with length V and spaced $\frac{\pi}{3}$ apart. We synthesize a space vector $V_\alpha + iV_\beta$ whose direction cannot be produced by a single switching pattern from the two neighboring ones whose direction can be reached, and the two zero patterns. The time t_l in the procedure SPACEVECTORPWM, Fig. 8.25, is the time the left neighbor must be active in one pulse width modulation cycle, the time t_r is the one for the right neighbor, and t_0 is the time for one of the zero switching patterns. In the switching pattern of one of the neighbors a single top switch is conducting, while in the switching pattern of the other two top switches are conducting. In order to minimize the number of switching actions a pulse width modulation cycle as

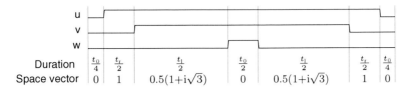

Fig. 8.24 One pulse width modulation cycle for a space vector with $0 \leq \theta < \pi/3$. A low logic level activates the *bottom* switch, a high level the *top* switch

1: **procedure** SPACEVECTORPWM(V_α, V_β)
2: $\quad \theta \leftarrow \arctan \frac{V_\beta}{V_\alpha}$
3: $\quad A \leftarrow \sqrt{V_\alpha^2 + V_\beta^2}$
4: $\quad t_\mathrm{l} \leftarrow \frac{2\hat{t}A}{\sqrt{3}V} \sin(\theta \bmod (\pi/3))$
5: $\quad t_\mathrm{r} \leftarrow \frac{2\hat{t}A}{\sqrt{3}V} \sin(\pi/3 - \theta \bmod (\pi/3))$
6: $\quad t_0 \leftarrow \hat{t} - t_\mathrm{l} - t_\mathrm{r}$
7: $\quad$ **if** $0 \leq \theta < \pi/3$ **then**
8: $\quad\quad (t_\mathrm{u}, t_\mathrm{v}, t_\mathrm{w}) \leftarrow (t_0/4, t_0/4 + t_\mathrm{r}/2, \hat{t}/2 - t_0/4)$
9: $\quad$ **else if** $\pi/3 \leq \theta < 2\pi/3$ **then**
10: $\quad\quad (t_\mathrm{u}, t_\mathrm{v}, t_\mathrm{w}) \leftarrow (t_0/4 + t_\mathrm{l}/2, t_0/4, \hat{t}/2 - t_0/4)$
11: $\quad$ **else if** $2\pi/3 \leq \theta < \pi$ **then**
12: $\quad\quad (t_\mathrm{u}, t_\mathrm{v}, t_\mathrm{w}) \leftarrow (\hat{t}/2 - t_0/4, t_0/4, t_0/4 + t_\mathrm{r}/2)$
13: $\quad$ **else if** $\pi \leq \theta < 4\pi/3$ **then**
14: $\quad\quad (t_\mathrm{u}, t_\mathrm{v}, t_\mathrm{w}) \leftarrow (\hat{t}/2 - t_0/4, t_0/4 + t_\mathrm{l}/2, t_0/4)$
15: $\quad$ **else if** $4\pi/3 \leq \theta < 5\pi/3$ **then**
16: $\quad\quad (t_\mathrm{u}, t_\mathrm{v}, t_\mathrm{w}) \leftarrow (t_0/4 + t_\mathrm{r}/2, \hat{t}/2 - t_0/4, t_0/4)$
17: $\quad$ **else**
18: $\quad\quad (t_\mathrm{u}, t_\mathrm{v}, t_\mathrm{w}) \leftarrow (t_0/4, \hat{t}/2 - t_0/4, t_0/4 + t_\mathrm{l}/2)$
19: $\quad$ **end if**
20: **end procedure**

Fig. 8.25 Procedure SPACEVECTORPWM. The mod operation computes $r = \theta + z\pi/3$ for some integer $z \in \mathbb{Z}$ such that $0 \leq r < \pi/3$

computed by SPACEVECTORPWM starts out with all bottom switches conducting, a zero pattern. At time $t_0/4$ it changes to the pattern with a single top switch conducting for half of the required time. In order to minimize the number of switching actions a single pulse width modulation cycle follows the pattern in Fig. 8.24.

When implementing the procedure SPACEVECTORPWM we again have to tabulate all used mathematical functions in order to meet the strict real-time requirements. We have to tabulate the sinusoid used for computing the phase voltages for one sixth period only.

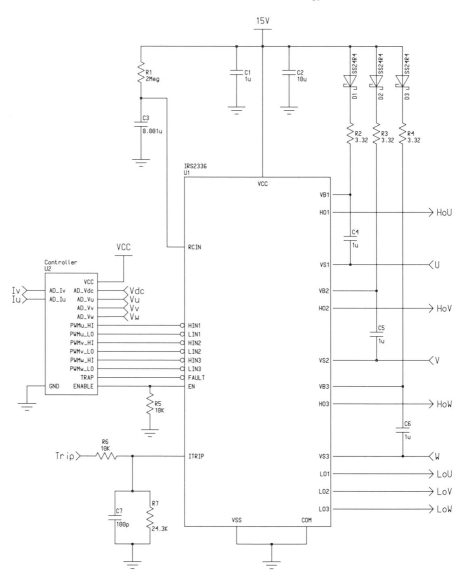

Fig. 8.26 Low voltage three phase power stage, gate driver

8.9 A Low-Voltage Power Stage

The low-voltage power stage in Figs. 8.26, 8.27 and 8.28 can be operated with supply voltages up to 36 V. It provides phase currents up to 10 A peak to peak. There is no isolation barrier between the logic portion of the circuit and the power part, the logic's ground potential is connected to the negative supply terminal instead.

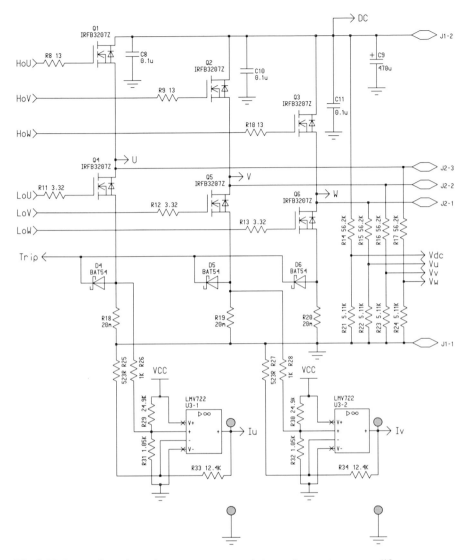

Fig. 8.27 Low voltage three phase power stage, switches and current sense amplifiers

The gate driver U1 in Fig. 8.26 shifts the logic levels output by the controller U2 to the voltages required to operate the power transistors in Fig. 8.27. Interestingly, it does not demand a separate supply for driving the gates of the top transistors but uses a clever circuit instead. While the bottom transistor Q4 for example connects the phase output U to the negative supply rail, the 15 V rail charges the capacitor C4 via the diode D1 and the resistor R2 to about 15 V. Note that the positive end of C4 is now 15 V above the source of the top transistor Q1. After the gate driver has switched off the bottom transistor it uses the charge stored in C4 to bias the top transistor into

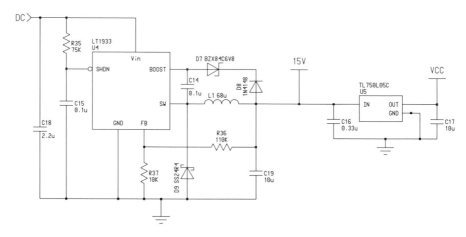

Fig. 8.28 Low voltage three phase power stage, power supplies for gate driver and logic

conduction. The voltage at the phase output U rises rapidly taking the negative end of
C4 with it. The diode D1 becomes reverse biased and disconnects the positive end of
C4 from the 15V rail. Therefore, the positive end of C4 stays 15 V above the source of
the top transistor Q1, exactly what the gate driver needs in order to keep biasing Q1
into conduction. In principle the switching transistors are controlled by voltage alone,
but eventually inevitable leakage currents will discharge the capacitor C4. Before this
happens the gate driver has to switch off the top transistor and switch on the bottom
transistor so that the 15V rail can charge C4 again. While being exceedingly simple
these so-called bootstrap circuits do not allow the top transistors to stay conducting
all the time. Therefore, the controlling software must not request such pulse width
modulation patterns.

The shunts R18, R19, and R20 sense the phase currents. The diodes D4, D5, and
D6 provide the maximum of these three current signals to the gate driver. When the
maximum of the three phase currents exceeds 10 A the gate driver switches off all
power transistors and activates the FAULT signal in order to notify the controller. The
two amplifiers U3-1 and U3-2 condition the current information for phases U and V. A
current of 0 A through a phase results in a voltage of 2.5 V at the output of the phase's
current sense amplifier, one of 10 A in a voltage of 5 V and one of −10 A in a voltage
of 0 V. Each amplifier has a bandwidth of 400 kHz, plenty enough for this application.
We expect that the accuracy of this circuit will be limited by voltage drops in the
ground connections due to the relatively large currents involved. We, therefore, allow
ourself to bias the positive inputs of the amplifiers with voltage dividers connected
directly to the imprecise VCC rail.

The power supply in Fig. 8.28 provides 15 V for the gate driver and 5 V for the
amplifiers and the logic. The supply for the gate driver is a step-down converter
built around U4. The logic supply is a linear voltage regulator. Although the dropout

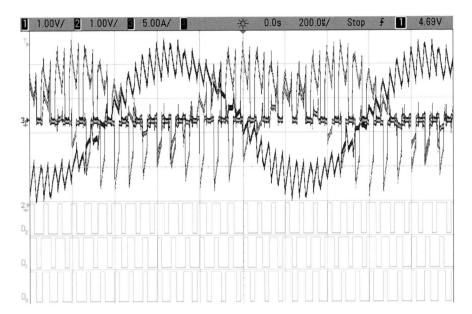

Fig. 8.29 Three phase currents and the pulse width modulation signals for switching the low-side power transistors. The currents for the phases U and V are represented by the outputs of the two current sense amplifiers, the current for phase W is measured with a separate current probe

voltage is 10 V we expect this rail to be loaded with less that 150 mA. Therefore, the power dissipation is bearable.

The measurement in Fig. 8.29 shows the situation when the motor is being loaded by hand and running with a velocity of 3,570 revolutions per minute. Whenever a bottom switch conducts, the respective current sense amplifier reproduces the current through that switch with acceptable accuracy. During the time the switch is not conducting the output of the respective current amplifier ideally is a constant 2.5 V. The observable raggedness results from resistive drops in the ground connections due to the relatively high currents.

8.10 Bibliographical Notes

Electrical machines and their control is an important topic, therefore the literature is vast. Schröder in the series of books (2006, 2009, 2012, 2013) gives a comprehensive overview of the field, covering the different kinds of electrical machines, the control of these machines, the power components, and the power circuits. He also provides a comprehensive bibliography. The book by Fischer (2009) provides another general introduction into the field of electrical machines. Hendershot and Miller in their two books (1994, 2010) provide the reference for the design of brushless DC and permanent magnet synchronous machines. El-Refaie (2010) gives an overview over permanent magnet synchronous motors. Liwschitz (1943) studies winding patterns and the resulting distributions of the magnetic field. Zhu et al. (1993), Zhu and Howe (1993a, b, c), and Zhu et al. (2002) model the magnetic field of permanent magnet synchronous machines without using finite element analysis. The book by Mohan et al. (2003), Undeland, and Robbins is a good reference for power electronics in general, not just for electric drives.

In the classical articles Park (1929, 1933) derives the model we present here. El-Refaie et al. (2006) derive a model of a permanent magnet synchronous motor in stator coordinates, which includes cogging effects and non-sinusoidal field distributions.

Karl Hasse (1969) and Felix Blaschke (1974) invented field-oriented control. When cheap power electronics components became available in the 1990s the field took off.

Starting an electrical motor from standstill at full torque without relying on position sensors is still open to research. One approach exploits the dependence of the inductances of the windings on the position of the rotor. Schrödl et al. measure the inductances of the windings by injecting suitable low-frequency measurement patterns into the normal switching sequence and, from measurements of the resulting change of the phase currents, compute these inductances (Schrödl and Lambeck 2003; Rieder and Schrödl 2005; Schrödl et al. 2006, 2007, 2009; Staffler and Schrödl 2010). Wisniewski and Koczara (2008a, b) sense the change of the inductances of the windings due to saturation effects using low-frequency test signals. Linke et al. (2003) and Perassi et al. (2005) apply a high-frequency signal to measure the drop in the winding's inductances due to saturation caused by the permanent magnets.

8.11 Exercises

Exercise 8.1 In Sect. 8.8.1 we claim that adding a sinusoidal voltage at three times the base frequency to the sinusoidal phase voltages does not affect the machine. Why?

Exercise 8.2 The LT1933 in Fig. 8.28 uses an integrated NPN transistor as top switch. In order to switch the transistor fully on the base drive circuit of the LT1933 requires a voltage larger than the supply. Can you spot a bootstrap circuit?

References

Blaschke F (1974) Method for controlling asynchronous machines
Braess H, Seiffert U (2005) Handbook of automotive engineering. SAE International, Warrendale
El-Refaie A (2010) Fractional-slot concentrated-windings synchronous permanent magnet machines: Opportunities and challenges. IEEE Trans Industr Electron 57(1):107–121. doi:10.1109/TIE.2009.2030211

El-Refaie A, Jahns T, Novotny D (2006) Analysis of surface permanent magnet machines with fractional-slot concentrated windings. IEEE Trans Energy Convers 21(1):34–43. doi:10.1109/TEC.2005.858094

Fischer R (2009) Elektrische Maschinen. Hanser, Munich

Hasse K (1969) Zur Dynamik drehzahlgeregelter Antriebe mit stromrichtergespeisten Asynchron-Kurzschlußläufermaschinen. Ph.D. thesis, Technische Hochschule, Darmstadt

Hendershot J, Miller T (1994) Design of brushless permanent magnet motors. Magna Physics Publishing, Oxford

Hendershot J, Miller T (2010) Design of brushless permanent-magnet machines. Motor Design Books LLC, Oxford

Linke M, Kennel R, Holtz J (2003) Sensorless speed and position control of synchronous machines using alternating carrier injection. In: Electric machines and drives conference, IEMDC'03. IEEE International, vol 2, pp 1211–1217. doi:10.1109/IEMDC.2003.1210394

Liwschitz MM (1943) Distribution factors and pitch factors of the harmonics of a fractional-slot winding. Trans Am Inst Electr Eng 62(10):664–666. doi:10.1109/T-AIEE.1943.5058623

Mohan N, Undeland T, Robbins W (2003) Power electronics: converters, applications and design. Wiley, Australia

Park R (1929) Two-reaction theory of synchronous machines generalized method of analysis - part I. Trans Am Inst Electr Eng 48(3):716–727. doi:10.1109/T-AIEE.1929.5055275

Park R (1933) Two-reaction theory of synchronous machines - II. Trans Am Inst Electr Eng 52(2):352–354. doi:10.1109/T-AIEE.1933.5056309

Perassi H, Berger G, Petzoldt J (2005) Practical implementation of the sensorless field oriented control of a PMSM for wide speed range. In: European conference on power electronics and applications. doi:10.1109/EPE.2005.219552

Rieder UH, Schrödl M (2005) A simulation method for analyzing saliencies with respect to enhanced INFORM-capability for sensorless control of PM motors in the low speed range including standstill. In: European conference on power electronics and applications. doi:10.1109/EPE.2005.219744

Schröder D (2006) Leistungselektronische Bauelemente. Springer, Berlin

Schröder D (2009) Elektrische Antriebe - Regelung von Antriebssystemen. Springer, Berlin

Schröder D (2012) Leistungselektronische Schaltungen: Funktion Auslegung und Anwendung. Springer, Berlin

Schröder D (2013) Elektrische Antriebe - Grundlagen: Mit durchgerechneten Übungs- und Prüfungsaufgaben. Springer, Berlin

Schrödl M, Lambeck M (2003) Statistic properties of the INFORM method for highly dynamic sensorless control of PM motors down to standstill. In: Industrial electronics society, IECON '03. The 29th annual conference of the IEEE, vol 2, pp 1479–1486, doi:10.1109/IECON.2003.1280276

Schrödl M, Hofer M, Staffler W (2006) Combining INFORM method, voltage model and mechanical observer for sensorless control of PM synchronous motors in the whole speed range including standstill. e & i Elektrotechnik und Informationstechnik 123(5):183–190. doi:10.1007/s00502-006-0340

Schrödl M, Hofer M, Staffler W (2007) Extended EMF- and parameter observer for sensorless controlled PMSM-machines at low speed. In: European conference on power electronics and applications. pp 1–8, doi:10.1109/EPE.2007.4417536

Schrödl M, Staffler W, Hofer M (2009) Accuracy of the sensorless determined rotor position for industrial standard drives in the whole speed range. In: Power electronics and applications, EPE '09. 13th European conference on, pp 1–6

Staffler W, Schrödl M (2010) Extended mechanical observer structure with load torque estimation for sensorless dynamic control of permanent magnet synchronous machines. In: 14th international power electronics and motion control conferences (EPE/PEMC), pp S1-18–S1-22, doi:10.1109/EPEPEMC.2010.5606516

Wisniewski J, Koczara W (2008a) Control of axial flux permanent magnet motor by the PIPCRM
    method at standstill and at low speed. In: 13th Power electronics and motion control conference,
    EPE-PEMC 2008, pp 2254–2260. doi:10.1109/EPEPEMC.2008.4635599
Wisniewski J, Koczara W (2008b) The sensorless rotor position identification and low speed oper-
    ation of the axial flux permanent magnet motor controlled by the novel PIPCRM method. In:
    Power electronics specialists conference, PESC 2008, IEEE, pp 1502–1507. doi:10.1109/PESC.
    2008.4592149
Zhu Z, Howe D (1993a) Instantaneous magnetic field distribution in brushless permanent magnet
    DC motors. II. armature-reaction field. IEEE Trans Magn 29(1):136–142. doi:10.1109/20.195558
Zhu Z, Howe D (1993b) Instantaneous magnetic field distribution in brushless permanent magnet
    DC motors. III. effect of stator slotting. IEEE Trans Magn 29(1):143–151. doi:10.1109/20.195559
Zhu Z, Howe D (1993c) Instantaneous magnetic field distribution in permanent magnet brushless
    DC motors. IV. magnetic field on load. IEEE Trans Magn 29(1):152–158. doi:10.1109/20.195560
Zhu Z, Howe D, Bolte E, Ackermann B (1993) Instantaneous magnetic field distribution in brushless
    permanent magnet DC motors. I. open-circuit field. IEEE Trans Magn 29(1):124–135. doi:10.
    1109/20.195557
Zhu Z, Howe D, Chan C (2002) Improved analytical model for predicting the magnetic field distrib-
    ution in brushless permanent-magnet machines. IEEE Trans Magn 38(1):229–238. doi:10.1109/
    20.990112

Chapter 9
Data Recovery from Noise—Lock-In Detection

Practically all signals have a random component added to their true value, the so-called noise. Noise is unavoidable, resulting from the aleatory behavior of nature at the thermodynamical and quantum-mechanical levels.[1] Noise is characterized by the probability density of its power with respect to frequency. White noise has a constant power density. Approximately resistors show white noise because of thermally induced movement of the charge carriers. Semiconductor junctions exhibit noise whose spectral power density is proportional to $1/f$, so-called $1/f$ noise, see e.g. Allan (1987). This form of noise disturbs the measurement of constant and slowly varying quantities, e.g., the weight of some object. When amplifying the very small voltages or currents produced by detectors, the $1/f$ noise produced in the amplifier's semiconductors might overpower the relevant information. Lock-in detection, also called lock-in amplification, is the standard method for the extraction of information which is buried in overpowering noise, Dicke (1946). Lock-in detection based approaches take many measurements and average these in order to minimize the influence of the added noise on the answer. While keeping the very narrow bandwidth achievable at low frequencies using low-order filters, lock-in detection moves the frequencies at which the detector's amplifier operates away from regions dominated by $1/f$ noise.

After introducing lock-in methods, we study the spectral sensitivity of different lock-in detectors. Doing so we use the Fourier series representations of the signals involved. We will consider software and hardware realizations of the different lock-in methods. Signal conditioning for weight and torque transducers shows how lock-in techniques help suppress thermally induced disturbances in the measurements. We lock at generalized lock-in methods, which allow for the simultaneous measurement of several parameters using a single detector only. Using one of these schemes we measure moisture using infrared light.

---

[1] In short, noise is a fact of life, like taxes.

© Springer International Publishing Switzerland 2015

P. Hintenaus, *Engineering Embedded Systems*, DOI 10.1007/978-3-319-10680-9_9

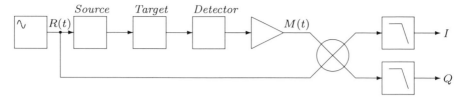

Fig. 9.1 Lock-in amplifier for measuring some physical property of the target. The multiplier multiplies the measurement $M(t)$ both with the reference $R(t)$ and the phase-shifted version $R'(t)$. The amplitude of the part of the target's answer that is in-phase with the reference is I, the one that is out of phase by $-\pi/2$ is Q

9.1 Operational Principle

In the most basic configuration of a lock-in amplifier, Fig. 9.1, an oscillator produces a sinusoidal reference signal $R(t)$ at the working frequency of the amplifier. The source converts the electrical signal $R(t)$ into a form fitting the target and impinges $R(t)$ onto the target. The detector captures the target's response. After amplification, the measurement $M(t)$ is multiplied with the reference $R(t)$ and with $R'(t)$, which is derived from $R(t)$ by shifting $R(t)$ by a quarter period. Both products are lowpass filtered using identical filters in order to produce the output I, the amplitude of the part of $M(t)$ that is in phase with $R(t)$, and the output Q, the amplitude of the part of $M(t)$ that is in quadrature (out of phase by $-\pi/2$) with $R(t)$, of the lock-in detector.

To understand the operation of such a lock-in, we assume that the combination of the source, the target and the detector is a linear time-invariant system, see Sect. 2.4.1. Let the reference $R(t)$ and $R'(t)$ be

$$R(t) = A \cos 2\pi f t$$
$$R'(t) = A \sin 2\pi f t.$$

The measurement $M(t)$ is

$$M(t) = B \cos (2\pi f t - \psi),$$

for some amplitude B and some phase-shift ψ as the combination of source, target and detector is linear and time-invariant. The products $M(t)R(t)$ and $M(t)R'(t)$ are

$$M(t)R(t) = \frac{AB}{2} (\cos \psi + \cos (4\pi f t - \psi))$$
$$M(t)R'(t) = \frac{AB}{2} (\sin \psi + \sin (4\pi f t - \psi)).$$

Assuming that the lowpass filters attenuate the contents with frequency $2f$ completely we get

$$I = \frac{AB}{2}\cos\psi$$

$$Q = \frac{AB}{2}\sin\psi.$$

If the working frequency f has been chosen prudently, such that the attenuation of the composition of the source, the target and the detector is negligible, B contains all information about the quantity to be measured. This information can be recovered by computing

$$B = \frac{2}{A}\sqrt{I^2 + Q^2}.$$

Furthermore, if the phase ψ is close to zero, one can dispense with the computation of Q and use $\frac{2I}{A}$ instead of B.

A sinusoidal signal $D(t)$ with frequency f', phase-shift ψ' and amplitude B',

$$D(t) = B'\cos\left(2\pi f't - \psi'\right),$$

interfering with the measurement $M(t)$ produces the products

$$D(t)R(t) = \frac{AB'}{2}\left(\cos\left(2\pi(f + f')t - \psi'\right) + \cos\left(2\pi(f - f')t + \psi'\right)\right)$$

$$D(t)R'(t) = \frac{AB'}{2}\left(\sin\left(2\pi(f + f')t - \psi'\right) + \sin\left(2\pi(f - f')t + \psi'\right)\right).$$

Provided the frequency f' has enough distance from the working frequency f of the lock-in, such that $|f - f'|$ falls into the stop bands of the lowpass filters, both frequency components in the products will be attenuated by the lowpass filters. In essence, the lock-in scheme converts the lowpass filters into a bandpass filter centered around the working frequency f but with the shape of the lowpass filters. A lock-in detector with first-order lowpass filters with a corner frequency of 1 Hz each, and a working frequency of 1 kHz will attenuate an interfering signal at 1.004 Hz by more than 12 dB. A bandpass filter alone without the multipliers would require an extreme order to achieve the same attenuation, rendering such a design impractical. The $1/f$ noise of the amplifier in the band between 0.999 and 1.001 kHz will not distort the measurement results at all.

The settling time of the filters determines the measurement time of a lock-in detector. The small bandwidth typically chosen makes lock-in-based approaches slow but precise.

9.2 Square-Wave Reference

Using nonsinusoidal signals, square waves, in particular, can simplify the implementation of a lock-in. In a digital implementation, driving the source with a sinusoid will
require a periodically converting digital-to-analog converter together with a suitable
reconstruction filter to produce the waveform. Driving the source with a square wave
requires a couple of switching transistors only. A second opportunity for simplifi-
cation appears when the multiplier is fed with a square wave instead of a sinusoid
as the reference. As the only factors used are 1 and -1, no multiplication operation is required. To analyze both cases we assume that the reference $R(t)$ and the
measurement $M(t)$ are band limited and consider their Fourier series,

$$R(t) = \sum_{n=1}^{k} A_n \cos(2\pi f n t - \phi_n)$$

$$R'(t) = \sum_{n=1}^{k} A_n \sin(2\pi f n t - \phi_n)$$

$$M(t) = \sum_{n=1}^{k} B_n \cos(2\pi f n t - \psi_n),$$

for some k indicating the highest nonzero harmonic in $R(t)$ and $M(t)$, some ampli-
tudes $A_1, \ldots, A_k$ and $B_1, \ldots, B_k$, and some phase-shifts $\phi_1, \ldots, \phi_k$ and $\psi_1, \ldots, \psi_k$.
The constant term I of the product $M(t)R(t)$ and the constant term Q of the product
$M(t)R'(t)$ are

$$I = \frac{1}{2} \sum_{n=1}^{k} A_n B_n \cos(\psi_n - \phi_n)$$

$$Q = \frac{1}{2} \sum_{n=1}^{k} A_n B_n \sin(\psi_n - \phi_n).$$

When the source is driven with an arbitrary waveform while the reference $R(t)$ is a
sinusoid, the second and higher harmonics in $M(t)$ will be attenuated by the lowpass
filters, therefore driving the source with an arbitrary waveform does not influence
the result.

In order to analyze the influence of a nonsinusoidal reference, we consider the
interference $D(t) = B' \cos(2\pi f' t - \psi')$. Its product with the reference is

$$D(t)R(t) = \frac{B'}{2}\sum_{n=1}^{k} A_n \cos\left(2\pi(nf - f')t + \psi' - \phi_n\right)$$

$$+ \frac{B'}{2}\sum_{n=1}^{k} A_n \cos\left(2\pi(nf + f')t - \psi' - \phi_n\right).$$

The second sum cannot contribute any nonzero constants, but the first sum contributes—if $f' = nf$ for some n with $A_n \neq 0$—to either $D(t)R(t)$ or $D(t)R'(t)$ or both. In short, a nonsinusoidal $R(t)$ will transform the lowpass filters not just into a bandpass filter centered around the working frequency of the lock-in f but also into bandpass filters centered around each nonzero harmonics of $R(t)$, and the lock-in is sensitive to interference at the frequencies of these harmonics too.

9.3 External Reference

Thermal sources, like incandescent lamps, are hard to modulate electrically, but the light they emit can be chopped easily by using a rotating disk made out of an opaque material into which radial slits are cut. In such a setup, the lock-in amplifier has to synthesize a reference internally that is synchronized with the passing of the slits. An auxiliary photodetector picks up the light shining through each passing slit in order to provide a signal R_e for the lock-in to synchronize to. When one uses the signal R_e as reference directly, the lock-in becomes sensitive to interference at unknown frequencies because of the unknown harmonic content of R_e. Instead one employs a phase-locked loop for producing a purely sinusoidal internal reference R, which is synchronized to the external reference R_e. A phase-locked loop, see Fig. 9.2, is a feedback system that locks the frequency of its output signal R to the frequency of its input signal R_e. It consists of a phase detector or phase frequency detector that produces the error signals, the loop filter, which acts as the controller and produces the signal to detune a variable frequency oscillator, which produces the output signal of the phase-locked loop. The phase frequency detector drives its output to $+1$ if the signal on its first input leads the signal on its second input, to -1 if the signal on its first input lags the signal on its second, and to 0 if both signals are in phase. Figure 9.3 shows an implementation of a phase frequency detector in hardware. In Fig. 9.4,

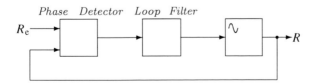

Fig. 9.2 Phase-locked loop for generating a synchronized purely sinusoidal internal reference

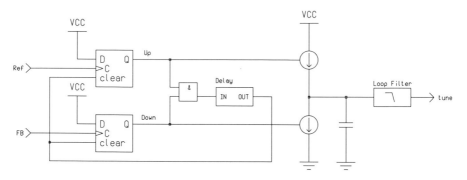

Fig. 9.3 Hardware implementation of a phase frequency detector. The two D flip-flops are triggered by the rising edge each. The output of the *upper* flip-flop controls a current source which deposits some charge in the capacitor, while the *lower* flip-flop controls a current sink which withdraws charge from the capacitor. The delay in the clear path of the two flip-flops prevents them from generating control pulses which are too short for the source and the sink

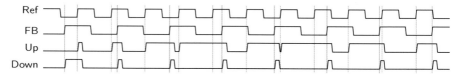

Fig. 9.4 Behavior of the phase frequency detector when the frequencies of the two input signals differ. The Up signal is active more often than the Down signal, therefore the voltage at the tune output increases

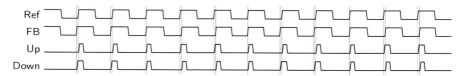

Fig. 9.5 Behavior of the phase frequency detector when the frequencies of the signals at the inputs are the same. The delay block elongates the Up and the Down pulses so that the current source and the current sink can deposit and withdraw minute charges

the frequency of the signal at the Ref input is larger than that of the signal at the feedback input FB. The phase-locked loop is unlocked. By driving the Up signal active for a longer time than the Down signals the phase frequency detector tunes the variable frequency oscillator toward higher frequencies to bring the loop into lock. In Fig. 9.5, the phase-locked loop has locked. The delay in Fig. 9.3 elongates the Up and Down pulses to give the current source and the current sink enough time to react. This prevents dead band behavior, when the loop has locked. Without the delay, the control pulses would be too short to fully activate the source and the sink, and the capacitor would not get charged or discharged until the two input signals are out of phase by an appreciable amount.

The loop filter integrates the output of the phase frequency detector. In the simplest case, it is a lowpass filter. For better performance, it can consist of a proportional-integral-differential structure. Like any other control system, the loop filter of a phase-locked loop has to be tuned for stability and step response.

9.4   Implementation

Commercially available lock-in amplifiers offer flexible choices in terms of synchronization to external references, the frequency of operation, the type of lowpass filter and the amplification of the signal captured by the detector. Older instruments performed the multiplication operations and the lowpass filtering using analog circuits, nowadays these operations are performed using digital logic or microprocessors. In order to be able to extract minute information buried in overpowering noise, the analog elements processing the detector's signal must have a very large dynamic range. Coupled with the bandwidth required to support a large range of frequencies for the reference, this poses a formidable design challenge. The analog-to-digital converter that samples the amplified output of the detector must be fast and have high resolution.

When designing a dedicated sensor, for example the optical sensor sketched in Fig. 9.6, we do not need most of the flexibility universal lock-in amplifiers offer.

When choosing the source, we will look for a source, like a light emitting diode (LED), which can be modulated electrically. Doing so avoids mechanical modulation devices resulting in a compact and robust device. Furthermore, one does not need a phase-locked loop for synchronizing to an external reference. The sensor will operate at a fixed frequency f. As designers we will chose f far enough away from the frequencies stemming from the most prominent sources of interference we expect. For optical sensors, for example, we expect to see interference from fluorescent lights[2] at twice the mains frequency. Placing f far enough from the expected interferences allows for the introduction of a bandpass filter, which attenuates these interferences and makes the circuit narrow-band. The amplifier following the filter can operate at a higher gain without clipping than would be possible without the filter. This

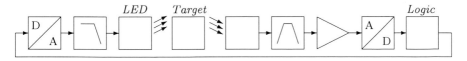

Fig. 9.6 Block diagram of an optical sensor. The bandpass filter in the signal path after the detector is designed to attenuate expected interference. This allows for high gain in the amplifier following it, and for using an analog-to-digital converter of moderate resolution

---

[2] Fluorescent lights flicker whenever their supply voltage becomes zero momentarily. In Europe, therefore they flicker with 100 Hz, in the USA the frequency is 120 Hz.

amplifier with its high gain brings the signal of interest within the dynamic range of an analog-to-digital converter of moderate, lets say about 12 bit, resolution.

The sampling rate f_s will be set up as an integer multiple of f, $f_s = pf$, where p is the number of samples taken during one period of the reference. In order to be able to shift the phase of the reference by $\frac{\pi}{2}$, we set p to an integer multiple of four. The upper corner frequency and the order of the bandpass, the gain of the amplifier, the sampling rate, and the resolution of the analog-to-digital converter must match in order to satisfy the design rule for an anti-aliasing filter, see Sect. 6.4. Instead of computing the required values of the reference over and over again, we tabulate these in the table r, $r_l = A \cos \frac{2\pi l}{p}$ for $0 \le l < p$ and an amplitude A. The simplest lowpass filter possible is an averaging filter whose output is the sum of the inputs of one measurement period. We set the length of the measurement period by choosing the number of cycles m of the reference used for one measurement. The procedure DOSAMPLE in Fig. 9.7 is the interrupt service routine, which is activated whenever the analog-to-digital converter has acquired a new sample M. Initially, we have to set the counters $i \leftarrow 0$, $q \leftarrow \frac{3p}{4}$ and $c \leftarrow 0$. We have to set the accumulators I for the in-phase part and Q for the part in quadrature to $I \leftarrow Q \leftarrow 0$. The procedure works well with integer arithmetic as long as we choose the amplitude A so that all bits in a word are used while integer overflows during the computation are avoided.

When we decide to use a square-wave reference, we can simplify both hardware and software. In the hardware, we can replace the digital-to-analog converter that provides the signal for driving the source with a simple switch. We can eliminate the reconstruction filter altogether. In the software, we can replace the tabulated reference and the multiplications with a case analysis, see the procedure DOSAMPLESQ in Fig. 9.8. This solution allows us to use processors which do not have an instruction for multiplication, without putting undue burden on the processor.

The elimination of the tabulated reference and of the multiplication operations pays off when we implement a lock-in using digital logic, Fig. 9.9. The strobe s indicates that a new sample M is ready to be processed. The two arithmetic blocks add or subtract the new sample to or from the accumulated samples. The inverted outputs of a two-bit gray-code counter control the operations of the two arithmetic blocks. The gray-counter increments every time a quarter of the period of the reference

Fig. 9.7 Procedure DOSAMPLE for processing one sample M

```
 1: procedure DOSAMPLE(M)
 2:     (I, Q) ← (I + r_i M, Q + r_q M)
 3:     i ← (i + 1) mod p
 4:     q ← (q + 1) mod p
 5:     write r_i to the LED
 6:     c ← c + 1
 7:     if c = mp then
 8:         report I, Q
 9:         c ← 0
10:         I ← Q ← 0
11:     end if
12: end procedure
```

```
 1: procedure DOSAMPLESQ(M)
 2:     if p/4 ≤ i < 3p/4 then
 3:         write 0 to LED
 4:             I ← I − M
 5:     else
 6:         write 1 to LED
 7:             I ← I + M
 8:     end if
 9:     if i < p/2 then
10:         Q ← Q + M
11:     else
12:         Q ← Q − M
13:     end if
14:     i ← (i + 1)  mod p
15:     c ← c + 1
16:     if c = mp then
17:         report I, Q
18:             c ← 0, I ← Q ← 0
19:     end if
20: end procedure
```

Fig. 9.8 Procedure DOSAMPLESQ for processing one sample M using square-wave references

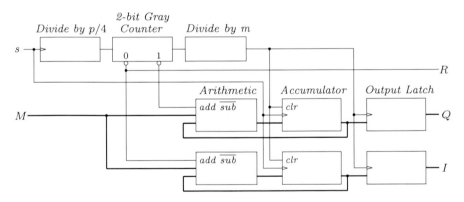

Fig. 9.9 Sketch of a lock-in amplifier using a square-wave reference, realized in digital logic

has elapsed. When a whole measurement time has elapsed, the contents of the accumulators are transferred to the output latches and the accumulators are cleared.

9.5 Signal Conditioning for Strain Gauges

Strain gauges are used for measuring deformations, see e.g. Schrüfer et al. (2012). When a strain gauge is stretched in its sensitive direction, the electrical resistance between its two terminals increases, when it is compressed the resistance decreases.

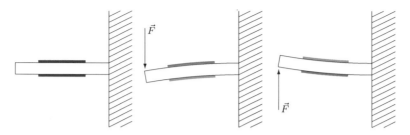

Fig. 9.10 Load-bearing structure with strain gauges glued to it. *Blue-colored* gauges are not stressed, *orange-colored* ones are stretched and *green-colored* ones are compressed

Strain gauges are glued to load-bearing structures, Fig. 9.10. They consist of a thin foil of metal deposited on a plastic carrier. The metal is etched into a series connection of long and thin structures oriented in the sensitive direction. The structures perpendicular to the gauge's sensitive direction are made wide. Such a geometry maximizes the change of the gauge's resistance when the gauge is deformed in its sensitive direction and minimizes it for deformations perpendicular to the sensitive direction. For small strains and at constant temperature, the change of resistance ΔR of a strain gauge with nominal resistance R is

$$\frac{\Delta R}{R} = K \frac{\Delta l}{l},$$

where Δl is the change in length, l is the unstrained length of the gauge, and K is the strain factor which describes the gauge. For the metals typically used for constructing strain gauges, the strain factor K has a value between 1 and 2. The strains $\frac{\Delta l}{l}$ to be measured range between 10^{-6} and 10^{-3}. Therefore, any circuit processing the response of a strain gauge has to deal with minute changes of the gauge's resistance.

When a strain gauge is used as one of the two resistors in a voltage divider, the output of this divider consists of a large, constant voltage superimposed with a small voltage variation, which carries the information. When amplifying such a signal the constant voltage limits the gain achievable before the amplifier reaches one of its supply rails. In a Wheatstone bridge, see Figs. 9.11, 9.12, and 9.13, the voltage difference between two points in the bridge circuit carries the information. If no strain is applied to the gauges in the bridge, this difference is zero. An instrumentation amplifier can amplify this difference without hitting the supply voltages.

In the quarter-bridge configuration, see Fig. 9.11, a Wheatstone bridge consists of three resistors and a strain gauge. The three resistors and the strain gauge all have the same (nominal) resistance. The output voltage V_o of the bridge is

$$V_o = -V_i \frac{\Delta R}{4R + 2\Delta R}.$$

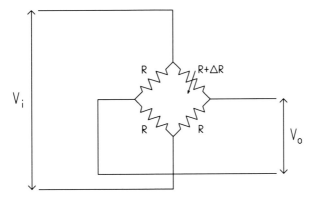

Fig. 9.11 Resistive sensor in quarter-bridge configuration

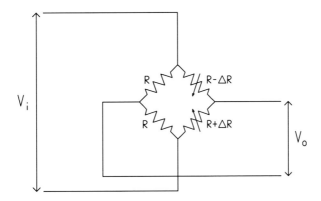

Fig. 9.12 Resistive sensor in half-bridge configuration

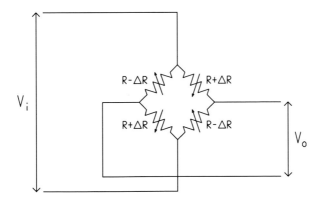

Fig. 9.13 Resistive sensor in full-bridge configuration

The output voltage V_o is proportional to the excitation V_i of the bridge. Such configurations are called ratiometric. Furthermore, V_o is approximately linear to ΔR as long as ΔR is sufficiently small. In order to match the temperature characteristics of the resistor in the right branch of the quarter-bridge with that of the strain gauge, a second strain gauge, placed close to the first to provide good thermal matching but not stressed, is used for the resistor.

For the half-bridge configuration, see Fig. 9.12,

$$V_o = -V_i \frac{\Delta R}{2R}.$$

The configuration is ratiometric. The output voltage V_o depends linearly on ΔR. Compared with the quarter-bridge, the voltage swing of the output has doubled. The half-bridge requires a place for mounting the second gauge such that the second gauge sees a strain opposite to the strain the first gauge sees. Given such a place the half-bridge is preferable.

For the full-bridge configuration, see Fig. 9.13, the output voltage V_o is

$$V_o = -V_i \frac{\Delta R}{R}.$$

The full-bridge is ratiometric again. The output voltage swing is double that of the half-bridge. For an excitation voltage V_i of 5 V and a strain of $\pm 10^{-3}$ a full-bridge sensor composed of four strain gauges with a strain factor of 2 each produces a voltage differential of $\pm 10\,\text{mV}$. The full-bridge is the preferable configuration for dedicated sensors, like the torque sensor in Fig. 9.14.

The thermoelectric effect disturbs the measurement. Pairings of dissimilar metals in the output path of the bridge in Fig. 9.15 form parasitic thermocouples. These thermocouples constitute a voltage source in series with the bridge, whose voltage V_t depends on the temperatures of the contact points between the dissimilar metals and on the kinds of metals involved.

A lock-in approach, also called AC excitation, eliminates both the $1/f$ noise of the amplifiers and the influence of the parasitic thermocouples in the output path of the bridge. The AD7195 analog-to-digital converter, designed by Analog Devices, embodies an elegant implementation of a lock-in, see Fig. 9.16. Owing to its Σ-Δ architecture, it over-samples its input. It uses a square-wave signal as reference. For the first half of a measurement cycle, the AD7195 applies the positive supply voltage to the top of the bridge and ground to the bottom by making the transistors Q2 and Q3 conduct. In the second half, it reverses the polarity of the bridge's excitation by making Q1 and Q4 conduct. The measurement input as well as the reference voltage input are differential, for producing a measurement the converter compares the voltage between AIN3 and AIN4 with the voltage between the reference inputs REFINp and REFINn. Wiring the reference inputs to the top and the bottom of the bridge achieves two goals. First the circuit becomes ratiometric, therefore it requires no precise voltage reference. Second changing the polarity of the bridge's excitation amounts to changing the sign of the reference signal. The down-conversion filter in the converter completes the lock-in scheme. Together with a microcontroller and a power supply, preferably a linear one in order to keep the noise low, this circuit forms a complete instrument for strain-gauge-based measurements.

9.6 Simultaneous Measurements

In many situations, the response of a target to a single type of stimulus is not enough to measure one or several of the target's properties. We can subject the target to the different stimuli and acquire the responses one after the other. In order to keep the time for taking a measurement reasonable and to be able to extract information about targets whose state evolves over time, we prefer to subject the target to the stimuli simultaneously. As long as each detector used is sensitive to the target's response to a single stimulus only, all we have to do is run several measurements in parallel. When one detector picks up the target's responses to several different stimuli, we must compute the individual responses from the combined detector signal.

In Fig. 9.17, the sources $S_1, \ldots, S_n$ impose the references $R_1, \ldots, R_n$ simultaneously onto the target. The single detector picks up the target's combined response. The signal conditioning electronics amplifies and filters this response and produces the measurement $M(t)$. For analyzing this setup, we assume that the combination of the source S_i, the target, the detector, and the signal conditioning electronics

Fig. 9.14 Torque sensor using four strain gauges in a full-bridge (ME-Meßsysteme GmbH). The inner cross of the sensor is mounted on some foundation. The outer ring of the sensor holds the housing of a motor. The torque the motor produces at its shaft twists the cross against the ring. Four strain gauges, each sitting on one of the connecting arms and wired in a full-bridge circuit convert this twist into a small voltage difference

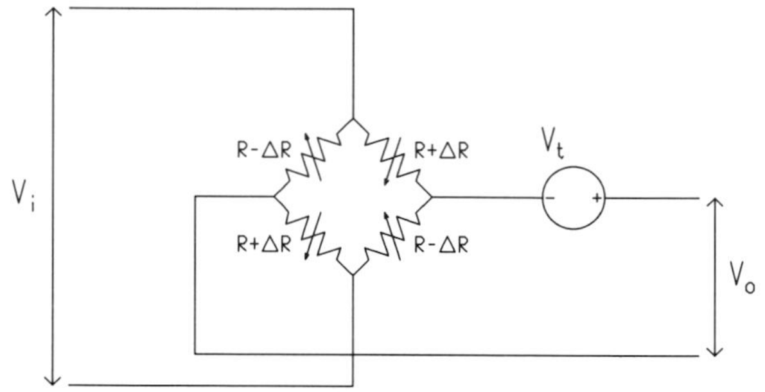

Fig. 9.15 Wheatstone bridge with parasitic thermocouples

form a linear, time-invariant system with impulse response $P_i(t)$ for $1 \leq i \leq n$. The response of this system to the reference $R_i(t)$, the single channel measurement $M_i(t)$, is the convolution of the impulse response $P_i(t)$ and $R_i(t)$,

$$M_i(t) = (P_i * R_i)(t).$$

The single channel measurement $M_i(t)$ conveys information about the reaction u_i of the target to the physical stimulus delivered by source S_i. Alas all we can measure in this setup is the measurement $M(t)$. We assume that $M(t)$ is the sum of the single channel measurements,

$$M(t) = \sum_{i=1}^{n} M_i(t).$$

We attack the extraction of estimates for the reactions $u_i(t)$ from $M(t)$ with a lock-in for each channel.

9.6.1 Sinusoidal References

We choose the references $R_i = A_i \cos(2\pi f_i t)$ for some amplitudes $A_1, \ldots, A_n$ and some commensurate frequencies $f_1, \ldots, f_n$ to be sinusoidal. The single channel measurement $M_i(t)$ is sinusoidal with frequency f_i because the combination of the source S_i, the target, the detector, and the signal conditioning electronics form a linear time-invariant system. For each channel, we feed the product $R_i(t)M(t)$ to the lowpass filter F_i to arrive at I_i and the product $R_i'(t)M(t)$ to F_i' to arrive at Q_i, where $R'(t) = A_i \sin(2\pi f_i t)$. The filters F_i and F_i' shall have the same frequency response. We estimate the reaction u_i with

$$u_i \approx \sqrt{I_i^2 + Q_i^2}.$$

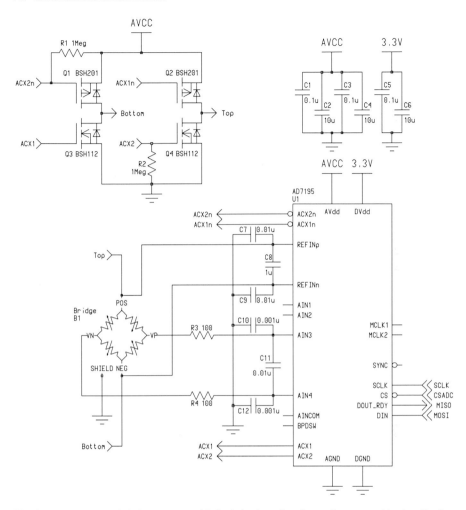

Fig. 9.16 Analog-to-digital converter with lock-in detection for strain gauges (Analog Devices 2010)

When designing the filters $F_1, \ldots, F_n$ and $F'_1, \ldots, F'_n$ and when choosing the frequencies $f_1, \ldots, f_n$, we must avoid settings, which result in crosstalk between channels. If we let all nonzero frequencies in the products $R_i(t)M(t)$ and $R'_i(t)M(t)$ fall onto zeros of the frequency responses of the filters in all channels we avoid crosstalk completely. Of course, each channel has to have its own frequency of operation different from those of all the other channels. Finding appropriate filters and frequencies looks hard to achieve. But consider an averaging filter, which computes the arithmetic mean of its input over a time-span τ. The frequency response of this filter has zeros at nonzero integer multiples of $\frac{1}{\tau}$. For designing such a scheme, we have to choose the time τ, which elapses between two answers of the lock-in ampli-

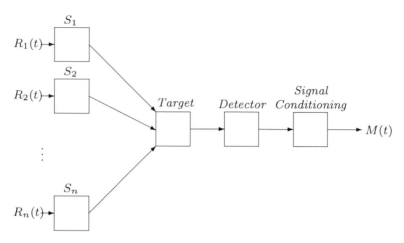

Fig. 9.17 Measurement setup with several sources $S_1, \ldots, S_n$, a single target and a single detector

fiers. Assuming that we accept averaging filters we have to use frequencies f_i only, for which r_i perform an integral number of periods during τ.

The sample rate f_s must be a multiple of all f_i, furthermore, f_s/f_i must be a multiple of four. We must obey the design rule for anti-aliasing filters when building the signal conditioning electronics. By running n copies of the procedure DOSAMPLE, Fig. 9.7, each copy with different setting of the variable p, we can implement this scheme.

Looking at the choice of the filters $F_1, \ldots, F_n$ and the frequencies $f_1, \ldots, f_n$ from a different vantage point, we observe that I_i and $-Q_i$ are the real and imaginary parts of a point of the discrete Fourier transform of

$$m(0), m(1/f_s), \ldots, m(\tau/f_s - 1).$$

For a system with many sources, it is computationally more efficient to compute a fast Fourier transform once the samples are taken.

9.6.2 Square-Wave References

Using square-wave references promises to simplify a multichannel lock-in considerably. For avoiding crosstalk however, we not only have to consider the fundamentals but the higher harmonics of the references too. For reasons of simplicity, we restrict the discussion to filters averaging over a time-span τ only. The reference R_i has to perform an integral number of periods in the time-span τ. According to Sect. 9.2, the ith lock-in is not only sensitive at the fundamental frequency of R_i but at the higher harmonics of R_i too. Therefore, no two references may share a common harmonic frequency, or else the two channels will exhibit crosstalk.

Again we sample the measurement with rate f_s. When we restrict ourself to references with odd harmonics only, which luckily includes square waves, the following number theoretic observation tells us how to chose the period of each reference in terms of number of samples.

Lemma 9.1 *Let $p = 2^e q$ for a positive integer e and a positive odd integer q. For $0 \leq i \leq e$ let $p_i' = 2^i q_i$, where q_i is a positive odd integer that divides q, and let P_i' be the set $\left\{ p_i', \frac{p_i'}{3}, \frac{p_i'}{5}, \ldots \right\}$. Then the sets P_i' and P_k' are disjoint for $i \neq k$.*

Proof Assume $i > k$. Consider the integer $s = 2^i q_i q_k$. Then the set P_i' equals $\left\{ \frac{s}{q_k}, \frac{s}{3q_k}, \frac{s}{5q_k}, \ldots \right\}$, and the set $P_k' = \left\{ \frac{s}{2^{i-k}q_i}, \frac{s}{2^{i-k}3q_i}, \frac{s}{2^{i-k}5q_i}, \ldots \right\}$. As the denominators of all fractions in P_i' are odd, and the denominators of all fractions in P_k' are even, the two sets are disjoint.

In the lemma, the p_i' for $i \geq 2$ are the candidates for the periods of the references. The first two, p_0' and p_1' cannot be used, because they are not divisible by four. We first choose a range for the rate f_s so that the hardware at hand will not be overstressed by the amount of data it has to deal with. Next we look for the number of samples p that are used in the computation of a single answer. Together with the estimate for f_s this gives us an idea how much time will pass between answers. Finally, we try combinations of the candidates until satisfied with our choice. Choosing periods which are in the same range allows us to design a narrow band signal conditioning electronics.

When we start out with a sampling rate $f_s = 100\,\mathrm{kHz}$ and an approximate measurement rate of 20 Hz, the number of samples used for a single answer is $p \approx 5,000$. The suitable powers of 2 less than p are 4, ..., 4,096. For a sensor with three sources and a working frequencies slightly above 2 kHz, the powers of two 4, 8, 16 and 32 remain. Some possible choices for the periods are 36, 44, 24, 40 and 48, resulting in the nominal frequencies 2.78, 2.27, 4.17, 2.5 and 2.08 kHz. The combination with the smallest bandwidth requirement are the periods 44, 40, and 48 with nominal frequencies 2.27, 2.5 and 2.08 kHz.

9.7 A Moisture Sensor

Drying materials is part of many industrial production processes. Typically a maximum water content is specified. In order to be certain that their product exceeds such a specification manufacturers often reduce the moisture content to much lower values than the specified ones. For running drying operations efficiently, one needs sensors measuring continuously the water content of the material to be dried. Because of the high temperatures involved optical sensors, which can make use of optical fibers, are preferable. The moisture sensor at hand exploits the fact that water absorbs infrared light with wavelengths around 1.45 μm strongly. The sensor illuminates its target with the light of three LEDs, one with a peak emission wavelength of 1.3 μm, one with a peak emission wavelength of 1.45 μm and one with a peak emission wavelength of 1.65 μm. A single band of wavelengths is insufficient for distinguishing the moisture content from other influences such as the distance between the target and the sensor. A photodiode picks up the light reflected from the target. In order to be able to separate the response to the light emitted by an individual LED from the light emitted by the others and from the ambient illumination, the sensor employs a multichannel lock-in scheme with square-wave references.

The circuit in Fig. 9.18 is responsible for controlling the LEDs. The amount of light, which is emitted by a single LED and reflected by the target, is proportional to the power radiated in the form of light by this LED. In order to achieve reproducible measurements, the sensor either has to measure the emitted power for each individual LED or keep these powers constant by some other means. Measuring the light output of each individual LED continuously is expensive, therefore the sensor drives each LED with its own programmable constant current supply. By keeping these currents constant the sensor achieves constant light output for each LED, as the amount of light emitted by each LED is almost proportional to the current through the LED. Furthermore, constant current drive prevents thermal runaway and destruction of the diodes, see Sect. 7.3. The amplifier U1-1 together with the transistor Q1 and the shunt R3 form the constant current supply for the first LED D1. By driving the base of transistor Q2 positive, the software can make the transistor Q2 short-circuit D1 and turn it off without interrupting the current through the source. Therefore, the source stays in regulation and is able to provide the correct current to the LED once the software asks Q2 to stop conducting by pulling the signal LedA to ground. The digital-to-analog converter U2 programs the current for each of the three current sources. The ferrite beads L1 and L2 keep high-frequency interference from leaking into the sensitive analog electronics via the power supply lines. The beads L3 to L11 clean up the signal lines.

Before we can discuss the circuit for signal conditioning the target's response in detail, we have to specify the frequencies for the LEDs and the sampling frequency for the analog-to-digital conversion. As mentioned before, we have to expect severe interference from fluorescent lights at twice the mains frequency. While the photodiode used in the sensor is not really sensitive to visible light, enough signal is picked up optically[3] to become a problem. Introducing a highpass filter as soon as possible of reasonable, lets say second order, will help. A second-order filter provides 40 dB per decade of attenuation in the stop-band, see Sect. 6.2, therefore a corner frequency of 1 kHz for the filter is acceptable. Choosing frequencies above 2 kHz for the LEDs gets us well away from the filter's knee. Choosing $p = 5280$, $p_1 = 40$, $p_2 = 44$, and $p_3 = 48$ and a sample rate of about 100 kHz places the frequencies for the LEDs between 2 and 2.5 kHz. A second-order lowpass filter with a corner frequency of 5 kHz gives us 40 dB attenuation at the Nyquist frequency of 50 kHz. As we do not

---

[3] The coupling path being optical and not electrical can be verified by observing the output of the first amplifier while the sensor is turned away from and while it is pointed toward the lights.

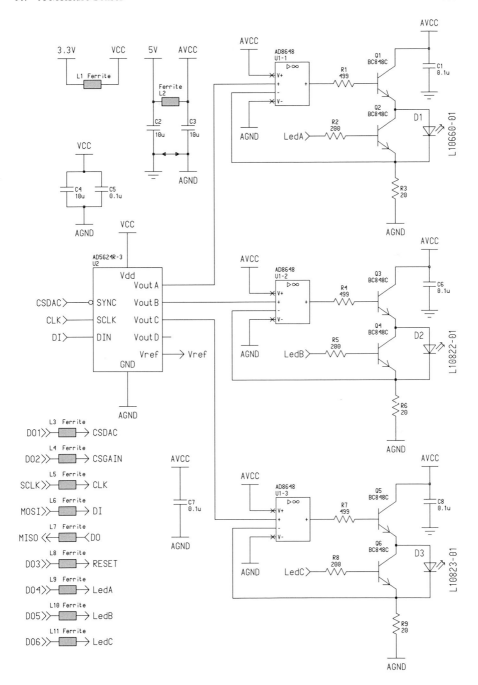

Fig. 9.18 Moisture sensor, LED control

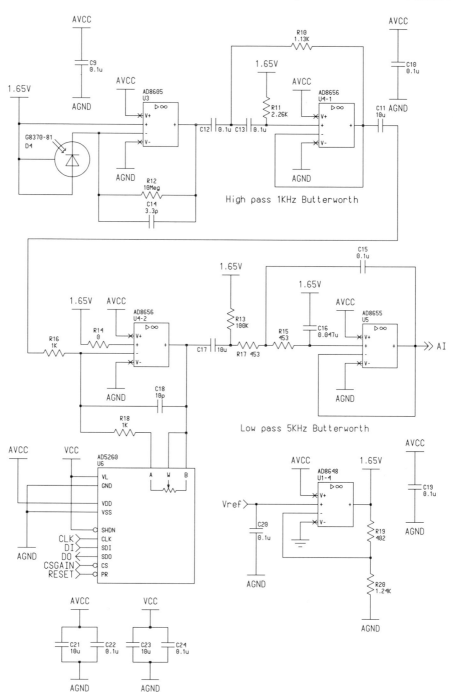

Fig. 9.19 Moisture sensor, photodiode amplifier

expect much signal content at that frequency, we deem the attenuation adequate for a 12-bit converter.

The photodiode in Fig. 9.19 produces a current, which is proportional to the radiant flux striking the diode for a wide range of fluxes. The amplifier U3 acts as transimpedance amplifier, which converts the diode's current into a voltage with a gain of $10\,M\Omega$. The circuit around U4-1 is a highpass filter of second order in Sallen-Key topology. Next follows an inverting amplifier with gain between 1 and 200 programmable in 256 steps. The digital potentiometer U6 wired into the feedback path of the amplifier U4-2 allows the software to set the gain. The last stage around the amplifier U5 is the anti-aliasing filter. All stages work with a single supply only. Therefore the stages with capacitors to block DC. The amplifier U1-4 buffers a voltage of 1.65 V, which is used as virtual ground. Together with a microcontroller and a power supply, these circuits form a complete sensor system.

The sensor's software consists mainly of the interrupt service routine DOSAMPLEMSQ, Fig. 9.20, for serving the "conversion complete" interrupt of the analog-to-digital converter. Upon startup, the software configures the hardware so that the analog-to-digital converter is triggered at the sampling rate. Initially we have to set the counters $c_1, \ldots, c_n$ and c and the accumulators $I_1, \ldots, I_n$ and $Q_1, \ldots, Q_n$ to zero. While a hundred thousand interrupts per second seem excessive, measurements indicate a load of about 50 % on a 32-bit processor running at 80 MHz, which is quite doable. In order to achieve accurate timing, the priority of the conversion complete interrupt must be highest. The program estimates how much of the ith LED's light is reflected by computing $\sqrt{I_i^2 + Q_i^2}$. Using these estimates, a linear regression model

Fig. 9.20 Procedure DOSAMPLEMSQ for processing one sample M using n sources and square-wave references

```
1: procedure DOSAMPLEMSQ(M)
2:     for all i ∈ {1,...,n} do
3:         if c_i < p_i/2 then
4:             I_i ← I_i + M
5:         else
6:             I_i ← I_i − M
7:         end if
8:         if p_i/4 ≤ c_i < 3p_i/4 then
9:             Q_i ← Q_i + M
10:        else
11:            Q_i ← Q_i − M
12:        end if
13:        c_i ← (c_i + 1)  mod p_i
14:        prepare to write c_i < p_i/2 to source S_i
15:    end for
16:    update sources S_1,...,S_n simultaneously
17:    c ← c + 1
18:    if c = p then
19:        report I_1,...,I_n, Q_1,...,Q_n
20:        c ← 0
21:        I_1 ← Q_1 ← ··· ← I_n ← Q_n ← 0
22:    end if
23: end procedure
```

provides the moisture of the target. For different types of targets specific models have to be built, see e.g. Hintenaus and Trinker (2013).

9.8 Bibliographical Notes

Cosens in (1934) introduced an instrument based on lock-in. Dicke used lock-in detection when designing a radiometer (Dicke 1946). Our description of a lock-in in Sects. 9.1 and 9.2 follow the presentation by Blair and Sydenham in (1975).

Lock-in amplifiers, which generate the reference internally, are used with sources like light emitting diodes that can be modulated rapidly (Wang 1990; Barragn et al. 2001; Oh et al. 2003; Carrato et al. 1989; Pogue et al. 1997; Dixon and Wu 1989; Pei et al. 2008; Liu et al. 2012). Restelli et al. describe an implementation in logic using an internally generated sinusoidal reference (Restelli et al. 2005).

For lock-in amplifiers which allow an external reference see e.g. Alonso et al. (2003), Dixon and Wu (1989), Barone et al. (1995), Josephs et al. (1987). For an implementation using a digital signal processor, see Han et al. (2012).

Early digital lock-in amplifiers (Morris and Johnston 1968; Cova et al. 1979; Carrato et al. 1989) employ a voltage-to-frequency converter and a counter for analog-to-digital conversion, see e.g. (Kester and Analog Devices Inc. Engineering 2005). Such schemes lend themselves both to architectures with internal as well as external reference signals.

Lock-in amplifiers for simultaneous measurements of several parameters use either a dedicated detector for each single source or a single detector for all sources. For lock-in amplifiers which impose a single reference onto a target via several sources and which use a separate detector for each source, see Probst and Jaquier (1994), Albertini and Kleeman (1997) for systems with internal reference and (Machel and von Ortenberg 1995) for a system with external reference. Hielscher et al. use a lock-in with several sources and a single detector for optical tomography of human tissues (Schmitz et al. 2002). In optical tomography systems Flexman, Hielscher et al. use a combination of both approaches resulting in a system that applies several references to several sources and picks up the responses with several detectors (Flexman et al. 2011, 2012).

The presentation in Sect. 9.6.1 follows (Masciotti et al. 2008). There Masciotti, Lasker, and Hielscher introduce a lock-in scheme in which a single detector picks up the target's responses to several sinusoidal references. Hintenaus and Trinker (2013) use nonsinusoidal references, square-wave ones in particular. They state the rule (due to Trinker) for picking each reference's frequency, which allows one to avoid crosstalk.

9.9 Lab Exercises

Exercise 9.1 On a microcontroller with integrated analog-to-digital converter implement a phase-locked loop producing a digital output that is synchronized to an analog input signal. Determine the range of input frequencies the loop locks onto reliably.

Exercise 9.2 Implement a lock-in with several sources and a single detector on a microcontroller with an integrated analog-to-digital converter. Simulate the target and the detector with the circuit in Fig. D.15. Use digital outputs as sources. Implement both sinusoidal and square-wave references. Demonstrate crosstalk between the channels by choosing the observation time and the frequencies of the references in violation of the design rules. Show that crosstalk will be eliminated by following the design rules.

References

Albertini A, Kleeman W (1997) Analogue and digital lock-in techniques for very-low-frequency impedance spectroscopy. Meas Sci Tech 8:666–672

Allan D (1987) Should the classical variance be used as a basic measure in standards metrology? IEEE Trans Instrum Meas IM–36(2):646–654. doi:10.1109/TIM.1987.6312761

Alonso R, Villuendas F, Borja J, Barragan LA, Salinas I (2003) Low-cost, digital lock-in module with external reference for coating glass transmission/reflection spectrophotometer. Measur Sci Technol 14:551–557

Analog Devices (2010) Precision weigh scale design using a 24-bit sigma-delta ADC with internal pga and AC excitation, circuit note CN-0155. http://www.analog.com/static/imported-files/circuit_notes/CN0155.pdf

Barone F, Calloni E, DiFiore L, Grado A, Milano L, Russo G (1995) High-performance modular digital lock-in amplifier. Rev Sci Instrum 66:3697–3702

Barragn LA, Artigas JI, Alonso R, Villuendas F (2001) A modular, low-cost, digital signal processor-based lock-in card for measuring optical attenuation. Rev Sci Instrum 72:247–251

Blair DP, Sydenham PH (1975) Phase sensitive detection as a means to recover signals buried in noise. J Phys E: Sci Instrum 8:621–827

Carrato S, Paolucci G, Tommsini R, Rosei R (1989) Versatile low-cost digital lock-in amplifier suitable for multichannel phase-sensitive detection. Rev Sci Instrum 60:2257–2259

Cosens CR (1934) A balance-detector for alternating-current bridges. Proc Phys Soc 46(6):818

Cova S, Longoni A, Freitas I (1979) Versatile digital lock-in detection technique: application to spectrofluorometry and other fields. Rev Sci Instrum 50:296–301

Dicke RH (1946) The measurement of thermal radiation at microwave frequencies. Rev Sci Instrum 17:275–286

Dixon PK, Wu L (1989) Broadband digital lock-in amplifier techniques. Rev Sci Instrum 60:3329–3336

Flexman ML, Khalil MA, Kim HK, Fong CJ, Desperito E, Hershman DL, Barbour RL, Hielscher AH (2011) Digital optical tomography system for dynamic breast imaging. J Biomed Opt 16:1–16

Flexman ML, Kim HK, Stoll R, Khalil MA, Fong CJ, Hielscher AH (2012) A wireless handheld probe with spectrally constrained evolution strategies for diffuse optical imaging of tissue. Rev Sci Instrum 83(3):033108. doi:10.1063/1.3694494

Han X, Ding P, Xie J, Shi J, Li L (2012) Precise measurement of the inductance and resistance of a pulsed field magnet based on digital lock-in technique. IEEE Trans Appl Supercond 22

Hintenaus P, Trinker H (2013) Multifrequency lock-in detection with nonsinusoidal references. IEEE Trans Instrum Meas 62(4):785–793. doi:10.1109/TIM.2013.2240095

Josephs RM, Compton DS, Krafft CS (1987) Application of digital signal processing to vibrating sample magnetometry. IEEE Trans Magn MAG–23:241–244

Kester (ed) W, Analog Devices Inc Engineering (2005) Data Conversion Handbook. Elsevier-Newnes, Amsterdam-London, http://www.analog.com/library/analogDialogue/archives/39-06/data_conversion_handbook.htm

Liu Z, Zhu L, Koffman A, Waltrip BC, Wang Y (2012) Digital lock-in amplifier for precision audio frequency bridge. In: Conference on Precision Electromagnetic Measurements (CPEM), pp 586–587.10.1109/CPEM.2012.6251065

Machel G, von Ortenberg M (1995) A digital lock-in-technique for pulsed magnetic field experiments. Physica B 211:355–359

Masciotti J, Lasker J, Hielscher A (2008) Digital lock-in detection for discriminating multiple modulation frequencies with high accuracy and computational efficiency. IEEE Trans Instrum Meas 57:182–189

Morris ED, Johnston HS (1968) Digital phase sensitive detector. Rev Sci Instrum 39:620–621

Oh TI, Lee JW, Kim KS, Lee JS, Wu EJ (2003) Digital phase-sensitive demodulator for electrical impedance tomography. In: Engineering in Medicine and Biology Society, 2003. Proceedings of the 25th Annual International Conference of the IEEE, Cancun, Mexico, pp 1070–1072

Pei R, Velichko A, Jiang Y, Hong Z, Katayama M, Coombs T (2008) High-precision digital lock-in measurements of critical current and ac loss in hts 2g-tapes. In: SICE Annul Conference 2008. pp 3147–3150. doi:10.1109/SICE.2008.4655206

Pogue BW, Testorf M, McBride T, Osterberg U, Paulsen K (1997) Instrumentation and design of a frequency-domain diffuse optical tomography imager for breast cancer detection. Opt Express 1:391–403

Probst PA, Jaquier A (1994) Multiple-channel digital lock-in amplifier with PPM resolution. Rev Sci Instrum 65:747–750

Restelli A, Abbiati R, Geraci A (2005) Digital field programmable gate array-based lock-in amplifier for high-performance photon counting applications. Rev Sci Instrum 76(9):1–8. doi:10.1063/1.2008991

Schmitz CH, Locker M, Lasker JM, Hielscher AH, Barbour RL (2002) Instrumentation for fast functional optical tomography. Rev Sci Instrum 73:429–439

Schrüfer E, Reindl L, Zagar B (2012) Elektrische Messtechnik, Messung elektrischer und nichtelektrischer Größen. Hanser, Berlin

Wang X (1990) Sensitive digital lock-in amplifier using a personal computer. Rev Sci Instrum 61:1999–2001

Chapter 10
Short-Range Radar

Radar uses radio waves to detect distant targets and typically measures their distance and/or velocity. A radar consists of a transmitter that emits radio waves and a receiver that receives the radio waves that are reflected off of distant objects. There are two kinds of radar systems: pulse radars, which transmit the radio energy in short pulses, and continuous-wave radars, which transmit continuously. The radar systems that are used for air-traffic control are of the former type, the ones used for measuring the velocity of motor vehicles are of the latter. Short-range radars of the frequency-modulated continuous-wave type increasingly find their applications in automobiles for collision avoidance.

While studying the Doppler radar for measuring the velocity of some object, we stumble upon a complex-valued signal whose frequency encodes the velocity information we are after. The Fourier transforms recommend themselves as the methods of choice in this situation. We argue why we can apply the discrete Fourier transform to a sampled version of part of this complex-valued signal to arrive at the velocity information. The frequency-modulated continuous-wave radar allows us to measure the distance and the velocity of an object simultaneously. Finally, we discuss the design of a simple continuous-wave radar.

10.1 The Unmodulated Continuous-Wave Radar

When a moving target reflects a radio wave, the frequency of the reflected wave is shifted by the so-called Doppler shift. Unmodulated continuous-wave radars employ this Doppler shift to measure the speed and the direction of the target.

10.1.1 The Doppler Shift

The transmitter in a continuous-wave (CW) radar generates a radio-frequency signal X with constant frequency f in order to emit a radio wave with the same frequency;

© Springer International Publishing Switzerland 2015
P. Hintenaus, *Engineering Embedded Systems*, DOI 10.1007/978-3-319-10680-9_10

thus, the wave's wavelength is $\lambda = \frac{c}{f}$, where c is the speed of light in air. A target T that approaches the radar with speed v observes this wave with the shorter wavelength

$$\lambda_T = \frac{c}{f} - \frac{v}{f},$$

therefore the frequency at the target is $f_T = \frac{cf}{c-v}$. The difference $f_d = f_T - f$ is called the Doppler shift. Substituting f_T and simplifying results in

$$f_d = \frac{fv}{c-v}.$$

For targets moving at speeds that are small when compared with the speed of light f_d can be approximated by

$$f_d = \frac{fv}{c}.$$

By applying this twice, once from the radar to the target and once the other way round, we compute the Doppler shift at the radar,

$$f_d + \frac{(f + f_d)v}{c} = \frac{2fv}{c} + \frac{fv^2}{c^2}.$$

As v^2/c^2 is very small for a small speed v, the Doppler shift observed by the radar is $2f_d$. For a receding target, the Doppler shift is negative.

A target that approaches a radar transmitting at 24 GHz with a speed of $1\,\mathrm{ms}^{-1}$, about the speed of a slow pedestrian, produces a Doppler shift at the radar of

$$2\frac{24\,\mathrm{GHzms}^{-1}}{300{,}000\,\mathrm{kms}^{-1}} = 160\,\mathrm{Hz},$$

which is well within the audible range.

10.1.2 Separating the Doppler Shift from the Carrier

The velocity information contained in the signal received by a CW Doppler radar is contained in a small frequency shift $2f_d$ on top of the frequency f the radar transmits at. To separate $2f_d$ from f one uses a mixer. A mixer, Fig. 10.1, is a circuit with two inputs. The input IN is fed with the received signal r, the input LO is fed with the output signal l of a so-called local oscillator, which is part of the circuit around the mixer. An ideal mixer multiplies the two signals r and l. The output OUT carries the product. If both r and l are sinusoidal with frequencies f_r and f_l their product is the sum of two sinusoidal signals, with frequencies $f_r - f_l$ and $f_r + f_l$. In most circuits,

Fig. 10.1 A mixer with a single output

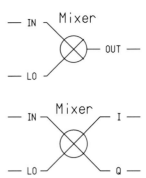

Fig. 10.2 A mixer with in-phase and quadrature outputs

a bandpass (or lowpass) filter follows a mixer to attenuate the unwanted summand. A simple CW Doppler radars uses the signal X it generated for transmission and a lowpass filter in order to move the received signal into the audible frequency range where it can be processed easily.

After lowpass filtering the sign of the frequency of the resulting signal cannot be recovered anymore, therefore the information whether the target is approaching or receding has been lost. In order to preserve this information, the actually employed mixer, Fig. 10.2, produces two outputs, the in-phase part $I = rl$ at the output I and the quadrature part $Q = rl'$ at the output Q, where the signal l' is derived from l by shifting the phase of l by a quarter period.

10.1.3 Down-Conversion of the Doppler Shift

To simplify the following discussion we introduce the angular frequencies $\omega_X = 2\pi f$ and $\omega_d = 2\pi f_d$, thus the transmitted signal $X(t)$ is $A_X \cos(\omega_X t)$ with amplitude A_X and the received signal $R(t)$ is $A_R \cos((\omega_X + 2\omega_d)t + \phi)$ for some phase shift ϕ and amplitude A_R. Note that the information we are interested in is contained in ω_d, the phase shift ϕ is an artifact introduced by the distance to the target, the target itself and the receiving circuit. The product of the transmitted signal and the received signal is

$$X(t)R(t) = A_X A_R \cos(\omega_X t) \cos((\omega_X + 2\omega_d)t + \phi)$$
$$= \frac{A_X A_R}{2} \left(\cos(2(\omega_X + \omega_d)t + \phi) + \cos(2\omega_d t + \phi)\right).$$

After a lowpass filter has removed the content at approximately twice the transmit frequency, the in-phase part remains

$$I(t) = \frac{A_X A_R}{2} \cos(2\omega_d t + \phi).$$

The product of the phase-shifted version $X'(t) = X(t - \frac{\pi}{2\omega_X})$ of $X(t)$ and the received signal $R(t)$ is

$$X'(t)R(t) = A_X A_R \sin{(\omega_X t)} \cos{((\omega_X + 2\omega_d)t + \phi)}$$
$$= \frac{A_X A_R}{2} (\sin{(2(\omega_X + \omega_d)t + \phi)} - \sin{(2\omega_d t + \phi)}).$$

After lowpass filtering, the quadrature part remains

$$Q(t) = -\frac{A_X A_R}{2} \sin{(2\omega_d t + \phi)}.$$

For further processing, we interpret I and Q as the real and imaginary parts of a complex signal $y(t)$,

$$y(t) = I(t) + iQ(t)$$
$$= \frac{A_X A_R}{2} (\cos{(2\omega_d t + \phi)} - i \sin{(2\omega_d t + \phi)})$$
$$= \frac{A_X A_R}{2} e^{-i(2\omega_d t + \phi)}. \tag{10.1}$$

For a target that approaches the radar ω_d is greater than zero, therefore y rotates clockwise in the complex plane as t increases. For a receding target, ω_d is less than zero and y rotates counterclockwise.

10.1.4 Extracting Speed Information

Suggested by (10.1) we represent the signal $y(t)$ in the frequency domain. Then each target corresponds to an amplitude peak in the spectral representation $Y(\omega)$ of $y(t)$. A peak at an angular frequency ω_T corresponds to a target T moving at a velocity v_T of

$$v_T = c\frac{\omega_T}{2\omega_X}$$

toward the radar.

For a finite number of targets n_T, $y(t)$ is the sum of n_T sinusoidal signals of unknown frequencies. Therefore, the complex signal $y(t)$ is continuous, band-limited and periodic, but the period is unknown. In this case, the mathematically correct transform is the continuous-time Fourier transform

$$Y(\omega) = (\mathcal{F}(y))(\omega) = \int_{-\infty}^{\infty} y(t)e^{-i\omega t} \, dt.$$

Most probably the signal $y(t)$ is not absolutely integrable so we have to accept Dirac deltas in the transform. The problem with this transform is, however, that observing $y(t)$ for an infinite time span is physically impossible. Therefore, we choose a time τ and approximate $Y(\omega)$ by

$$
\begin{aligned}
Y(\omega) \approx \tilde{Y}(\omega) &= \int_{-\tau}^{\tau} y(t) e^{-i\omega t} \, dt \\
&= \int_{-\infty}^{\infty} y(t) b_\tau(t) e^{-i\omega t} \, dt \\
&= (\mathcal{F}(y b_\tau))(\omega) \\
&= \frac{1}{2\pi} (\mathcal{F}(y) * \mathcal{F}(b_\tau))(\omega)
\end{aligned}
\tag{10.2}
$$

where $b_\tau(t)$ is the so-called boxcar signal of length 2τ,

$$
b_\tau(t) = \begin{cases} 1 & \text{if } -\tau \le t \le \tau \\ 0 & \text{otherwise,} \end{cases}
$$

and $(F * G)(\omega)$ is the convolution of the signals F and G.

In order to understand the errors we introduce by considering $\tilde{Y}(\omega)$ instead of $Y(\omega)$ we study the situation for

$$
\hat{y}(t) = e^{i\omega_0 t},
$$

where $\omega_0 \in \mathbb{R}$ is some angular frequency. The continuous-time Fourier transform of $\hat{y}(t)$ is the single spectral line $(\mathcal{F}(\hat{y}))(\omega) = 2\pi \delta(\omega - \omega_0)$, $\delta(\omega)$ is the Dirac delta. The continuous-time Fourier transform of $b_\tau(t)$ is

$$
(\mathcal{F}(b_\tau))(\omega) = 2\frac{\sin(\omega\tau)}{\omega},
$$

and the transform of the product $\hat{y}(t) b_\tau(t)$ is

$$
2\frac{\sin((\omega - \omega_0)\tau)}{\omega - \omega_0}.
$$

Note that this function has zeros at $\omega_0 + \frac{k\pi}{\tau}$ for $k \in \mathbb{Z}$, $k \ne 0$. Approximating the continuous Fourier transform by observing the input for the finite time span $[-\tau, \tau]$ only replaces a line at ω_0 with the Fourier transform of b_τ shifted to ω_0. It is possible to use a different function $w(t)$, a so-called window function, being 0 outside of $[-\tau, \tau]$ as multiplier in (10.2) for cutting a piece of finite length out of $\hat{y}(t)$. This choice determines the error introduced by observing $\hat{y}(t)$ for only a finite amount of time.

In order to be able to compute with the signal $y(t)$, we have to sample it using two synchronously operating analog-to-digital converters. Let $y_i = y(-\tau + i\frac{2\tau}{n})$ for integers i with $0 \le i < n$. For satisfying the Nyquist criterion, the number of samples n, taken during the observation time, has to be chosen such that $\frac{n}{2\tau}$ is bigger than twice the highest frequency contained in $y(t)$. To be mathematically correct, we have to treat the series $\langle y_0, \ldots, y_{n-1} \rangle$ as part of an aperiodic discrete signal. The discrete-time Fourier transform of the signal $y_i' = y_i$ for $0 \le i < n$, $y_i' = 0$ otherwise, is a periodic continuous function $Y'(\omega)$, which does not lend itself easily to programmatic treatment. The continuous-time transform $Y(\omega)$ is zero outside $-\frac{\pi n}{2\tau} < \omega < \frac{\pi n}{2\tau}$, as the signal $y(t)$ is band-limited accordingly. The discrete-time Fourier transform $Y'(\omega)$ is related to the continuous-time transform by

$$Y'(\omega) = \frac{n}{2\tau} Y\left(\frac{n\omega}{2\tau}\right), \tag{10.3}$$

for $-\pi < \omega < \pi$.

When we construct a discrete periodic signal by extending the sequence $\langle y_0, \ldots, y_{n-1} \rangle$ periodically, we can compute the discrete Fourier transform coefficients Y_i for $i \in \mathbb{Z}$. The discrete Fourier transform coefficients are related to the discrete-time transform $Y'(\omega)$ by

$$Y_k = Y'\left(k\frac{2\pi}{n}\right). \tag{10.4}$$

Combining (10.3) and (10.4) we get

$$Y_k = Y'\left(k\frac{2\pi}{n}\right) = \frac{n}{2\tau} Y\left(\frac{k\pi}{\tau}\right).$$

Subsequent discrete Fourier transform coefficients are mapped to frequencies in the continuous-time transform, which are separated by an angular frequency of $\frac{\pi}{\tau}$. Converting back from angular frequencies to frequencies, we observe that the frequency resolution attainable by using the discrete Fourier coefficients is the inverse of the observation time. The speed resolution Δv of the CW Doppler radar is

$$\Delta v = \frac{c}{4\tau f}.$$

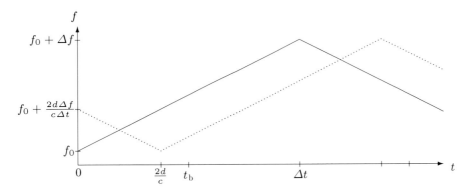

Fig. 10.3 Triangle modulation and the return of a target at distance d. The *solid line* is the frequency of the transmitted wave. The *dotted line* is the frequency of the reflected wave when the wave is received by the radar

10.1.5 Spectral Leakage

The signal $e^{i\frac{k\pi}{\tau}t}$ for $k \in \mathbb{Z}$ exhibits exactly k periods during the observation time τ. If we use a boxcar function for making the observation time finite, then it contributes to the lth discrete Fourier transform coefficient for some $0 \le l \le n - 1$. All other coefficients in this period are mapped to frequencies in the continuous transform, which are zeros of the shifted transform of the window function. Therefore, these coefficients deliver no contribution. Frequency content in $y(t)$ on the other hand, which does not exhibit full periods during the observation time, contributes to all discrete Fourier transform coefficients according to the shifted transform of the window. This effect is called spectral leakage. Periodic signals which exhibit full periods during the observation time obey the no-leakage condition.

10.2 The Frequency-Modulated Continuous-Wave Radar

A CW Doppler radar cannot detect stationary objects as their echoes do not exhibit a frequency shift and, therefore, mingle with the returns of the surrounding. To tackle this problem, a frequency-modulated CW radar modulates the frequency of the transmitted wave to not only gain information about the velocity of the objects in sight of the radar, but also measure their distance. Most often triangle-shaped modulation is used.

Let us assume that the radar sweeps the frequency of the transmitted wave upward from f to $f + \Delta f$ in the time interval from 0 to Δt and downward in the interval from Δt to $2\Delta t$, Fig. 10.3. The round-trip time for an electromagnetic wave between the radar and a stationary target at distance d is $\frac{2d}{c}$. Therefore, the reflected wave arrives at the radar with a frequency offset o_d of

$$o_d = \pm \frac{2d\Delta f}{c\Delta t}, \tag{10.5}$$

where the offset is negative during an up-sweep and positive during a down-sweep. Right after the radar has switched from an up-sweep to a down-sweep or vice versa, it has to skip data for a time t_b, depending on the radar's maximum range, to avoid combined processing of reflections stemming from different sweeps.

Like in the CW Doppler radar, an FMCW radar down-converts and samples the received signal. For short-range radars with a sweep time Δt of about 10 ms and a range of several hundred meters, we can ignore the time t_b and use Δt as observation time. When we equate the frequency resolution of $\frac{1}{\Delta t}$ with the absolute value of the frequency offset (10.5) and solve for the distance d, the distance resolution Δd turns out to be

$$\Delta d = \frac{c}{2\Delta f}.$$

To compute distance and velocity of a target, we combine an up-sweep and the following down-sweep. While the frequency shift in the received wave due to the frequency modulation changes sign between these sweeps the Doppler shift does not. Assuming that we are able to assign a peak in the spectrum computed during the up-sweep and a peak in the spectrum of the down-sweep to the same target, the doubled FMCW shift is the difference of the frequencies of the two peaks, and the doubled Doppler shift is their sum.

10.3 Implementation of a Continuous-Wave Radar

Our radar can be operated both in the CW Doppler and the FMCW mode. It transmits in the 24 GHz ISM[1] band. Operating equipment in this frequency band does not require licensing by the authorities. The legal bandwidth of 250 MHz allows for a resolution not better than 0.6 m in the FMCW mode.

10.3.1 Overview

A modulation block, *Mod* in Fig. 10.4, optionally provides modulation to the voltage-controlled oscillator, *Vco*, which generates the high-frequency signal. The power amplifier *Pa* amplifies this signal, and the directional transmitting antenna transmits it. The receiving antenna picks up the signal that has been bounced back from the targets. The down-converter consists of a mixer and two lowpass filters, one for the in-phase part *I* and one for the quadrature part *Q*. The Fourier transform block *Ft*

---

[1] Industrial, Scientific, and Medical.

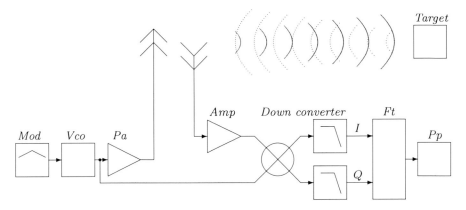

Fig. 10.4 Block diagram of a simple continuous-wave radar. The target is approaching the antennae

converts I and Q to digital and performs a complex fast Fourier transform to produce a spectrum. The optional peak processor Pp analyzes the spectrum for peaks, identifies targets, and computes distance and velocity for each target.

10.3.2 The Analog and High-Frequency Signal Path

The schematic in Fig. 10.5 shows the analog and high-frequency signal path. The heart of the radar is the module U1.[2] It contains the voltage-controlled oscillator, a power amplifier, the transmitting antenna, the receiving antenna, a high-frequency amplifier, the mixer, and two audio-frequency amplifiers. As the radar module is operating from a single 5 V supply, and does not tolerate voltages outside the 0 to 5 V range, we designed the analog circuitry around it to operate from the same supply.

The modulation amplifier U2A amplifies the modulation signal to provide enough voltage at the tune input of the radar module. The resistor R3 serves as protection. It limits the current from the amplifier, should the modulation signal become negative because of a software problem. The gain of the modulation amplifier is 1.65. The amplifiers U3A and U3B amplify the signals I and Q. Each of them forms a first-order lowpass filter to prevent aliases in the subsequent analog-to-digital conversion with a corner frequency of 10 kHz and a gain of -33 in the pass band.[3] As usual in single supply circuits, direct current to these amplifiers is blocked by capacitors (C3 and C5) to prevent the amplifiers from clipping permanently at one of their supply rails. The resistors R10 and R11, the capacitor C8 and the amplifier U2B provide 2.5 V at a low impedance for the other amplifiers to use as a virtual ground.

---

[2] IVS-148, InnoSent GmbH.

[3] In Salzburg's wet weather this gain is fine, in dryer climates it has to be reduced by increasing the values of the resistors R4 and R7.

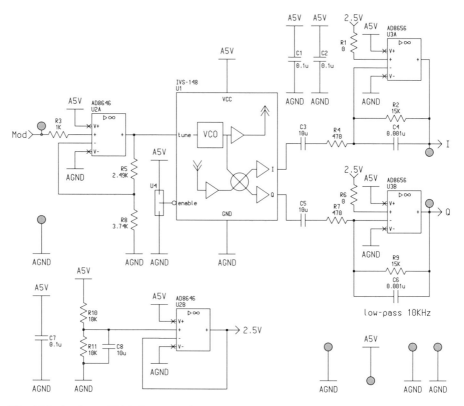

Fig. 10.5 Analog and high-frequency signal path

10.3.3 The Digital Signal Processor and its Support Circuits

The digital signal processor U4 in Fig. 10.6 is an ADSP-21363, a floating-point processor manufactured by Analog Devices. It controls the radar, provides the modulation, computes the spectra, and communicates the results to the outside world. The clock oscillator X1 provides a square-wave clock at 24.576 MHz to the processor and the data converters. As the processor chip uses an internal phase-locked loop, which is an analog block itself, supplying the processor will not disturb the clock signal so much as to render it unusable for the data converters. The JTAG interface U6 is a connector, which is used during software development to connect an emulator, which in turn connects to the computer running the development software.

The reset generator U5 monitors both the 3.3 and 1.2 V supplies of the processor. It activates the processor once both supplies are stable and within the range required for correct operation of the circuit. The push-button J1 resets the processor too—aiding

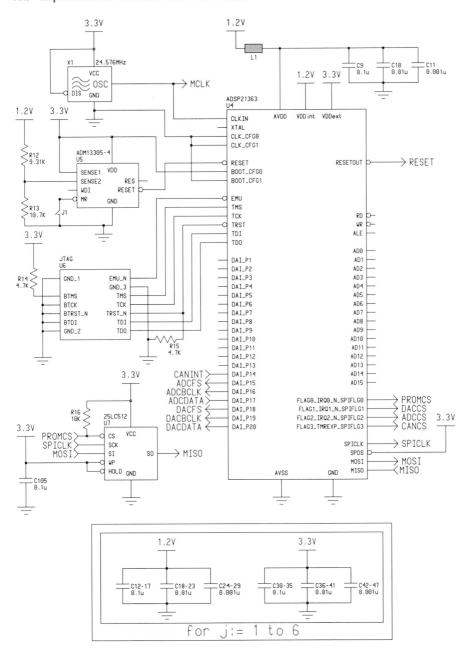

Fig. 10.6 Digital signal processor and support circuits

in software development. The processor itself provides a system-wide signal RESET, to set all peripheral components to a well-defined state on startup.

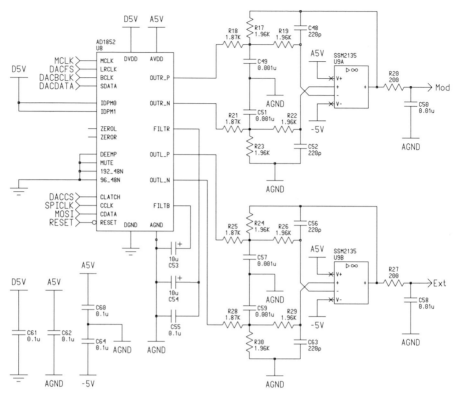

Fig. 10.7 Dual channel digital-to-analog converter with reconstruction filters

The processor is strapped to boot from the serial EEPROM U7 via its SPI interface, consisting of the signals SPICLK, MISO, and MOSI. Pullup resistors on the chip-select signals ADCCS and CANCS are crucial, as the processor does not drive these pins to a high level right after reset. As the data converters, the CAN interface and the EEPROM share the same SPI bus, forgetting these resistors results in contention on signal MISO after reset, which prevents the processor from booting.

10.3.4 The Digital-to-Analog Converter

The digital-to-analog converter U8 converts both channels synchronously and periodically, Fig. 10.7. Because of its Σ-Δ architecture it introduces a group delay of $903.8\,\mu s$ at a sampling rate of $48\,kHz$ at its digital inputs. This group delay has to be taken into account by the radar's software. The sample rate at its analog outputs is $6.144\,MHz$. We use one of its channels for converting the modulating waveform, the other is left unused. The digital data are streamed continuously over one of the digital

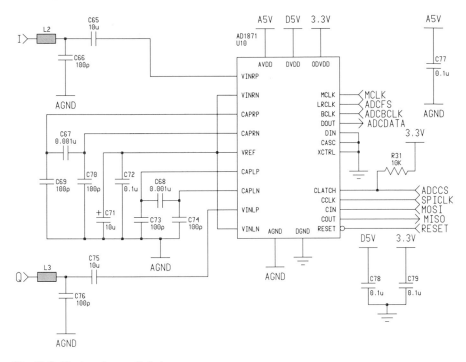

Fig. 10.8 Dual analog-to-digital converter

signal processor's serial ports using the signal DACFS to indicate the start of a new pair of words for the two channels, the signal DACBCLK as clock for the individual bits, and the DACDATA for transferring the data bits. After reset, the digital-to-analog converter is configured via SPI.

The design of the reconstruction filters, U9A and U9B, is taken from the converter's data sheet. They are third-order lowpass filters with Bessel characteristics, a corner frequency of 75 kHz and a group delay of about 3.5 µs. At the Nyquist frequency of 3.072 MHz, the filters attenuate by more than 90 dB.

10.3.5 The Analog-to-Digital Converter

The analog-to-digital converter U10 periodically and synchronously converts both of its channels, Fig. 10.8. Owing to its Σ-Δ architecture, it introduces a group delay of 910 µs, which has to be allowed for when writing the software. Its digital output rate is 48 kHz, its analog input rate is 6.144 MHz. The converter streams the digital data continuously to one of the signal processor's serial ports. The serial port provides the signals ADCFS, which indicates the start of a new pair of words, and ADCBCLK, the

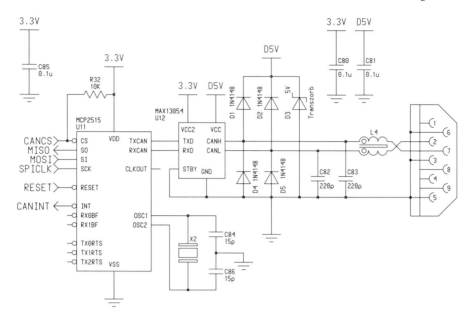

Fig. 10.9 CAN bus interface

bit clock for the data bits. The converter transfers the data to the serial port using the
signal ADCDATA. After reset, the signal processor initializes the converter using SPI.

The first-order lowpass filters in the analog signal path prevent aliasing during the
analog-to-digital conversion. Their corner frequency of 10 kHz limits the maximum
detectable speed to more than 225 km h^{-1} in the Doppler mode and to a maximum
range larger than 300 m for a sweep time of 50 ms in FMCW mode. These filters
provide an attenuation of about 48 dB at the Nyquist frequency of 3.072 MHz, which
is adequate for the radar module's signal to noise ratio. The passive network at the
analog inputs of the converter is taken from its data sheet. While the resolution of the
converter is not required in this application, its synchronous dual channel architecture
makes it a viable choice.

10.3.6 The Communication Interface

The radar communicates with the outside world using a CAN bus, Fig. 10.9. The
signal processor communicates with the CAN interface U11 using SPI. The CAN
transceiver U12 provides the electrical interface to the bus. The diode network D1 to
D5 clamps the bus lines between 0 and 5 V. The common-mode choke, L4, passes the
differential CAN signals but blocks common-mode interference.

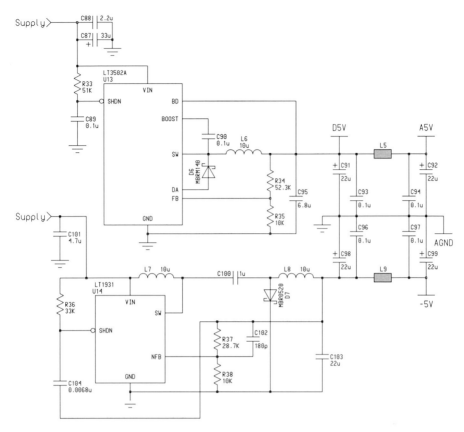

Fig. 10.10   +5 V and −5 V power supply

10.3.7 The Power Supplies

The radar system requires an external supply of 12 V and about 200 mA at the connectors J2 and J3, Figs. 10.10 and 10.11. The switching regulator U13 steps down the input voltage to 5 V. It operates at a switching frequency of 2.2 MHz. The regulator U14 is a Ćuk converter operating at 1.2 MHz. It converts the input voltage to −5 V. The network consisting of the capacitors C91 to C93 and the ferrite bead L5 removes unwanted high-frequency content to produce the positive analog supply A5 V. The network consisting of the capacitors C96 to C99 and the ferrite bead L9 filters the output of the Ćuk converter in order to provide the negative analog supply −5 V.

The dual step-down converter U15, operating at a switching frequency of 575 kHz, produces the digital supplies of 3.3 and 1.2 V. It is configured to first let the 1.2 V rail reach regulation, before powering the 3.3 V rail. This behavior satisfies the power-up sequencing requirements of the signal processor.

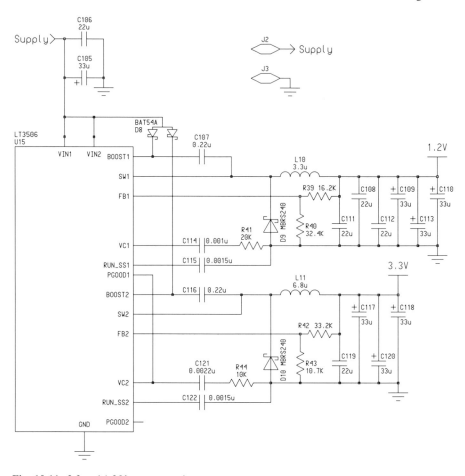

Fig. 10.11 3.3 and 1.2 V power supply

Fig. 10.12 Interrupt service
routine for handling the data
of one sweep

```
1: procedure HANDLEADC
2:    if state = idle then
3:       for i ← 0, n − 1 do
4:          Iᵢ ← d₍₂ᵢ₊₂ₛ₎Wᵢ
5:          Qᵢ ← d₍₂ᵢ₊₂ₛ₊₁₎Wᵢ
6:       end for
7:       state ← processing
8:    end if
9: end procedure
```

10.3.8 Software

The radar's software consists of a MAIN procedure, Fig. 10.13, and the interrupt
service procedure HANDLEADC, Fig. 10.12, for processing the data acquired during

```
 1: procedure MAIN
 2:     for i ← 0, n/2 − 1 do
 3:         (ρ_i, ι_i) ← (cos (2iπ/n), − sin (2iπ/n))
 4:     end for
 5:     for i ← 0, l − 1 do
 6:         r_i ← i
 7:     end for
 8:     initialize window W
 9:     start input-output processor and converters
10:     enable interrupts
11:     loop
12:         while state = idle do
13:             IDLE
14:         end while
15:         transfer I and Q via CAN
16:         FFT(I, Q, ρ, ι, n)
17:         for i ← 0, n − 1 do
18:             (I_i, Q_i) ← (√(I_i^2 + Q_i^2), arctan (Q_i/I_i))
19:         end for
20:         transfer portions of I and Q around frequency 0 via CAN
21:         state ← idle
22:     end loop
23: end procedure
```

Fig. 10.13 Main procedure of the radars software

a single sweep. The software delegates the provision of data to the digital-to-analog converter to the input-output processor, which is part of the ADSP-21363 chip. The input-output processor traverses the ramp r for FMCW mode that has been tabulated in memory back and forth using direct memory access and feeds this data to the converter via a serial port. No more intervention is required after the software has initialized and started this process. For CW Doppler mode, in which no modulation is required, the ramp is replaced with a horizontal line.

The input-output processor is responsible for transferring the data produced by the analog-to-digital converter into memory too. Again it uses direct memory access to transfer the data it received from the converter via a serial port into the array d. In this process, the in-phase part ends up on even indexes and the quadrature part on odd ones. Every time the data produced during a complete sweep has been transferred, the input-output processor delivers an interrupt to the processor in order to wake up the software. To compensate for the group delays of the converters, the length l of each sweep is somewhat longer than the number of samples n that is required for computing the spectra. The software skips the first s samples in order to process data from a single sweep only. The transfers to the digital-to-analog converter and the transfers from the analog-to-digital converter stay synchronized automatically, as both data streams are governed by the same clock signal and operate at the same rate.

The procedures MAIN and HANDLEADC communicate through three shared variables. The variable $state$ keeps HANDLEADC from corrupting the other two variables I and Q while MAIN processes their content. If the software is in the state *idle* when data of a complete sweep have been collected, HANDLEADC converts the sampled data into floating point, windows it, and distributes it into the two arrays I and Q. There is a logic race between HANDLEADC and the input-output processor, but the software wins this race every time, because the signal processor is more than fast enough.

The procedure MAIN initializes the arrays ρ and ι with the twiddles for the fast Fourier transform, the window W and the ramp r. In the main loop, it sits idle till it receives data from HANDLEADC. In its most simple form, the software transmits this data via the CAN bus for inspection, fast Fourier transforms it, and computes magnitude and phase of the spectrum. Before becoming *idle* again, it transfers the relevant portions of the magnitude and the phase spectrum.

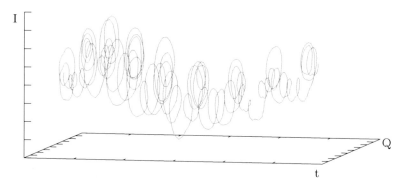

Fig. 10.14   The complex signal $I + jQ$ versus time, recorded for the situation in Fig. 10.15

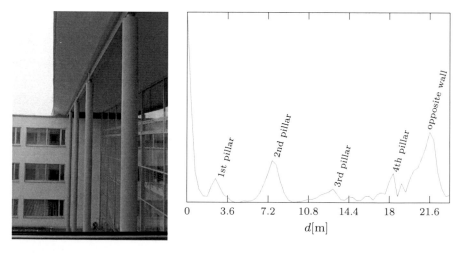

Fig. 10.15   Sample measurement with the radar in FMCW mode. The situation is shown to the *left*, the magnitude spectrum to the *right*. The four pillars and the opposite wall are featured prominently in the spectrum. On the horizontal axis, d is the distance

10.4 Sample Measurements

For the sample measurements, we operated the radar in FMCW mode. We pointed it across the court in Fig. 10.15 at an angle, so that all four pillars were in view of the radar. The data straight from the analog-to-digital converters in Fig. 10.14 are impossible to interpret by hand. In the magnitude spectrum in Fig. 10.15, on the other hand the first three pillars are resolved nicely. The fourth one can be recognized, but the unclear situation in the far corner makes interpretation of the returns hard. The distance between two subsequent pillars is 5.1 m, the distances between the near wall and the first pillar, and between the fourth pillar and the opposite wall are 3.1 m each.

10.5 Bibliographic Notes

The books (Richards et al. 2010) and (Melvin et al. 2012) cover radar technology comprehensively. Bentley in 1928 invented the frequency-modulated continuous-wave radar (Bentley 1935). Radar altimeter used in aircraft employ the FMCW principle for measuring the height above ground. Automotive radars use the FMCW principle for determining velocity and position of several targets. By sweeping the beam horizontally, they produce two-dimensional information. Frequency-modulated continuous-wave radars are also used for probing the environment, see for example (Okorn et al. 2014) for radar-based avalanche prediction.

10.6 Lab Exercises

Exercise 10.1 Change the radar's software so that it operates as continuous-wave Doppler radar. Observe the signals I and Q and the amplitude spectrum for an approaching target and for a receding one.

Exercise 10.2 Familiarize yourself with Kalman Filters. Extend the radar's FMCW mode so that it computes range and velocity of the targets simultaneously. In order to increase the robustness of the measurements use Kalman filters for updating hypotheses about the distance and the velocity of each target.

References

Bentley JO (1935) Airplane altitude indicating system
Melvin WL, Scheer JA, Holm WA et al (2012) Principles of modern radar: advanced techniques.
  SciTech Publishing, New Jersey

Okorn R, Brunnhofer G, Platzer T, Heilig A, Schmid L, Mitterer C, Schweizer J, Eisen O
(2014) Upward-looking L-band FMCW radar for snow cover monitoring. Cold Reg Sci Technol
103(0):31–40
Richards MA, Scheer JA, Holm WA et al (2010) Principles of modern radar: basic principles.
SciTech Publishing, New Jersey

Chapter 11
Infrared Spectrometry

Infrared spectrometers are instruments which measure the intensity distribution of a beam of infrared light with respect to the "colors", more precisely the frequencies, of the components of the beam. As many chemicals absorb infrared light of different frequencies in a way that is characteristic for the chemical in question, that is the chemical species delivers a "fingerprint" in the light spectrum that passes through it, infrared spectrometers are used routinely for chemical analyses.

11.1 Light Waves

In order to be able to understand the working of these instruments, we state some properties of light waves without deeper discussion. Light is an electromagnetic wave. The electric field $\vec{E}(\vec{r}, t)$ and the magnetic field $\vec{B}(\vec{r}, t)$ of a light wave are perpendicular to each other at any point $\vec{r}$ in space and at any time t. The direction of propagation is $\vec{E}(\vec{r}, t) \times \vec{B}(\vec{r}, t)$ and, therefore, perpendicular to both fields. A light wave is linearly polarized if its electric field oscillates in the given direction, the so-called polarization, at any point in space and at any time. The electric field of a monochromatic light wave propagating in the z-direction in vacuum that is polarized in the x-direction has the vector representation

$$\vec{E}(z, t) = \begin{pmatrix} E \cos(2\pi f t - 2\pi \vec{k} \cdot \vec{e}_z z) \\ 0 \\ 0 \end{pmatrix}.$$

The wave's magnetic field oscillates in the y-direction. The temporal frequency of the wave is f, $c = 299{,}792{,}458 \text{ ms}^{-1}$ is the speed of light in vacuum. The vector $\vec{k}$ is the wave vector, it points in the direction of propagation of the wave and its magnitude is $v = \frac{f}{c}$, its spacial frequency, or wavenumber, in vacuum. The vector $\vec{e}_z$ is the unit vector in the z-direction and determines propagation in the z-direction.

© Springer International Publishing Switzerland 2015
P. Hintenaus, *Engineering Embedded Systems*, DOI 10.1007/978-3-319-10680-9_11

The wave's wavelength in vacuum is $\lambda = \frac{c}{f}$. The frequency f of an electromagnetic wave can be stated either by f itself, or by the wavelength λ in vacuum or by the wavenumber v in vacuum. In spectrometry, λ and v are commonly used. A wave is called monochromatic when the oscillation of the electric field at any point in space is purely sinusoidal and of a single, constant frequency. A wave packet is a spatially confined light field, which propagates as one piece. Light emitted by thermal radiators is composed of wave packets: When an electron drops from an excited state to it's ground state, it emits the energy difference in form of a wave pulse. We call a mixture of a number of monochromatic waves (or wave packets) with a distribution of frequencies as "white" light.

The points in space at which the wave oscillates at a constant phase define a wave front. In vacuum and in isotropic media, that is in media in which the speed of propagation is independent of the direction of propagation and of the direction of polarization, the wavefronts are perpendicular to the wave vector $\vec{k}$. The wavefronts of monochromatic parallel beams are planes, the wavefronts produced by a monochromatic point source are spherical.

Light is not just an electromagnetic wave, it is at the same time a beam of particles, the photons, which travel with the speed of light. In vacuum a photon in a light beam with frequency f has the energy

$$E = hf,$$

where $h = 6.62606957 \times 10^{-34}$ J s is the Planck constant. The momentum of this photon is

$$p = \frac{hf}{c}.$$

The momentum of a light beam has been experimentally verified by observing the change of the momentum of an object from which the beam is deflected.

11.1.1 Coherence

For two light waves to interfere constructively, their fields must be of similar phase. The notion of coherence describes what this similarity means. Two light waves with electric fields $\vec{E}_1$ and $\vec{E}_2$ are temporally and spatially coherent at points $\vec{r}_1$ and $\vec{r}_2$ if and only if the electric field of the second wave at the point $\vec{r}_2$, $\vec{E}_2(\vec{r}_2, t)$, and the electric field of the first wave at the point $\vec{r}_1$, $\vec{E}_1(\vec{r}_1, t - \tau)$, exhibit a defined phase relation $\phi(\tau)$, which must depend on τ only, for each time t and each time shift τ. The requirements for coherence can be relaxed in two ways by either looking at the two fields at the same time or by looking at the two fields at the same place. The two waves at points $\vec{r}_1$ and $\vec{r}_2$ respectively are spatially coherent if and only if $\vec{E}_2(\vec{r}_2, t)$ and $\vec{E}_1(\vec{r}_1, t)$ exhibit a constant phase difference ϕ for all times t. The two waves are temporally coherent at a point $\vec{r}$ if and only if the electric field of the second wave at point $\vec{r}$, $\vec{E}_2(\vec{r}, t)$, and the time-shifted electric field of the first wave at the same point $\vec{E}_1(\vec{r}, t - \tau)$, exhibit a phase relation $\phi(\tau)$ for each time t and each time shift τ. Temporally coherent light is monochromatic.

Two light waves show some type of coherence if they both originate from the same physical process. The light emitted by a point source in vacuum is spatially coherent with itself at points which are equidistant with respect to the source. Depending on the source, light is also approximately temporally coherent. The largest time shift τ_0 for which the two electric fields of two different emission events resemble each other is called the coherence time, the distance $c\tau_0$ the coherence length. The coherence length of a thermal source is a few micrometers. Some lasers emit temporally and spatially coherent light with a coherence length of several hundred meters to several kilometers. Two coherent light waves interfere with each other. Their interaction produces a static pattern of bright and dark areas, which is easily observable. Dependent on the phase between their electric fields at a point $\vec{r}$, they either amplify or extinct each other. When both are in phase the amplitude of the resulting field at $\vec{r}$ is the sum of the amplitudes of the individual fields at $\vec{r}$, when they are out of phase by half a cycle, they interfere destructively.

11.1.2 Intensity

Light waves transport energy. The intensity I of a light wave is the energy which is transferred per unit time through the unit area perpendicular to the direction of propagation, therefore its unit is $[W/m^2]$. A polarized monochromatic beam in vacuum has an intensity of

$$I = \frac{c\varepsilon_0}{2} E^2,$$

where ε_0 is the dielectric constant of vacuum and E is the amplitude of the electric field of the beam.

11.1.3 Refraction

A parallel beam of monochromatic light in vacuum that hits the boundary of a medium at an angle α to the surface normal propagates with an angle β to the surface normal inside the medium. The ratio

$$n = \frac{\sin \alpha}{\sin \beta},$$

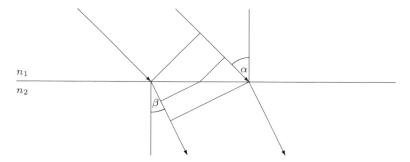

Fig. 11.1 A beam of light is refracted when it passes from a medium with index of refraction n_1 into a medium with index of refraction n_2. The wavefronts of the outgoing beam have to match the wavefronts of the incoming one, although the propagation speeds of the wavefronts are different, therefore they have to bend

the index of refraction, is a property of the medium. The index of refraction of air is almost exactly 1, for optical glasses and visual light it ranges between about 1.5 and 2. When the light beam enters the medium its temporal frequency f remains constant, but the speed at which the wavefronts of the beam propagate, the phase velocity $v_p = \frac{c}{n}$, changes. As the wavefronts have to stay continuous, at the boundary they have to bend. If the beam passes from a medium with a refractive index n_1 into a medium with refractive index n_2, Fig. 11.1, Snell's law of refraction

$$n_1 \sin \alpha = n_2 \sin \beta$$

determines the angle β at which the beam propagates from the boundary.

The refractive index of a medium and, therefore, the phase velocity v_p in the medium depend on the temporal frequency f of the wave. This effect is called dispersion. For very high frequencies, the refractive index becomes <1 for some materials. As the wavefronts of a monochromatic wave do not carry information or energy, a phase velocity greater than c is possible.

When a dispersive material is introduced into a beam of white light, a beam of light containing waves of many frequencies, the retardation each constituent of the beam experiences depends on the frequency of the constituent. In our application we have to compensate for this effect. In optical instruments built out of lenses, the dispersion of the glasses used is responsible for color fringes around images of objects. By combining appropriate glasses for the different lenses of an optical system, a lens designer to some extent compensates the dispersion in the individual lenses.

11.1.4 Reflection

Electromagnetic waves are reflected by boundaries between media of different indexes of refraction. If the wave is reflected by a boundary of an optically thinner medium (a medium with a smaller index of refraction), the reflected wave has the same phase as the incoming one. If the incoming wave is reflected at a boundary to an optically denser medium (a medium with a larger index of refraction), then the phase of the reflected wave is shifted by π (= half a period) with respect to the phase of the incoming wave.

11.2 Spectrometer Types

There exist two classes of instruments, diffractive ones and Fourier transform instruments.

11.2.1 Diffractive Instruments

Diffractive instruments use a diffraction grating to separate spatially the monochromatic constituents of a beam of white light that enters the instrument. The monochromatic constituents are focused onto individual pixels of a linear sensor array where their intensities are measured. Diffractive instruments can be very fast, as the whole spectrum is recorded simultaneously and there are no moving parts involved, but they suffer from low sensitivity as the light entering the instrument is spread over the pixels of the sensor array. In order to be insensitive to variations in the intensity of the input beam, the sensitivities of the individual pixels in the sensor array have to be matched closely. Furthermore, the ticks of wavelength partitions on the spectral axis will drift when the optical setup changes slightly due to temperature influences. For diffractive instruments, the scale on the abscissas of the spectra labels the wavelength.

11.2.2 Fourier Transform Instruments

Fourier transform spectrometers use a clever arrangement of mirrors and beam splitters in order to produce an output beam whose intensity is the Fourier transform of the intensity distribution of the beam of light fed into the instrument. The instrument measures the intensity of the output beam and computes the intensity distribution as a function of the spatial position of a mirror. In the rest of this case study, we present a Fourier transform spectrometer of our own design.

11.3 The Michelson Interferometer

A Michelson interferometer consists of two mirrors, one of them is movable, at exactly 90° and a beam splitter set at 45° facing the mirrors. A parallel beam of infrared light entering the interferometer, at the bottom of Fig. 11.2, is split into two halves, one is reflected horizontally toward the fixed mirror, the other is transmitted vertically toward the movable mirror. After being reflected the two beams meet again at the beam splitter, a partially transmissive/reflective plate. The vertical beam has to pass the beam splitter twice and, therefore, suffers from the frequency dependent retardation due to the substrate of the beam splitter, while the reflected horizontal beam does not pass the beam splitter at all. A compensator plate made out of the same material and of the same dimensions as the beam splitter is introduced into the horizontal path for phase compensation. Each returning beam is split again, with one quarter each directed horizontally toward the detector. The two other quarters travel back to the source.

In the space between the beam splitter and the detector's optics, the two horizontal beams interfere with each other. Assuming that the beam entering is perfectly parallel, the wavefronts are kept plane. Moving the movable mirror changes the retardation of one interfering beam with respect to the other, i.e. it changes the phase relation between the two interfering beams. So, for monochromatic light, shifting the movable mirror modulates the intensity of the superimposed beam. Either a lens or an off-axis paraboloid mirror concentrates the light on the detector aperture.

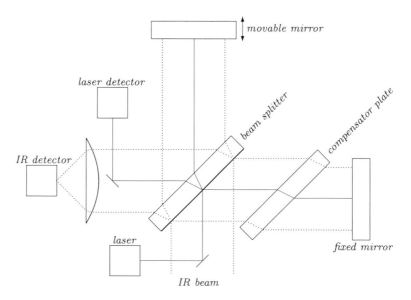

Fig. 11.2 Michelson interferometer. The *red lines* indicate the paths of the reference laser, the *dotted lines* indicate those of white, infrared light. The surface drawn bold in the beam splitter is a layer subdividing the beam into a transmitted and reflected part at equal intensities

Fourier transform spectrometers record the interference pattern with respect to the retardation between the two interfering beams. For measuring this retardation with high precision, an additional laser is shown feeding it's beam through the optics, and the interference pattern of this laser beam is detected separately.

11.3.1 Monochromatic Light

After entering the Michelson interferometer, Fig. 11.3, a polarized monochromatic parallel beam b, the laser beam, with spatial frequency $\tilde{v}_0$ is split into two beams, one, b_r, is reflected toward the fixed mirror, the other, b_t, is transmitted toward the moving mirror. When b_r arrives at the beam splitter again after being reflected by the fixed mirror, it is split into a beam b_{rr} that is reflected back into the source and into the beam b_{rt} that is transmitted toward the detector. The beam b_t is split a second time into b_{tt} that is transmitted back into the source and b_{tr} that is reflected toward the detector. The phase relations at beam splitters between the incoming beam and the outgoing beams are rather subtle and depend on the optical layout of the beam splitter. In the sequel, we disregard the thin layer that is responsible for splitting the incoming beam into almost equal parts, one that is transmitted through the layer and the other that is reflected by the layer and argue using the index of refraction of the layer's carrier only.

The beam b_{rr} is reflected three times at a boundary to an optically denser medium, therefore its phase has jumped by 3π ($= \frac{3}{2}$ periods) due to reflection. The phase of b_{rt} has jumped by one period, the phase of b_{tt} by half a period, and the phase of b_{tr} by half a period, as b_{tr} is reflected by the boundary of the beam splitter with air, which is optically thinner than the beam splitter. The beams b_{rt} and b_{tr} interfere with each other as do the beams b_{rr} and b_{tt}.

Let E be the amplitude of the electric field of the beam b. The component of the electric field of the beam b_{rt} in the direction of polarization is

$$\frac{E}{2}\cos\left(2\pi f t - 2\pi\tilde{v}_0 x\right).$$

Fig. 11.3 The paths of a beam of monochromatic light through a Michelson interferometer

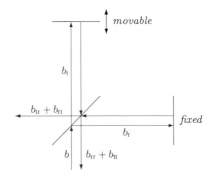

The one of the electric fields of b_{tr} is

$$-\frac{E}{2}\cos\left(2\pi ft - 2\pi \tilde{v}_0 x - 2\pi \tilde{v}_0 \delta\right),$$

at some distance x between the beam splitter and the detector's optics, where δ is the optical path difference, the difference of the lengths of the two optical paths in the interferometer. The sum of the two components is

$$\frac{E}{2}\left(\cos\left(2\pi ft - 2\pi \tilde{v}_0 x\right) - \cos\left(2\pi ft - 2\pi \tilde{v}_0 x - 2\pi \tilde{v}_0 \delta\right)\right)$$
$$= E\sin\left(\pi \tilde{v}_0 \delta\right)\sin\left(2\pi \tilde{v}_0 x + \pi \tilde{v}_0 \delta - 2\pi ft\right).$$

The amplitude of the electric field is $E\sin\left(\pi \tilde{v}_0 \delta\right)$ and the intensity $I'(\delta)$ of the beam hitting the laser detector is

$$I'(\delta) = \frac{c\varepsilon_0 E^2}{4}\left(1 - \cos\left(2\pi \tilde{v}_0 \delta\right)\right). \tag{11.1}$$

By a similar argument, the intensity of the beam reflected back into the source is

$$\frac{c\varepsilon_0 E^2}{4}\left(1 + \cos\left(2\pi \tilde{v}_0 \delta\right)\right),$$

and the sum of the two intensities is $\frac{c\varepsilon_0}{2} E^2$ demonstrating conservation of energy.[1]

Using (11.1) we can measure the position of the movable mirror to a high precision: When the mirror is moved the optical path difference and thus the intensity $I'(\delta)$ become functions of time. As long as the direction of movement stays the same, neighboring solutions t_i and t_{i+1} of the equation

$$\frac{c\varepsilon_0 E^2}{4}\left(1 - \cos\left(2\pi \tilde{v}_0 \delta(t)\right)\right) = \frac{c\varepsilon_0 E^2}{4} \tag{11.2}$$

have the property that $|\delta(t_i) - \delta(t_{i+1})| = \frac{1}{2v_0}$. By recording the laser interferogram $I'(t)$ and solving (11.2), the optical path difference can be measured with a resolution of $\frac{1}{2v_0}$. In practice, it is better to record either the even or the odd-numbered solutions only, to compensate for offsets introduced by the laser detector's electronics. Using interpolation the resolution can be increased below the reference laser's wavelength.

---

[1] Checking for conservation of energy or the other conservation laws relevant in the situation at hand helps to avoid mistakes in the calculation.

11.3.2 White Light

For the monochromatic constituents of a beam of white light (11.1) holds too, as long as δ does not exceed the beam's coherence length. Let $B(\tilde{\nu})$ be the intensity distribution with respect to the spatial frequency $\tilde{\nu}$ of the beam of light. When this beam passes through a Michelson interferometer with an optical path difference of δ, light of the intensity

$$S'(\delta) = \frac{1}{2} \int_0^\infty B(\tilde{\nu})(1 - \cos(2\pi\,\tilde{\nu}\delta))\,d\tilde{\nu} = \frac{1}{2} \int_0^\infty B(\tilde{\nu})\,d\tilde{\nu} - \frac{1}{2}S(\delta),$$

hits the detector. The integral $S(\delta)$,

$$S(\delta) = \int_0^\infty B(\tilde{\nu})\cos(2\pi\,\tilde{\nu}\delta)\,d\tilde{\nu} \tag{11.3}$$

is called the interferogram. The interferogram and the intensity distribution $B(\tilde{\nu})$ form a Fourier transform pair up to a constant factor, therefore $B(\tilde{\nu})$ can be recovered by applying the inverse Fourier transform to $S(\delta)$. A spectrum obtained with a Fourier transform spectrometer is the distribution $B(\tilde{\nu})$; its abscissa is labeled in units of the spacial frequency, usually in cm^{-1}. The maximum displacement of the movable mirror δ_x limits the resolution achievable to worse than $\frac{1}{\delta_x}$, see the detailed discussion in Chap. 10.

11.3.3 Imperfections in the Optics

The alignment of the two mirrors and of the beam splitter with respect to each other is absolutely critical. Accuracies in the range of a few arc seconds have to be achieved and maintained over the whole travel of the movable mirror. It is absolutely not easy to achieve such a precision particularly when the instrument is subject to thermal variations. The instrument discussed here uses a setup based on a cube-corner retroreflector, which ensures alignment by optical means.

Parts of the input beam, which are not hitting the beam splitter at exactly 45°, the divergent rays, experience a different optical path than the ones, which do. The difference in the lengths of the optical paths depends on the angle of divergence and on the displacement of the movable mirror. Once this difference becomes too large, so that different interference maximums will be focused onto the detector simultaneously, moving the mirror further does not provide new information. The divergence of the input beam can be limited by introducing a circular stop at a position in the optical path of the interferometer, where an image of the source is formed by the optics.

In real instruments, the beam splitter and the compensator plate will differ slightly. Therefore, some dispersion effects will be present in the recorded interferogram. To quantify these we split the optical path difference δ into a part δ_m, which is due to the displacement of the movable mirror, and one $\delta(\tilde{\nu})$, which is due to the residual dispersion, $\delta = \delta_m + \delta(\tilde{\nu})$; (11.3) becomes

$$S(\delta_m) = \int_0^\infty B(\tilde{\nu})\cos(2\pi\,\tilde{\nu}(\delta_m + \delta(\tilde{\nu})))\,d\tilde{\nu}$$

$$= \frac{1}{2}\int_{-\infty}^\infty B(\tilde{\nu})e^{-2\pi i\tilde{\nu}(\delta_m + \delta(\tilde{\nu}))}\,d\tilde{\nu}$$

$$= \frac{1}{2}\int_{-\infty}^\infty B(\tilde{\nu})e^{-2\pi i\tilde{\nu}\delta_m - i\phi(\tilde{\nu})}\,d\tilde{\nu}$$

$$= \frac{1}{2}\int_{-\infty}^\infty B(\tilde{\nu})e^{-i\phi(\tilde{\nu})}e^{-2\pi i\tilde{\nu}\delta_m}\,d\tilde{\nu}, \qquad (11.4)$$

where $B(-\tilde{\nu}) = B(\tilde{\nu})$, for all $\tilde{\nu} \geq 0$, and $\phi(\tilde{\nu}) = 2\pi\,\tilde{\nu}\delta(\tilde{\nu})$ is the phase shift due to dispersion for monochromatic light with wave number $\tilde{\nu}$. According to (11.4) $S(\delta_m)$, the intensity that hits the detector in an imperfect interferometer, and $B(\tilde{\nu})e^{-i\phi(\tilde{\nu})}$ form a Fourier transform pair up to a constant factor. The intensity distribution $B(\tilde{\nu})$ is proportional to the magnitude of the inverse Fourier transform of $S(\delta_m)$.

11.4 Design of a Fourier Transform Spectrometer

Fourier transform infrared spectrometer can operate in two modes: In step-scan instruments, the mirror moves stepwise from one sample position to the next. While the mirror rests at the sample position, the instrument takes any number of measurements. In continuous-scan instruments, the mirror moves continuously and the instrument records (or computes) a sample for the very instant the mirror passes a sample position. Step-scan instruments, can produce time-resolved spectral data with very high resolution when applied to repeatable experiments, but they suffer from long measurement times and low signal-to-noise ratios. Continuous-scan instruments[2] offer better signal-to-noise ratios but cannot record spectral data of short events.

---

[2] Also known as rapid-scan instruments.

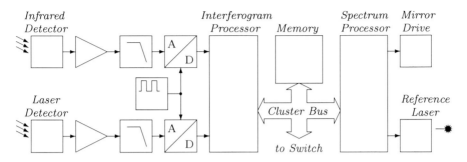

Fig. 11.4 Block diagram of the spectrometer electronics. A bus switch connects the cluster bus to the other processors in the instrument's electronics

The instrument described here performs automatized routine measurements in production processes in the chemical, food, and pharmaceutical industries. The instrument is of the continuous-scan type and produces spectra with a resolution of about $3.7\,\mathrm{cm}^{-1}$, and a signal-to-noise ratio better than 20,000 for a measurement time of 1 s.

In our design, we transfered all functions related to processing the interferograms into the instrument's software, Fig. 11.4. A cluster of two floating-point signal processors with large, shared off-chip memory produces the spectra and controls the spectrometer. The analog input stage consists of two channels, one for the laser interferogram, the other for the infrared light. The two channels work in lockstep. An indium-gallium-arsenide photodiode picks up the laser interferogram. The type of detector for the white-light interferogram depends on the spectral range of the instrument. In the near-infrared region of the spectrum this instrument uses indium-gallium-arsenide photodiodes, in the mid-infrared region it employs mercury-cadmium-telluride photodiodes.

Each channel consists of a transimpedance amplifier for amplifying the detector's signal, a lowpass anti-aliasing filter and a high resolution periodically converting analog-to-digital converter for converting the channel's signal to digital. The interferograms are oversampled considerably. The filters are matched to minimize noise due to speed variations of the mirror, so-called Zachor–Aaronson (1979) noise, in the computed spectra. They are second order filters with Bessel characteristic. Furthermore, the electronics centers the signal around 0. In order to preserve the temporal relation between the two channels, the analog-to-digital converters operate in lockstep. A separate clock generator with low jitter provides to the converters a clock signal that has not been polluted by digital electronics.

The interferogram processor picks up both streams of data. This instrument drives the mirror sinusoidally instead of operating it with constant velocity between the turning points. The interferograms both are sampled at equidistant times and, therefore, exhibit a frequency modulation. The interferogram processor demodulates the white-light signal with the help of the laser signal to arrive at a white-light interferogram that has been sampled at equally spaced increments of the optical path difference.

In order to keep synchronization efforts to a minimum, the interferogram processor waits until it received the data of at least one complete scan of the mirror before it starts to demodulate the white-light interferogram. The program the interferogram processor executes is fully pipelined as described in Chap. 5. This enables the interferogram processor to handle the large volume of data it is confronted with. The spectrum processor is responsible for compensating the imperfections in the optics by correcting the phase of the interferograms, for computing the spectra, for communicating the results to the rest of the instrument, and for controlling the mirror drive, the reference laser, and the cooling of the detector.

11.4.1 Recording the Interferograms

The instrument acquires the data of the two channels using direct memory access. It starts to process the data of a scan of the mirror after it has collected the data of the complete scan. Therefore, the software can process a scan's data without having to wait, synchronization with the data acquisition mechanism is only necessary between scans. While the software demodulates an interferogram, the direct memory access mechanism acquires data produced by the subsequent scan of the mirror.

11.4.2 Measuring the Mirror Movement

Errors in sampling the white-light interferogram at equally spaced positions along the travel of the mirror add artifacts to the whole spectrum. While a detailed study of these effects is beyond the scope of this treatise, we want to mention that a signal-to-noise ratio of 5,000 or above for a spectrum computed from a single scan of the mirror requires a random error of the sampling positions that is less than a few angstrom.[3]

Campbell (2008) describes a procedure for compensating variations in the speed of the moving mirror when the mirror moves with almost constant speed. He uses a mixer to compute the phase error in the laser interferogram. In the instrument described here, we find the solutions of (11.2) directly. We exploit that the analog electronics centers the laser signal around zero, and deal with the zero-crossings of the laser signal only. Initially, or after a restart from a stall the software has to search the beginning of a scan of the mirror. Between two scans the mirror has to change the direction of its travel, therefore its speed drops to zero momentarily. In the laser interferogram the time between two zeros becomes a (local) maximum at each turning point. To find such a maximum the software performs a coarse search considering every forth sample only. As the kinetic energy of the mirror assembly is low around the turning points, the assembly becomes susceptible to vibration. To counteract this effect, the software skips a configurable number of zeros of the

---

[3] We really mean 1×10^{-10} m, approximately the diameter of a single atom!

laser interferograms before using the data for the computation of spectra. When the software is in the middle of computing interferograms, it can start with processing a new scan where it detected the end of the previous one.

Typically, the software has to process more than 20,000 zero-crossings each second. Therefore, we exploit the special structure of the problem instead of resorting to generic numerical procedures. The software finds these estimates in two passes over the laser interferogram. The first pass performs a coarse search for the positive going zero-crossings by considering every fourth sample only. In the second pass, the software refines each zero-crossing found in the first pass. It considers only those samples around a zero-crossing found in the first pass that reside within a configurable distance of mirror travel around the zero. Therefore, for zeros that were recorded when the mirror was moving slowly, more samples are considered than for zeros for which the mirror was moving faster. In order to preserve the precision of the estimates, regardless of the position of a zero in the buffer for the laser interferogram, each zero is represented by the index of the sample closest to the zero and the floating-point offset from this index.

In order to find such an estimate we first search among the five candidates for the sample, which is the closest to zero. We temporarily shift indexes so that this sample is Y_0 and the samples $Y_{-k}, \ldots, Y_k$ are available. Next, we compute a polynomial

$$F = a_3 t^3 + a_2 t^2 + a_1 t + a_0$$

that approximates the points $(-k, Y_{-k}), \ldots, (k, Y_k)$ around the zero-crossing using the least-squares method. While computing such an approximation for arbitrary sets of points requires singular value decomposition for solving the resulting equations, see, e.g., Press et al. (2007), our special case is highly symmetric and allows for a direct solution. The sum E of squared errors is

$$E = \sum_{i=-k}^{k} \left(Y_i - a_3 i^3 - a_2 i^2 - a_1 i - a_0 \right)^2$$

and the partial derivatives of E with respect to the coefficients of F are

$$-\frac{1}{2}\frac{\partial E}{\partial a_0} = \sum_{i=-k}^{k} \left(Y_i - a_3 i^3 - a_2 i^2 - a_1 i - a_0 \right) = t - a_2 c_2(k) - a_0(2k+1)$$

$$-\frac{1}{2}\frac{\partial E}{\partial a_1} = \sum_{i=-k}^{k} i \left(Y_i - a_3 i^3 - a_2 i^2 - a_1 i - a_0 \right) = u - a_3 c_4(k) - a_1 c_2(k)$$

$$-\frac{1}{2}\frac{\partial E}{\partial a_2} = \sum_{i=-k}^{k} i^2 \left(Y_i - a_3 i^3 - a_2 i^2 - a_1 i - a_0 \right) = v - a_2 c_4(k) - a_0 c_2(k)$$

$$-\frac{1}{2}\frac{\partial E}{\partial a_3} = \sum_{i=-k}^{k} i^3 \left(Y_i - a_3 i^3 - a_2 i^2 - a_1 i - a_0 \right) = w - a_3 c_6(k) - a_1 c_4(k),$$

where $t = \sum_{i=-k}^{k} Y_i$, $u = \sum_{i=-k}^{k} i Y_i$, $v = \sum_{i=-k}^{k} i^2 Y_i$ $w = \sum_{i=-k}^{k} i^3 Y_i$ and $c_j(k) = \sum_{i=-k}^{k} i^j$, for $j \in \{2, 4, 6\}$. The sums with factor i or i^3 or i^5 vanish because the range of the sums is symmetric around 0. By setting the right-hand sides above to 0 and solving the resulting system for $\{a_3, a_2, a_1, a_0\}$ symbolically, we find expressions for the coefficients of the polynomial F that minimize E. The procedure FITCUBIC, Fig. 11.5, computes these coefficients.

To speed up the computation of the fitted polynomial the entries of the matrices $A(k)$ and $B(k)$,

$$A_{11}(k) = \frac{9k^2 + 9k - 3}{8k^3 + 12k^2 - 2k - 3}$$

$$A_{12}(k) = A_{21}(k) = \frac{-15}{8k^3 + 12k^2 - 2k - 3}$$

$$A_{22}(k) = \frac{45}{8k^5 + 20k^4 + 10k^3 - 5k^2 - 3k},$$

and

$$B_{11}(k) = \frac{75k^4 + 150k^3 - 75k + 25}{8k^7 + 28k^6 + 14k^5 - 35k^4 - 28k^3 + 7k^2 + 6k}$$

Fig. 11.5 Procedure FITCUBIC for computing the coefficients of a cubic polynomial fitted to the points $(-k, Y_{-k}), \ldots, (k, Y_k)$

```
1: function FITCUBIC(Y, k)
2:     if k > m then
3:         k ← m
4:     end if
5:     t ← Y_0
6:     u ← v ← w ← 0
7:     for i ← 1, k do
8:         t ← t + Y_i + Y_{-i}
9:         u ← u + i(Y_i - Y_{-i})
10:        v ← v + i²(Y_i + Y_{-i})
11:        w ← w + i³(Y_i - Y_{-i})
12:    end for
13:    a_0 ← A_{11}(k)t + A_{12}(k)v
14:    a_2 ← A_{21}(k)t + A_{22}(k)v
15:    a_1 ← B_{11}(k)u + B_{12}(k)w
16:    a_3 ← B_{21}(k)u + B_{22}(k)w
17:    return a
18: end function
```

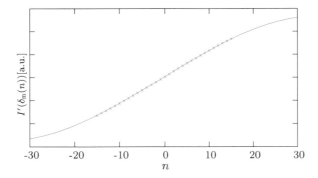

Fig. 11.6 Sampled laser interferogram versus time compared to the fitted polynomial, here $k = 15$. On the *horizontal* axis n is the relative sample number, which is proportional to time, on the *vertical* axis $I'(\delta_m(n))$ is the intensity of the laser interferogram in arbitrary units

$$B_{12}(k) = B_{21}(k) = \frac{-105k^2 - 105k + 35}{8k^7 + 28k^6 + 14k^5 - 35k^4 - 28k^3 + 7k^2 + 6k}$$

$$B_{22}(k) = \frac{175}{8k^7 + 28k^6 + 14k^5 - 35k^4 - 28k^3 + 7k^2 + 6k},$$

can be precomputed and tabulated up to the expected maximal number of points. Should a larger number of points be passed FITCUBIC can discard the superfluous points at the cost of a slightly reduced precision. In the derivation above we exploit that the samples are taken at equally spaced points in time. By making the fit symmetric around Y_0 the resulting linear system becomes sparse and is guaranteed to have a unique solution for $k \geq 2$. This allows us to use a fast procedure for computing the fits. The data in Fig. 11.6 demonstrates the quality of one such fit. To compute the floating-point offset we apply the Newton–Raphson method to the fitted polynomial and a starting value of zero. This converges rapidly to the desired offset.

11.4.3 Demodulating the White-Light Interferogram

The instrument's software demodulates the white-light interferogram in two passes. In the first pass it searches for the central burst. In a perfect instrument the minimum of the white-light interferogram coincides with a minimum of the laser interferogram. In the described instrument the two beams take slightly different paths and, therefore the interferograms are offset with respect to each other. Furthermore, there is a small random variation in the offset. This random variation is due to the index of refraction of the gas the beams pass through, which depends on the temperature of the gas. Slight drafts within the optic's housing create an uneven temperature distribution within the housing and, therefore, create the random variations. We compensate for this effect by computing the offset to align the two interferograms with respect to each other.

In the second pass the time between two zero-crossings of the laser interferogram is subdivided further into four or eight parts by linear interpolation. This increases the spectral range to twice or four times the wavenumber of the laser. Increasing the spectral range past the range set by the detector helps improving the signal-to-noise ratio of the measurements because of the processing gain of the discrete Fourier transform.

To interpolate between the samples of the white-light interferogram for computing estimates for the values at the times required one can use interpolation filters. In the situation at hand we do not know the time at which we need the estimate, therefore we either would have to compute the filter coefficients on the fly, or we would have to rely on precomputed tables of coefficients, where the filters were precomputed for a predetermined raster of times. The first approach is too taxing on the computational power of the interferogram processor, the second stresses the available memory bandwidth to its limits, as the tables of coefficients would have to be stored in off-chip memory because of their size and the required filter coefficients would have to be brought into the internal memory before being used in a computation. Instead we resort to the fitting procedure FITCUBIC again. Like Brault (1996) we let the software base the interpolation on the samples residing in a fixed interval of mirror travel around the desired position. In areas where the mirror is moving slowly more points are considered than in areas where the mirror is going fast. The software evaluates the fitted polynomial to compute the desired estimate. Letting the mirror go slow far away from the region of the central burst increases the vertical resolution in these regions and allows for the extraction of minute details. Figure 11.7 illustrates the interpolation for a favorable situation and a less favorable one.

11.4.4 Phase Correction and Computation of Spectra

The easiest way to eliminate the effects of residual dispersion is to acquire the white-light interferogram symmetrically around the central burst and to compute the discrete Fourier transform of the samples. From the resulting complex spectrum one extracts the amplitudes. For a given resolution however, the mirror has to travel twice as far as when recording a single-sided interferogram produced by a perfect interferometer. In instruments designed for high resolution the movement toward the interferometer might even be mechanically impossible.

As the residual dispersion of a practical interferometer changes slowly with wavenumber one usually pursues a different approach. Instead of acquiring a full double-sided interferogram one lets the mirror move past the central burst only a little, see Forman et al. (1966). From the data symmetric around the central burst one computes a coarse phase spectrum. Using this phase spectrum one corrects the

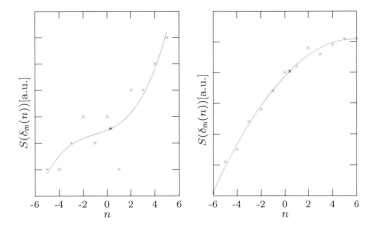

Fig. 11.7 Performance of the interpolation in two situations. On the *horizontal* axis n is the relative sample number, which is proportional to time, on the *vertical* axis $S(\delta_{\mathrm{m}}(n))$ is the intensity of the white-light interferogram in arbitrary units. The samples are colored *green*, the fitted polynomial *orange* and the interpolated point is indicated by a *blue asterisk*. Even in the situation on the *left*, where there is little variation in the measured points the interpolation performs satisfactorily

full-length interferogram. More precisely let

$$S_b, \ldots, S_{z-c/2+1}, \ldots, S_z, \ldots, S_{z+c/2}, \ldots, S_a$$

be the white-light interferogram sampled at equally spaced increments of the optical path difference. The sample S_z is the one taken at zero optical path difference. The samples $S_b, \ldots, S_{z-1}$ are the ones taken while the mirror moved on the near side of the point of zero retardation, the samples $S_z, \ldots, S_a$ are the ones sampled on the far side of the point of zero optical path difference, where near and far indicate the position of the movable mirror with respect to the beam splitter. The samples $S_{z-c/2+1}, \ldots, S_{z+c/2}$ are the ones used for estimating the phase spectrum of the interferometer, see Fig.11.8. Let k be the smallest power of 2 equal to or greater than $a - b + 1$. A fast Fourier transform program set up for length k computes all discrete Fourier transforms required in the sequel. We compute the discrete Fourier transform $\tilde{s}$ of the periodic extension of

$$\langle 0, \ldots, w_{-c/2+1} S_{z-c/2+1}, \ldots, w_{c/2} S_{z+c/2}, 0, \ldots, 0 \rangle,$$

for $w_{-c/2+1}, \ldots, w_{c/2}$ a sampled window function and with $k + z - a - c/2$ zeros added to the left and $a - z - c/2$ zeros added to the right. Its discrete Fourier transform $\tilde{s}$ is an interpolated version of the discrete Fourier transform of the periodic extension of $\langle w_{-c+1} S_{-c+1}, \ldots, w_c S_c \rangle$ in that sense that both transforms are composed of samples of the discrete-time Fourier transform of

$$\langle \ldots, 0, w_{-c+1} S_{-c+1}, \ldots, w_c S_c, 0, \ldots \rangle,$$

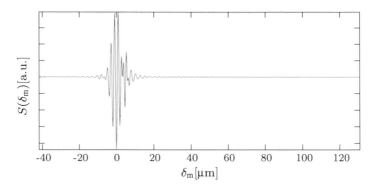

Fig. 11.8 Part of the white-light interferogram before phase correction. On the *horizontal* axis δ_m is the optical path difference, on the *vertical* axis $S(\delta_m)$ is the intensity of the white-light interferogram in arbitrary units. The interferogram is not symmetric around the point of zero retardation

see (2.15), with $\tilde{s}$ containing more samples. The estimate p for the phase spectrum is

$$p = \langle p_{-k} \ldots, p_{k-1} \rangle = \left\langle \frac{\tilde{s}_{-k}}{|\tilde{s}_{-k}|}, \ldots, \frac{\tilde{s}_{k-1}}{|\tilde{s}_{k-1}|} \right\rangle,$$

see Fig. 11.9. For correcting the phase the software computes the discrete Fourier transform $s = \langle s_{-k}, \ldots, s_{k-1} \rangle$ of

$$\langle 0, \ldots, S_b, \ldots, S_{z-c/2+1}, \ldots, S_z, \ldots, S_{z+c/2}, \ldots, S_a \rangle,$$

where $k - (a - b + 1)$ zeros have been added left of the white-light interferogram. By transforming the sequence $\langle s_i p_i^* \rangle$ for $-k \leq i \leq k - 1$ back, we arrive at the phase-corrected interferogram. To avoid the effects of the circular convolution, which we

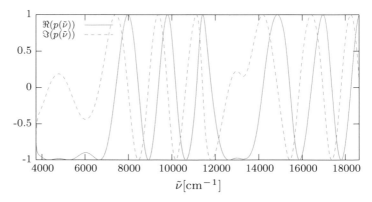

Fig. 11.9 Real part $\Re(p(\tilde{\nu}))$ and imaginary part $\Im(p(\tilde{\nu}))$ of the estimate p of the interferometer's phase spectrum. We used 512 points around the central burst for phase estimation. On the *horizontal* axis $\tilde{\nu}$ is the spatial frequency. The phase varies slowly with the spacial frequency

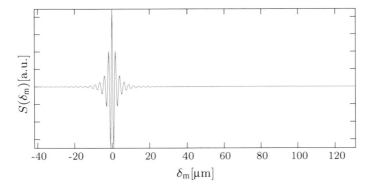

Fig. 11.10 Part of the white-light interferogram after phase correction. On the *horizontal* axis δ_m is the optical path difference, on *vertical* axis $S(\delta_m)$ is the intensity of the interferogram in arbitrary units. The corrected interferogram is essentially symmetric around the point of zero retardation

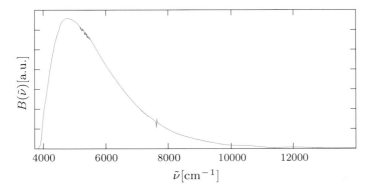

Fig. 11.11 The spectrum of the light emitted by a halogen lamp, corresponding to the interferogram in Fig. 11.10. On the *horizontal* axis $\tilde{\nu}$ is the spatial frequency, on the *vertical* axis $B(\tilde{\nu})$ is the intensity of the spectrum in arbitrary units. The peaks around $7{,}634\,\mathrm{cm}^{-1}$ result from stray light from the reference laser. Absorption by water vapor in the atmosphere creates the features between $5{,}160$ and $5{,}500\,\mathrm{cm}^{-1}$. The spectrum is not corrected for the wavenumber-dependent sensitivity of the detector

actually compute, we exploit that the phase spectrum changes slowly with wavenumber. Therefore, its inverse Fourier transform is concentrated around zero and only the ends of the interferogram are effected. We discard these parts, Fig. 11.10.

   In order to improve the signal-to-noise ratio we average a configurable number of corrected interferograms before computing the final spectrum, see Fig. 11.11. As the corrected interferograms are even functions we use for their computation a discrete cosine transform.[4]

---

[4] The discrete cosine transform is the specialization of the discrete Fourier transform to even functions. It can be computed more efficiently.

11.4.5 The Reference Laser

Many Fourier transform spectrometer use a Helium–Neon laser as reference. While being exceptionally stable in the long term these lasers are bulky, produce a lot of waste heat and require high voltage. To avoid these difficulties the instrument employs a semiconductor laser instead. The laser diode, of the distributed feedback type, emits at a wavelength of 1,310 nm. It is mounted in a housing together with a thermoelectric element and a thermistor. The thermoelectric element acts as controlling element, and the thermistor acts as sensor for a temperature control loop. The light emitted by the laser is coupled into an optical fiber within the housing. The wavelength of the light the laser emits depends on the laser's temperature. According to the laser's specification its wavelength changes by $0.09\,\text{nm}\,\text{K}^{-1}$. Using a PID controller for keeping the temperature of the laser diode constant the instrument achieves a temperature error of $<7\,\text{mK}$ over a period of one day. This translates into an uncertainty of the wavenumber scale of $0.004\,\text{cm}^{-1}$ at the wavenumber of the reference laser. This stability exceeds the requirements by a comfortable margin. By using pulse width modulation in the power stage of the PID controller, the efficiency of the electronics is kept high, reducing the waste heat introduced into the instrument.

The instrument monitors both the temperature of the laser and the power the laser emits. Should any of the two values leave predefined bounds, the instrument stops operation.

11.4.6 Detector Cooling

For reducing the noise obscuring the white-light signal the detector for the white-light interferogram is mounted on a thermoelectric cooler. We use a controller similar to the one for the laser to control the detector's temperature. The detector's temperature is configurable in the described instrument.

11.4.7 Driving the Movable Mirror

A voice coil actuator moves the movable mirror. Such an actuator consists of a magnetic coil that is mounted in the field of a permanent magnet. When an electric current flows through the coil, the coil exerts a force parallel to the axis of the permanent magnet. The electronics for driving the mirror consists of a switching power stage and a synthesizer. The synthesizer drives the power stage without burdening the software with tasks that have to be accomplished at a high rate. The software sets the frequency of the signals that drives the actuator in order to determine the rate at which interferograms are produced. It contains two proportional-integral controllers to first control the lengths of the interferograms by changing the amplitude of the driving signal, and to second control the position of the central burst by changing an offset added to the generated sinusoid.

11.4.8 Optical Setup

In the described instrument, we use an interferometer configuration introduced in Steel (1970). In this configuration a movable cube-corner retroreflector folds the beam to a fixed fold-back mirror compensating both for tilt and shear of the moving parts. A beam of light entering a retroreflector leaves it parallel to the direction of entry but with a lateral offset. Figure 11.12 shows the mechanical and optical layout of the instrument. Light enters the instrument through a window in the upper wall. The interferometer sits in the left upper corner. Located below is a porch-swing assembly, driven by a voice coil actuator that holds the retroreflector. To the right of the interferometer adjustable fixtures for coupling the reference laser's beam into the interferometer and for picking up the laser interferogram are placed. Both the laser and the detector for the laser interferogram sit outside the housing of the optics. We use optical fibers to guide the laser's light to the interferometer and back to the detector. An off-axis parabolic mirror concentrates the infrared radiation onto the detector for white light sitting below the mirror. The parabolic mirror has a central hole for letting the laser interferogram pass.

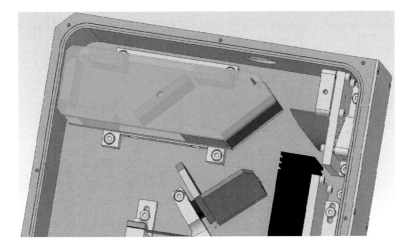

Fig. 11.12 Optical setup of the instrument. Courtesy of i-RED GmbH, with permission

11.4.9 Automation of Measurement Processes

For controlling industrial production processes uninterpreted spectra do not help at all. Multivariate regression models, so-called chemometric models, see, e.g., Varmuza and Filzmoser (2009), Brereton (2003), can extract information that is relevant for monitoring and controlling such processes. When light interacts with a material, chemical compounds in the material absorb constituents of the light at certain wavenumbers. The spectrum of the absorbed light is characteristic for each substance. Furthermore, the amount of light absorbed by a substance depends on the concentration of the substance in the material. Therefore, the spectra contain information about the concentrations of substances, known or unknown, in a material and models can extract this information automatically. The described instrument uses a second cluster of floating-point signal processors for running the numerical procedures these models require.

Many routine measurement situations require fully automatized measurements. Typically a measurement system has to handle several probing stations and process the acquired data without human intervention. In rare cases, the process automation system will change production parameters automatically based on the measurement results. To cope with such situations the described instrument is equipped with an additional embedded controller running programmable logic controller software that has been designed for handling spectral data. This controller is able to communicate with sensors and actuators attached to the process using a field bus. For communicating with reporting systems, it uses a LAN interface.

11.5 Bibliographical Notes

The Michelson interferometer was invented by Michelson and Morley (1995, 1961) in 1887, while producing evidence against the then prevalent theory that light waves need a medium, the so-called luminiferous aether, for propagation. Their discovery led Einstein to the formulation of the special theory of relativity. The book by Griffiths and de Haseth (2007) is a recent overview over Fourier transform infrared spectrometry. The article by Jackson (2006) surveys continuous-scan Fourier transform instruments for the mid-infrared region of the electromagnetic spectrum.

Our derivation of the influence of dispersion on the measured interferogram follows the one given by Forman et al. (1966).

Brault (1996) is the first using periodically converting analog-digital converters with very high resolution to sample the white-light interferogram. He demonstrates how to compensate the effects of variations in the speed of the mirror. Brasunas and Cushman (1997) sample both the laser and the white-light interferograms uniformly in time. By interpolating between zero-crossings of the laser interferogram they are able to extend the spectral range of their instrument into the visible region. Keens and Rapp (1999) sample the white-light interferogram uniformly in time while the mirror is moving with constant velocity. They derive a procedure for computing a filter that compensates for the phase response of the detector and the signal processing electronics. Simon et al. (2002) describe problems introduced by the frequency response of the detector and by fluctuations of the source and their elimination with the help of sampling the white-light interferogram uniformly in time. Yang et al. (2002) apply Brault's method to an ultra-rapid-scan Fourier transform spectrometer. Their optics uses a rotating mirror for varying the optical path difference sinusoidally with time. Moreover, they describe a method that fits cubic poynomials to the white-light interferogram, which is similar to the one described here. They use a constant number of points sampled uniformly in time for computing the fitted cubic polynomials, therefore the performance of this method is influenced by variations of the speed of the mirror. They compare Brault's approach, which compensates for variations in the speed of the mirror to their method, and conclude that the former is better. Roy et al. (2007) sample the infrared data of an imaging Fourier transformation spectrometer uniformly in optical path difference. They record the sampling times for each set of samples and use these times to compensate for variations in the speed of the mirror. Campbell (2008) samples both the laser and the white-light interferogram with two synchronously and periodically converting analog-digital converters. By down-conversion of the laser interferogram points equidistant in space are computed. Naylor et al. (2004) propose to use a nonuniform discrete Fourier transform to forgo the interpolation step when compensating for sampling unevenly in terms of the optical path difference.

Several authors have studied the influence of uncertainties in the sample positions on the quality of the resulting spectrum. Surh (1966) and Sakai (1970) study the influence on single spectral lines. Bell and Sanderson (1972) consider both single lines and spectra of white light. Zachor (1977) investigates the combined effect of the filters in the signal path for the laser and the white-light interferogram respectively and variations of the speed of the mirror. Zachor and Aaronson (1979) match the delay of the laser channel with the delay of the white-light channel to reduce errors introduced by variations of the speed of the movable mirror. Assuming several models of the frequency variation of the reference laser Meynart derives bounds on the laser's frequency stability for a desired signal-to-noise ratio in the measured spectra (Meynart 1992). He also derives estimates for the errors introduced by sampling at uneven increments of the optical path difference. Cohen (1999) studies the effects of uneven sampling on the noise of a spectrum that is compensated with the help of a background, i.e., a spectrum that covers the influences which are not pertinent to the observed scene. Learner et al. study ghosts of spectral lines due to amplitude and frequency modulation, and due to nonlinearities (Learner et al. 1996). Moreover, they state how phase correction is influenced by the ghosts. Palchetti and Lastrucci (2001) study the influence of variations of the speed of the movable mirror on the spectra when the interferogram is picked up with a slow detector. They describe compensation mechanisms piggybacked onto Brault's method (1996) for these errors. Saggin et al. (2007) study the errors introduced by mechanical vibrations theoretically. Comolli and Saggin (2010) investigate artifacts observed in an actual instrument numerically. Taurand et al. (2007) describe the errors introduced by difuse reflections at the mirror's surfaces.

Manning and Griffiths (1997) study sources on error in step-scan instruments. Their discussion of errors introduced by temperature variations and convection applies to rapid-scan instruments as well.

Methods for the correction of asymmetric interferograms where introduced by Forman et al. (1966), the approach we follow, and by Mertz (1967).

Norton and Beer (1976) and Bretzlaff and Bahder (1986) study artifacts introduced in the spectra by the applied window function.

Steel (1970) describes several basic configurations for the optical setup. The setup we use in our instrument is the shear and tilt compensated one described there. Ultra-high resolution instruments are being built, which allow a mirror travel of several meters, see Keens (2004). For the analysis of trace gases in the atmosphere several instruments have been built. Friedl-Vallon et al. (2004) describe an instrument operated at cryogenic temperatures for the mid-infrared region of the spectrum. Palchetti et al. (1999, 2004, 2005, 2006a, 2006b) report on an instrument for the far infrared. The instrument presented here is described in Hintenaus et al. (2007, 2010).

References

Bell EE, Sanderson RB (1972) Spectral errors resulting from random sampling-position errors in Fourier transform spectroscopy. Appl Opt 11(3):688–689. doi:10.1364/AO.11.000688

Bianchini G, Castagnoli F, Pellegrini M, Palchetti L (2006a) Frictionless mirror drive for intermediate resolution infrared Fourier transform spectroscopy. Infrared Phys Technol 48(3):217–222. doi:10.1016/j.infrared.2005.09.002

Bianchini G, Palchetti L, Carli B (2006b) A wide-band nadir-sounding spectroradiometer for the characterization of the earth's outgoing long-wave radiation. In: Proceedings of SPIE 6361. doi:10.1117/12.689260

Brasunas JC, Cushman GM (1997) Uniform time-sampling Fourier transform spectroscopy. Appl Opt 36(10):2206–2210. doi:10.1364/AO.36.002206

Brault JW (1996) New approach to high-precision Fourier transform spectrometer design. Appl Opt 35(16):2891–2896. doi:10.1364/AO.35.002891

Brereton R (2003) Chemometrics: data analysis for the laboratory and chemical plant. Wiley, Chichester

Bretzlaff RS, Bahder TB (1986) Apodization effects in Fourier transform infrared difference spectra. Revue de Physique Applquée (Paris) 21(12):833–844. doi:10.1051/rphysap:019860021012083300

Campbell J (2008) Synthetic quadrature phase detector/demodulator for Fourier transform spectrometers. Appl Opt 47(36):6889–6894. doi:10.1364/AO.47.006889

Cohen DL (1999) Performance degradation of a Michelson interferometer due to random sampling errors. Appl Opt 38(1):139–151. doi:10.1364/AO.38.000139

Comolli L, Saggin B (2010) Analysis of disturbances in the planetary Fourier spectrometer through numerical modeling. Planet Space Sci 58(5):864–874. doi:10.1016/j.pss.2010.01.011

Forman ML, Steel WH, Vanasse GA (1966) Correction of asymmetric interferograms obtained in Fourier spectroscopy. J Opt Soc Am 56(1):59–61. doi:10.1364/JOSA.56.000059

Friedl-Vallon F, Maucher G, Seefeldner M, Trieschmann O, Kleinert A, Lengel A, Keim C, Oelhaf H, Fischer H (2004) Design and characterization of the balloon-borne michelson interferometer for passive atmospheric sounding (MIPAS-B2). Appl Opt 43(16):3335–3355. doi:10.1364/AO. 43.003335

Griffiths PR, de Haseth JA (2007) Fourier transform infrared spectrometry. Wiley, Hoboken

Hintenaus P, Kvas G, Märzinger W (2007) An infrared spectrometer for process monitoring I, spectroscopy. In: Industrial electronics society, 2007. IECON 2007. 33rd annual conference of the IEEE, pp 2576–2579. doi:10.1109/IECON.2007.4459989

Hintenaus P, Märzinger W, Pöll H (2010) An infrared spectrometer for process monitoring II, chemometry and automatization. In: 2010 IEEE workshop on environmental energy and structural monitoring systems (EESMS), pp 9–13. doi:10.1109/EESMS.2010.5634173

Jackson RS (2006) Continuous scanning interferometers for mid-infrared spectrometry. Handbook of vibrational spectroscopy. Wiley, Chichester. doi:10.1002/0470027320.s0204

Keen A, Rapp N (1999) Method of obtaining an optical FT spectrum

Keens A (2004) A tribute to the Winnewissers or how science meets technology. J Mol Struct 695–696(0):379–384. doi:10.1016/j.molstruc.2003.12.040

Learner RCM, Thorne AP, Brault JW (1996) Ghosts and artifacts in Fourier-transform spectrometry. Appl Opt 35(16):2947–2954. doi:10.1364/AO.35.002947

Manning CJ, Griffiths PR (1997) Noise sources in step-scan FT-IR spectrometry. Appl Spectrosc 51(8):1092–1101

Mertz L (1967) Auxiliary computation for Fourier spectrometry. Infrared Phys 7(1):17–23. doi:10. 1016/0020-0891(67)90026-7

Meynart R (1992) Sampling jitter in Fourier-transform spectrometers: spectral broadening and noise effects. Appl Opt 31(30):6383–6388. doi:10.1364/AO.31.006383

Michelson AA (1961) Light waves and their uses. Phoenix science series, University of Chicago Press, Chicago

Michelson AA (1995) Studies in Optics. Dover books on astronomy. Dover, New York

Naylor DA, Fulton TR, Davis PW, Chapman IM, Gom BG, Spencer LD, Lindner JV, Nelson-Fitzpatrick NE, Tahic MK, Davis GR (2004) Data processing pipeline for a time-sampled imaging Fourier transform spectrometer. In: Proceedings of SPIE 5546, imaging spectrometry X, pp 61–72. doi:10.1117/12.560096

Norton RH, Beer R (1976) New apodizing functions for Fourier spectrometry. J Opt Soc Am 66(3):259–264. doi:10.1364/JOSA.66.000259

Palchetti L, Lastrucci D (2001) Spectral noise due to sampling errors in Fourier-transform spectroscopy. Appl Opt 40(19):3235–3243. doi:10.1364/AO.40.003235

Palchetti L, Barbis A, Harries J, Lastrucci D (1999) Design and mathematical modelling of the space-borne far-infrared Fourier transform spectrometer for refir experiment. Infrared Phys Technol 40(5):367–377. doi:10.1016/S1350-4495(99)00026-2

Palchetti L, Bianchini G, Pellegrini M, Esposito F, Restieri R, Pavese G (2004) Radiometric performances of the Fourier transform spectrometer for the radiation explorer in the far-infrared (REFIR) space mission. Proc SPIE 5570:433–444. doi:10.1117/12.565548

Palchetti L, Bianchini G, Castagnoli F, Carli B, Serio C, Esposito F, Cuomo V, Rizzi R, Maestri T (2005) Breadboard of a Fourier-transform spectrometer for the radiation explorer in the far infrared atmospheric mission. Appl Opt 44(14):2870–2878

Press WH, Teukolsky SA, Vetterling WT, Flannery BP (2007) Numerical recipes 3rd edition: the art of scientific computing, 3rd edn. Cambridge University Press, New York

Roy SA, Genest J, Giaccari P (2007) Hybrid sampling approach for imaging Fourier-transform spectrometry. Appl Opt 46(35):8482–8487. doi:10.1364/AO.46.008482

Saggin B, Comolli L, Formisano V (2007) Mechanical disturbances in Fourier spectrometers. Appl Opt 46(22):5248–5256. doi:10.1364/AO.46.005248

Sakai H (1970) Consideration of the signal-to-noise ratio in Fourier spectroscopy. In: Vanasse GA, Stair AT Jr, Baker DJ (eds) Aspen international conference on Fourier spectroscopy

Simon A, Metz W, Keens A (2002) Data acquisition and interferogram data treatment in FT-IR spectrometers. Vib Spectrosc 29(12):97–101. doi:10.1016/S0924-2031(01)00191-6

Steel WH (1970) Interferometers for Fourier spectroscopy. In: Vanasse GA, Stair AT Jr, Baker DJ (eds) Aspen international conference on Fourier spectroscopy

Surh MT (1966) The effect of time jitter in sampling of an interferogram. Appl Opt 5(5):880–881. doi:10.1364/AO.5.000880

Taurand G, Genest J, Cadotte M, Gibeault M (2007) Parasitic diffuse reflection in a Fourier transform spectrometer yielding subharmonic ghosts and line-shape distortion. Appl Opt 46(4):533–537. doi:10.1364/AO.46.000533

Varmuza K, Filzmoser P (2009) Introduction to multivariate statistical analysis in chemometrics. CRC Press, Boca Raton

Yang H, Griffiths PR, Manning CJ (2002) Improved data processing by application of Brault's method to ultra-rapid-scan FT-IR spectrometry. Appl Spectrosc 56(10):1281–1288

Zachor AS (1977) Drive nonlinearities: their effects in Fourier spectroscopy. Appl Opt 16(5):1412–1424. doi:10.1364/AO.16.001412

Zachor AS, Aaronson SM (1979) Delay compensation: its effect in reducing sampling errors in Fourier spectroscopy. Appl Opt 18(1):68–75. doi:10.1364/AO.18.000068

Appendix A
Block Diagrams

Engineers use block diagrams for communicating the basic structure of a device without reference to an implementation. At the level of block diagrams first design decisions, like setting corner frequencies and sampling rates, can be attempted. Table A.1 lists the symbols we use in the diagrams in this book. Whenever aesthetically pleasing we use arrows for connecting blocks, the direction of the arrow indicating the direction of the flow of information. Rarely, when there is no doubt about the direction, we use lines instead. We represent parallel buses by broad arrows. As block diagrams serve as a means of communication readability is paramount. Information shall flow from left to right, unless there is a good reason for the other direction, like when closing a feedback loop.

© Springer International Publishing Switzerland 2015
P. Hintenaus, *Engineering Embedded Systems*, DOI 10.1007/978-3-319-10680-9

Table A.1 Symbols used in block diagrams

Function	Symbol
Amplifier	
Highpass filter	
Lowpass filter	
Bandpass filter	
Sinus oscillator	
Triangle oscillator	
Clock generator	
Analog-to-digital converter	
Digital-to-analog converter	
Difference	
Differentiator	
Integrator	
PI controller	
PID controller	

(continued)

Table A.1 (continued)

Function	Symbol
Mixer	
Receiving antenna	
Transmitting antenna	
Photodetector	
Light emitter	
Laser	
Permanent magnet synchronous machine	
Squirrel cage fan	

Appendix B
Circuit Diagrams

When drawing circuit diagrams it is a good practice to have points in the upper portion of the page at a higher voltage than points in the lower portion. Signal flow should be from left to right, but for feedback signals. Table B.1 lists the symbols we use in this book's circuit diagrams.

© Springer International Publishing Switzerland 2015
P. Hintenaus, *Engineering Embedded Systems*, DOI 10.1007/978-3-319-10680-9

Table B.1 Symbols used in circuit diagrams

Function	Symbol
In-tag	A⟩
Out-tag	⟩ A
Ground	⏚
Ground	θ GND
DC voltage source	1V U
AC voltage source	U
Current source	1mA U
Battery	4.5V +B
Voltmeter	V U
Ammeter	A U
Resistor	R
Capacitor	C
Polarized capacitor	C +
Inductor	L
2-input And-gate	U &
2-input Or-gate	U ≥1
Inverter	U
Simple flip-flop	U P Q C
D flip-flop	U D clk

(continued)

Table B.1 (continued)

Function	Symbol
D flip-flop with clock enable	
D flip-flop with synchronous set	
D flip-flop with synchronous reset	
Operational amplifier	
Comparator	
Diode	
Schottky diode	
Zener diode	
Light emitting diode	
Photodiode	
PNP bipolar transistor	
NPN bipolar transistor	
P-channel enhancement mode MOSFET	

(continued)

Table B.1 (continued)

Function	Symbol
P-channel enhancement mode MOSFET with built-in diode	
N-channel enhancement mode MOSFET	
N-channel enhancement mode MOSFET with built-in diode	

Appendix C
The One-Sided Laplace Transform

Let $x: \mathbb{R} \to \mathbb{C}, t \mapsto x(t)$ be a continuous-time signal. The one-sided Laplace transform of x, $\hat{X}: \mathrm{roc}(x) \to \mathbb{C}$ is defined as

$$s \mapsto \hat{X}(s) = (\mathcal{L}(x))\,(s) = \int\limits_0^\infty x(t)\mathrm{e}^{-st}\mathrm{d}t.$$

The set $\mathrm{roc}(x)$ is the region of convergence of the Laplace transform of x (Table C.1).

© Springer International Publishing Switzerland 2015
P. Hintenaus, *Engineering Embedded Systems*, DOI 10.1007/978-3-319-10680-9

Table C.1 Some continuous-time signals and their one-sided Laplace transforms

$w(t)$	$\hat{W}(s) = (\mathcal{L}(w))(s)$	roc(w)
$ax(t) + by(t)$	$a\hat{X}(s) + b\hat{Y}(s)$	roc(w) $\supseteq$ roc(x) $\cap$ roc(y)
$x(t - \tau)u(t - \tau)$	$e^{-\tau s}\hat{X}(S)$	roc(x)
$\displaystyle\int_0^\infty x(\tau)y(t - \tau)d\tau$	$\hat{X}(s)\hat{Y}(s)$	roc(w) $\supseteq$ roc(x) $\cap$ roc(y)
$tx(t)$	$-\dfrac{d\hat{X}(s)}{ds}$	roc(x)
$e^{at}x(t)$	$\hat{X}(s - a)$	$\{s \in \mathbb{C}: s - a \in \text{roc}(x)\}$
$\displaystyle\int_0^t x(\tau)d\tau$	$\dfrac{\hat{X}(s)}{s}$	roc(w) $\supseteq \{s \in \text{roc}(x): \Re(s) > 0\}$
$\dfrac{dx(t)}{dt}$	$s\hat{X}(s) - f(0)$	roc(w) $\supseteq$ roc(x)
$\dfrac{d^n x(t)}{dt^n}$	$s^n \hat{X}(s) - \displaystyle\sum_{i=0}^{n-1} s^{n-1-i}\dfrac{d^i x}{dt^i}\Big|_{t=0}$	roc(w) $\supseteq$ roc(x)
$\delta(t - \tau)$	$e^{-s\tau}$	$\mathbb{C}$
1	$\dfrac{1}{s}$	$\{s \in \mathbb{C}: \Re(s) > 0\}$
t	$\dfrac{1}{s^2}$	$\{s \in \mathbb{C}: \Re(s) > 0\}$
e^{-at}	$\dfrac{1}{s + a}$	$\{s \in \mathbb{C}: \Re(s) > -\Re(a)\}$

The signals $x: \mathbb{R} \to \mathbb{C}$ and $y: \mathbb{R} \to \mathbb{C}$ are arbitrary; $\hat{X}: \text{roc}(x) \to \mathbb{C}$ and $\hat{Y}: \text{roc}(y) \to \mathbb{C}$ are their Laplace transforms. The numbers $a, b \in \mathbb{C}$ and $c, \tau \in \mathbb{R}$ and $n \in \mathbb{N}$ are arbitrary

Appendix D
Answers to Selected Exercises

D.1 Chapter 1

1.1 The point $\vec{r}$ travels an arc length of $s = r\theta$.

1.2 The angular velocity $\omega(t)$ of the body is the time-derivative of the body's angular displacement,

$$\omega(t) = \frac{d\theta(t)}{dt}.$$

The unit of the angular velocity is $[s^{-1}]$.

1.3 The angular acceleration $\alpha(t)$ of the body is the time-derivative of the body's angular velocity,

$$\alpha(t) = \frac{d\omega(t)}{dt} = \frac{d^2\theta(t)}{dt^2}.$$

The unit of the angular acceleration is $[s^{-2}]$.

1.4 We choose a coordinate system such that the body is located in the xy-plane. The axis of rotation shall coincide with the z-axis. Let us assume that the force $\vec{F} = \begin{pmatrix} F_x \\ F_y \end{pmatrix}$ acts on the body at the point $\begin{pmatrix} x \\ y \end{pmatrix}$. Then the torque τ the force $\vec{F}$ exerts on the body is

$$\tau = x F_y - y F_x.$$

1.5 Let the point-shaped body have mass m and be located at a distance r from the axis of rotation. Its moment of inertia I is

$$I = mr^2.$$

© Springer International Publishing Switzerland 2015
P. Hintenaus, *Engineering Embedded Systems*, DOI 10.1007/978-3-319-10680-9

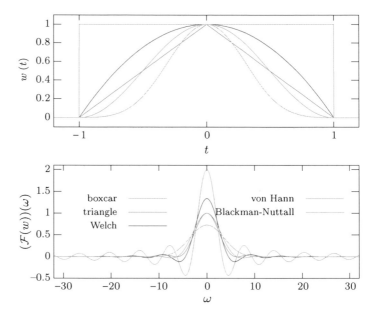

Fig. D.1 Some window functions and their Fourier transforms

The torque $\tau(t)$ necessary to achieve an angular acceleration $\alpha(t)$ of a body with moment of inertia I is

$$\tau(t) = I\alpha(t) = I\frac{\mathrm{d}^2\theta(t)}{\mathrm{d}t^2}.$$

D.2 Chapter 2

2.1 See Fig. D.1.

2.2 See Fig. D.1.

2.3 Fourier transform of the Welch window for $p = 1$,

$$(\mathcal{F}(w))(\omega) = \frac{4\sin\omega - 4\omega\cos\omega}{\omega^3}.$$

See Fig. D.1.

2.4 Fourier transform of the von Hann window for $p = 1$,

$$(\mathcal{F}(w))(\omega) = \frac{\pi^2 \sin\omega}{\pi^2\omega - \omega^3}.$$

See Fig. D.1.

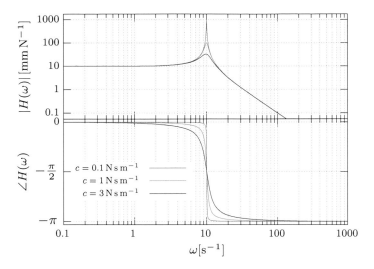

Fig. D.2 The resonant peak for three damping constants. The smaller the damping the higher and narrower the peak

2.5 Fourier transform of the Blackman–Nuttall window for $p = 1$,

$$(\mathcal{F}(w))(\omega) = \frac{\left(1814\omega^6 - 218143\pi^2\omega^4 + 6959771\pi^4\omega^2 - 65444742\pi^6\right)\sin\omega}{2500000\omega^7 - 35000000\pi^2\omega^5 + 122500000\pi^4\omega^3 - 90000000\pi^6\omega}.$$

See Fig. D.1.

2.6 See Fig. D.2. The smaller the damping constant the higher and narrower the peak.

2.7 The amplitude of the frequency response is

$$|H(\omega)| = \frac{1}{\sqrt{\left(k - m\omega^2\right)^2 + c^2\omega^2}}.$$

It has a maximum for $\omega = \sqrt{\frac{k}{m} - \frac{c^2}{2m^2}}$, which is real for

$$c \le \sqrt{2km}.$$

Negative damping constants would entail that the damper is adding energy to the system.

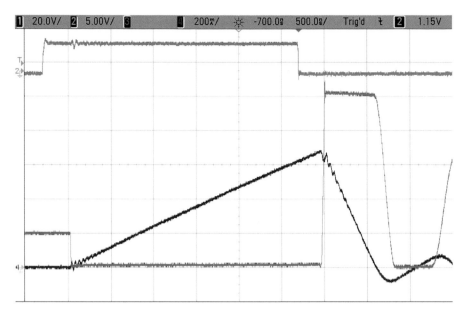

Fig. D.3 Behavior of the circuit in Fig. 3.33. Channel one shows the voltage at point A, channel two the signal operating the switch, and channel four shows the current through the inductor L

D.3 Chapter 3

3.1 See Fig. D.3. As long as the switch SW conducts the current through the inductor L ramps up linearly. In this first phase the voltage across L is 20 V. When the switch stops conducting this current breaks down instantaneously making the voltage at A jump high enough for the suppressor D to establish another current path. The current through L ramps then down. In the second phase the voltage across L is −80 V. The first phase is about four times as long as the second, demonstrating that the inductor stores energy with very little losses. What little energy is left stored in L when D stops conducting cycles between L and the combined parasitic capacitances of SW and of D. This causes the ringing visible at the end.

3.2 Let $I_C(t)$ be the current through the capacitor and $V_C(t)$ be the voltage across the capacitor. Then

$$V_C(t) = I_0 \sqrt{\frac{L}{C}} \sin \omega t + V_0 \cos \omega t$$

$$I_C(t) = -V_0 \sqrt{\frac{C}{L}} \sin \omega t + I_0 \cos \omega t$$

where $V_C(0) = V_0$, $I_C(0) = I_0$ and $\omega = \sqrt{\frac{1}{LC}}$. The energy stored in the capacitor is $W_C(t) = \frac{C}{2}(V_C(t))^2$, the energy stored in the inductor is $W_L = \frac{L}{2}(I_C(t))^2$. Therefore, the total energy $W(t)$ is

$$W(t) = W_C(t) + W_L(t) = \frac{C(V_C(t))^2 + L(I_C(t))^2}{2} = \frac{CV_0^2 + LI_0^2}{2},$$

which is the energy that was stored in the circuit at time 0.

3.3 The circuit can be understood as voltage divider. Amplitude and phase of the voltage at point B are

$$V_o = V_0 H(\omega) = V_0 \frac{\omega RC}{\omega RC - i},$$

when the circuit is excited at point A with the voltage $V_0(t) = V_0 \cos \omega t$. The circuit is a highpass filter. For a Bode diagram see Fig. 2.2 where $\beta = RC$.

3.4 Amplitude and phase of the voltage at point B are

$$V_o = V_0 H(\omega) = V_0 \frac{1 - \omega^2 LC}{i\omega RC - \omega^2 LC + 1},$$

when the circuit is excited at point A with the voltage $V_0(t) = V_0 \cos \omega t$. Figure D.4 shows the circuit's Bode plot for $C = 2.5\,\mu$F and $L = 1\,\mu$H. The frequency response $H(\omega)$ has a zero at $\omega = \frac{1}{\sqrt{LC}}$, the inductance together with the capacitor form a short at that frequency.

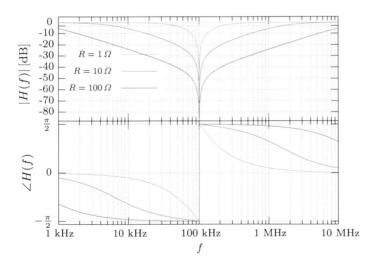

Fig. D.4 Bode plot for the circuit in Fig. 3.35, $C = 2.5\,\mu$F, $L = 1\,\mu$H

Fig. D.5 Edge detector
circuit

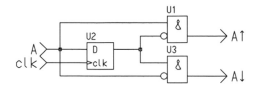

3.5 The root-mean-square voltage of $V_0 \cos \omega t$ is

$$\sqrt{\frac{\omega \int_0^{2\pi/\omega} (V_0 \cos \omega t)^2 dt}{2\pi}} = \frac{V_0}{\sqrt{2}}.$$

3.6 The voltage between L1 and L2 is

$$V_1(t) - V_2(t) = V(\cos(\omega t) - \cos(\omega t - 2\pi/3)) = \sqrt{3}V \cos(\omega t + \pi/6),$$

the voltage between L2 and L3 is

$$V_2(t) - V_3(t) = V(\cos(\omega t - 2\pi + 3) - \cos(\omega t + 2\pi/3)) = \sqrt{3}V \cos(\omega t - \pi/4),$$

and the voltage between L3 and L1 is

$$V_3(t) - V_1(t) = V(\cos(\omega t + 2\pi + 3) - \cos(\omega t)) = \sqrt{3}V \cos(\omega t + 5\pi/6).$$

D.4 Chapter 4

4.1 See Fig. D.5.

4.2 See Fig. D.6.

4.3 See Fig. D.7. The pulse p_1 has more high-frequency content than the pulse p_2.

D.5 Chapter 5

5.1 Our representation of an SPI transfer has the fields *initiate*, *process*, and *after*. The field *initiate* references a user supplied procedure to be called at the start of transfer. The procedure must configure the interface for the mode and the number of bits the involved slave needs, pull the slave's chip select low and load the data to be transmitted into the interface. The field *process* references a user supplied procedure to be called at the end of the transfer. The procedure must pull the slave's chip select

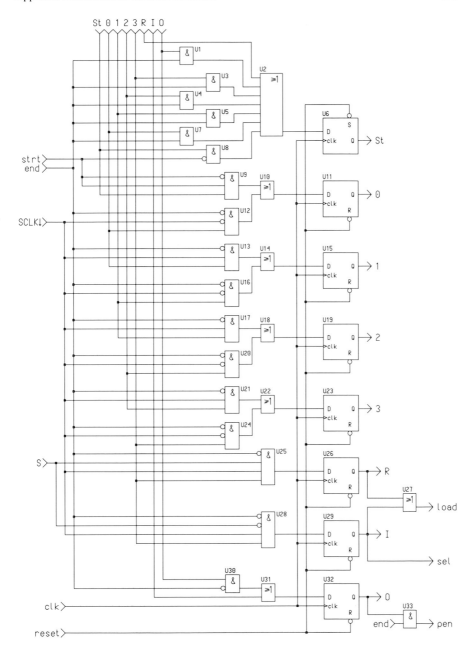

Fig. D.6 Realization of the state machine in Fig. D.13

high and read and evaluate the data received from the slave. We implement the state machine in Fig. D.8.

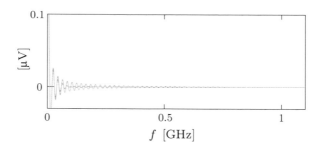

Fig. D.7 Frequency content of two pulses. Short rise and fall times introduce high-frequency content

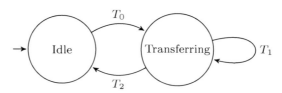

	Condition	Action	Thread
T_0	SCHEDULETRANSFER(x)	start transfer x	APPLICATION
T_1	transfer completed, transfers pending	start next transfer	HANDLESPI
T_2	transfer completed, no transfers pending	none	HANDLESPI

Fig. D.8 State transition diagram and transitions of a state machine for an SPI framework

The framework consists of the interrupt service routine HANDLESPI and the procedure SCHEDULETRANSFER, Figs. D.9 and D.10. When the framework is busy transferring data SCHEDULETRANSFER enqueues the transfer into a first-in first-out queue. The framework maintains three variables, *in*, *out* and *c*. The variables *in* and *out* reference the transfers at the entry and at the exit of the queue. The variable *c* references the transfer in progress. The framework does not represent its state explicitly in a separate state variable. The state Idle is represented by $c =$ **nil** and *out* = **nil**; the state Transferring is represented by any other combination of the two variables. The queue is a critical section. Therefore, SCHEDULETRANSFER executes with interrupts disabled.

5.2 Computing the first column of a block during the second pass requires 2^{2m-l} different twiddles; computing the $i + 1$st column requires 2^{2m-l+i} different ones. Therefore, the total number $t'/2$ of twiddles required for one block comprising $l - m$ columns is

$$t'/2 = \sum_{i=0}^{l-m-1} 2^{2m-l+i} = \sum_{i=0}^{m-1} 2^i - \sum_{i=0}^{2m-l-1} 2^i = 2^m - 2^{2m-l}.$$

```
1: procedure HANDLESPI
2:     c.process
3:     c.after ← nil
4:     if out ≠ nil then
5:         c ← out
6:         out ← out.after
7:         c.initiate
8:     else
9:         c ← nil
10:    end if
11: end procedure
```

Fig. D.9 Interrupt service routine of the SPI framework

```
1: procedure SCHEDULETRANSFER(x)
2:     x.after ← nil
3:     disable interrupts
4:     if out = nil then
5:         if c = nil then
6:             c ← x
7:             c.initiate
8:         else
9:             in ← out ← x
10:        end if
11:    else
12:        in.after ← x
13:        in ← x
14:    end if
15:    enable interrupts
16: end procedure
```

Fig. D.10 Procedure for scheduling an SPI transfer

We have to compute 2^{l-m} blocks during the second pass. The arrays referenced by T and T' have $2^l - 2^m$ entries each. The Fast Fourier Transform of length 2^l requires 2^{l-1} different twiddles. Arranging the twiddles in the order required by the second pass almost doubles the memory requirements.

5.3 See Fig. D.11.

5.4 The interrupt service routine HANDLEENCODER, Fig. D.12, triggered by a timer operating in continuous mode, tracks an incremental encoder. It keeps its state in the variable st, and the count accumulating the rotations since start in the global variable c. Right before starting the timer the two variables must be initialized with $(st, c) ← (\text{DECODEGRAY}(\text{READENCODER}), 0)$.

5.5 See Fig. D.13.

5.6 See Fig. D.14. The counter U1 counts how often a level different from the level at the output Q of the circuit has appeared in a row. When it raises its TC output

```
 1: procedure SCRAMBLETWIDDLES(l, T, T′)
 2:     (n_g, s_g, k) ← (2^{l−m}, 2^{2m−l}, 0)
 3:     for b ← 0, 2^{m+1}, 2(2^{m+1}), 3(2^{m+1}), …, 2^l − 2^{m+1} do
 4:         (d_{(tg)}, d_{(tw)}, f) ← (2^{2m−1}, 2^{m−1}, n_g b/2)
 5:         for n_{(tg)} ← 1, 2, 4, 8, …, 2^{l−m−1} do
 6:             for t ← 0, n_{(tg)} − 1 do
 7:                 for g ← 0, n_g − 1 do
 8:                     f′ ← f + d_{(tg)}t + d_{(tw)}g
 9:                     s′ ← f′ + d_{(tw)}n_g
10:                     (T_k, T′_k) ← (cos 2π f′/2^l, − sin 2π f′/2^l)
11:                     (T_{k+1}, T′_{k+1}) ← (cos 2π s′/2^l, − sin 2π s′/2^l)
12:                     k ← k + 2
13:                 end for
14:             end for
15:             (d_{(tg)}, d_{(tw)}, f) ← (d_{(tg)}/2, d_{(tw)}/2, f/2)
16:         end for
17:     end for
18: end procedure
```

Fig. D.11 Procedure for arranging the twiddles in the sequence as required by PIPELINEDFFT in Fig. 5.21

indicating it has reached its maximum value, the flip-flop U6 stores the new value. For the second channel the circuit has to be duplicated.

D.6 Chapter 6

6.1 Computing the continuous-time Fourier transform $W: \mathbb{R} \mapsto \mathbb{C}$ of

$$w(t) = \sum_{k=-\infty}^{\infty} s'_k \delta(t - kt_s)$$

we get

$$W(\omega) = \int_{-\infty}^{\infty} \left(\sum_{k=-\infty}^{\infty} s'_k \delta(t - kt_s) \right) e^{-i\omega t} \, dt$$

$$= \sum_{k=-\infty}^{\infty} s'_k \int_{-\infty}^{\infty} \delta(t - kt_s) e^{-i\omega t} \, dt$$

$$= \sum_{k=-\infty}^{\infty} s'_k e^{-i\omega k t_s}.$$

Fig. D.12 Timer Interrupt
service routine for tracking an
incremental encoder

```
 1: procedure HANDLEENCODER
 2:     (A, B) ← READENCODER
 3:     if st = 0 then
 4:         if (A, B) = (1, 0) then
 5:             (st, c) ← (1, c + 1)
 6:         else if (A, B) = (0, 1) then
 7:             (st, c) ← (3, c − 1)
 8:         end if
 9:     else if st = 1 then
10:         if (A, B) = (1, 1) then
11:             (st, c) ← (2, c + 1)
12:         else if (A, B) = (0, 0) then
13:             (st, c) ← (0, c − 1)
14:         end if
15:     else if st = 2 then
16:         if (A, B) − (0, 1) then
17:             (st, c) ← (3, c + 1)
18:         else if (A, B) = (1, 0) then
19:             (st, c) ← (1, c − 1)
20:         end if
21:     else
22:         if (A, B) = (0, 0) then
23:             (st, c) ← (0, c + 1)
24:         else if (A, B) = (1, 1) then
25:             (st, c) ← (2, c − 1)
26:         end if
27:     end if
28: end procedure
```

Comparing this with the definition of the discrete-time Fourier transform 2.13 we recognize $S'(\omega) = W(\omega/t_s)$. Together with 6.4 this finishes the argument.

6.2 In order to best maintain temporal relationships in the signal we choose Bessel characteristics. We set the filter's corner frequency to 100 kHz so that the filter does not disturb the signal. In order to achieve a resolution of 16 bits the frequency content of the signal at 10 MHz and above has to be down by about 96 dB. A second order Bessel filter gives us an attenuation of about 76 dB, a third order filter gives us about 110 dB. The third order filter is the save bet. The second order filter might do, provided the signal's frequency content between DC and 50 kHz dominates.

6.3 Kirchhoff's current law applied to the summing junction gives us

$$\frac{V_1}{R_1} + \frac{V_2}{R_2} + \frac{V_o}{R_3} = 0.$$

Solving for V_o produces the result.

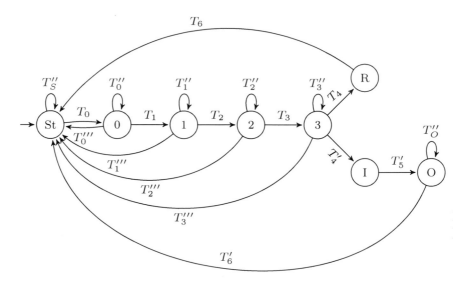

Fig. D.13 State transition diagram, transitions, and output function of the state machine for decoding command input in Fig. 5.32

	Condition		Condition		Condition
T_0	strt $= 1$	T_5'	**default**	T_O''	end $= 0$
T_1	(SCLK $\downarrow$, end) $= (1,0)$	T_6'	end $= 1$	T_0'''	end $= 1$
T_2	(SCLK $\downarrow$, end) $= (1,0)$	T_S''	strt $= 0$	T_1'''	end $= 1$
T_3	(SCLK $\downarrow$, end) $= (1,0)$	T_0''	(SCLK $\downarrow$, end) $= (0,0)$	T_2'''	end $= 1$
T_4	(SCLK $\downarrow$, end, S) $= (1,0,1)$	T_1''	(SCLK $\downarrow$, end) $= (0,0)$	T_3'''	end $= 1$
T_6	**default**	T_2''	(SCLK $\downarrow$, end) $= (0,0)$		
T_4'	(SCLK $\downarrow$, end, S) $= (1,0,0)$	T_3''	(SCLK $\downarrow$, end) $= (0,0)$		

State	sel	load	pen	State	sel	load	pen
St	0	0	0	3	0	0	0
0	0	0	0	R	0	1	0
1	0	0	0	I	1	1	0
2	0	0	0	O	0	0	end

6.4 Kirchhoff's current law applied to the summing junction gives us

$$\frac{V_i}{R_1} + 2i\pi f C_1 V_o + \frac{V_o}{R_2} = 0.$$

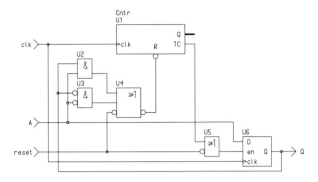

Fig. D.14 Circuit for filtering spurious transitions from a digital signal

Solving for V_o and taking the absolute value gives us

$$|V_o| = \frac{R_2 |V_i|}{\sqrt{4\pi^2 C_1^2 R_1^2 R_2^2 f^2 + R_1^2}}.$$

Equating

$$|V_o| = |V_i| \frac{R_2}{\sqrt{2} R_1}$$

and solving for f produces the result.

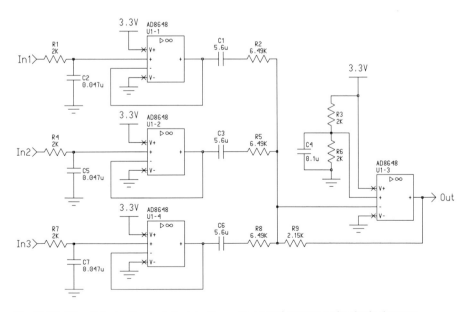

Fig. D.15 Circuit for testing lock-in detection with several sources and a single detector

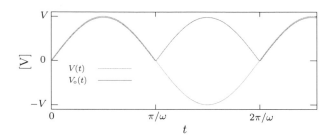

Fig. D.16 Input voltages $V(t)$ of a single phase rectifier and output voltage $V_o(t)$

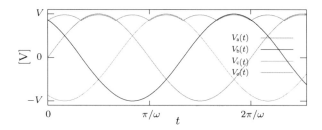

Fig. D.17 Input voltages $V_a(t)$, $V_b(t)$ and $V_c(t)$ of a three-phase rectifier and output voltage $V_o(t)$

6.5 See Fig. D.15. The amplifiers can be substituted with any other type as long as these have rail-to-rail inputs and rail-to-rail outputs. The circuit runs off a single 3.3 V supply, therefore the amplifiers are AC coupled.

6.6 Let us assume the output is at 0 V and the input In1 is at V_X. In order to produce the voltage V_X at the input Y of the comparator In2 must be driven to $V_0 = V_X \frac{R_1+R_2}{R_1}$. Now let us assume that the output sits at a voltage of V_S. In order to drive Y to the voltage V_X In2 must be driven to $V_1 = (V_X - V_S)\frac{R_1+R_2}{R_2}$. Therefore, the hysteresis is

$$V_H = V_0 - V_1 = V_S \frac{R_1 + R_2}{R_2}.$$

6.7 See Fig. D.16.

6.8 See Fig. D.17

D.7 Chapter 7

7.1 The duration of the first part of a switching cycle is $t_c V_o / V_i$, the voltage across the inductor during this period is $V_i - V_o$. Therefore, the current ripple through an inductor with inductance L is

$$\hat{I}_L = \frac{t_c V_o (V_i - V_o)}{L V_i}.$$

Rearranging yields the answer.

7.2 The voltage across the capacitor C2 reaches its minimum after half of the first part of a switching cycle, it reaches its maximum after half of the second part of a switching cycle. Integrating the current ramp during the first half of the first part of a switching cycle and dividing by C_2 gives us a voltage variation of

$$\frac{t_c^2 \left(V_o^3 - V_i V_o^2 \right)}{8 L C_2 V_i^2}.$$

Integrating the current ramp during the first half of the second part and dividing by C_2 gives us a voltage variation of

$$\frac{t_c^2 \left(V_o^3 - 2 V_i V_o^2 + V_i^2 V_o \right)}{8 L C_2 V_i^2}.$$

Therefore, the total variation during one cycle is

$$\hat{V}_o = \frac{t_c^2 V_o (V_i - V_o)}{8 L C_2 V_i}.$$

Rearranging yields the answer.

7.3 The duration of the first part of a switching cycle is $t_c (V_o - V_i)/V_o$. Therefore, the current ripple through the inductor is

$$\hat{I}_L = \frac{t_c V_i (V_o - V_i)}{L V_o}.$$

Rearranging yields the answer.

7.4 During the first part of the switching cycle the capacitor C2 solely has to supply the load. Therefore, the output voltage ripple is

$$\hat{V}_o = \frac{t_c I_o (V_o - V_i)}{C_2 V_o}.$$

Rearranging yields the answer.

7.5 The duration of the first part of a switching cycle is $t_c V_o/(V_i + V_o)$, the voltage across the inductors L1 as well as L2 during this period is V_i. Therefore, the current ripple through an inductor with inductance L_1 is

$$\hat{I_1} = \frac{t_c V_i V_o}{L_1(V_i + V_o)}.$$

Rearranging yields the answer.

7.6 During the first part of a switching cycle the capacitor C3 solely has to supply the load. Therefore, the output voltage ripple is

$$\hat{V_o} = \frac{t_c V_o I_o}{C_3(V_i + V_o)}.$$

Rearranging yields the answer.

7.7 During the first part of the switching cycle the current through C2 equals the current $-I_2(t)$. Integrating the ripple current waveform during the first part of a switching cycle and dividing by C_2 gives a voltage variation of

$$\hat{V_C} = \frac{t_c V_o I_o}{C_2(V_i + V_o)}.$$

Rearranging yields the answer.

7.8 The duration of the first part of a switching cycle is $t_c V_o/(V_o - V_i)$, the voltage across the inductor L during this period is V_i. Therefore, the current ripple is

$$\hat{I_L} = \frac{t_c V_i V_o}{L(V_o - V_i)}.$$

Rearranging yields the answer.

7.9 During the first part of a switching cycle the capacitor C2 solely has to supply the load. Therefore, the output voltage ripple is

$$\hat{V_o} = \frac{t_c V_o I_o}{C_2(V_o - V_i)}.$$

Rearranging yields the answer.

7.10 The duration of the first part of a switching cycle is $t_c V_o/(V_o - V_i)$, the voltage across the inductors L1 as well as L2 during this period is V_i. Therefore, the current ripple through an inductor with inductance L is

$$\hat{I_L} = \frac{t_c V_i V_o}{L(V_o - V_i)}.$$

Rearranging yields the answer.

7.11 The voltage across the capacitor C3 reaches its minimum after half of the first part of a switching cycle, it reaches its maximum after half of the second part of a switching cycle. Integrating the current ramp during the first half of the first part of a switching cycle and dividing by C_3 gives us a voltage variation of

$$\frac{t_c^2 V_i V_o^2}{8L_2 C_3 (V_o - V_i)^2}.$$

Integrating the current ramp during the first half of the second part and dividing by C_3 gives us a voltage variation of

$$\frac{t_c^2 V_i^2 V_o}{8L_2 C_3 (V_o - V_i)^2}.$$

Therefore, the total variation during one cycle is

$$\hat{V}_o = \frac{t_c^2 V_i V_o}{8L_2 C_3 (V_o - V_i)}.$$

Rearranging yields the answer.

7.12 During the first part of the switching cycle the current through C2 equals the current through L2. Integrating the current waveform during the first part of a switching cycle and dividing by C_2 gives a voltage variation of

$$\hat{V}_C = \frac{t_c I_o V_o}{C_2 (V_o - V_i)}.$$

Rearranging yields the answer.

7.13 The duration of the first part of a switching cycle is $t_c n V_o / (n V_o + m V_i)$, the voltage across L1 during this period is V_i. Therefore, the current ripple through L1 is

$$\hat{I}_1 = \frac{t_c n V_i V_o}{L_1 (n V_o + m V_i)}.$$

Rearranging yields the answer.

7.14 During the first part of the switching cycle the capacitor C2 solely supplies the load. Therefore, the ripple voltage is

$$\hat{V}_o = \frac{t_c V_i m I_o}{C_2 (n V_o + m V_i)}.$$

Rearranging yields the answer.

7.15 For stepping down from Vi to Vo the top switch of S2 must conduct continuously. During the first part of a switching cycle the top switch of S1 must conduct, during the second part the bottom switch of S1 must conduct. For stepping up from Vi to Vo the top switch of S1 must conduct continuously. During the first part of a switching cycle the bottom switch of S2 must conduct, during the second part the top switch of S2 must conduct.

D.8 Chapter 8

8.1 The three phases are

$$V_u = A \left(\cos \theta - \frac{1}{6} \cos 3\theta \right)$$

$$V_v = A \left(\cos \left(\theta + 2\pi/3 \right) - \frac{1}{6} \cos 3\theta \right)$$

$$V_w = A \left(\cos \left(\theta + 4\pi/3 \right) - \frac{1}{6} \cos 3\theta \right).$$

The machine sees the differences $V_u - V_v$, $V_v - V_w$ and $V_w - V_u$ so the third harmonic content cancels.

8.2 The capacitor C14 and the diodes D7 and D8 form a bootstrap circuit. The Zener diode limits the voltage applied to the BOOST pin.

Index

© Springer International Publishing Switzerland 2015
P. Hintenaus, *Engineering Embedded Systems*, DOI 10.1007/978-3-319-10680-9